The Vienna Series in Theoretical Biology
Gerd B. Müller, Günter P. Wagner, and Werner Callebaut, editors

The Evolution of Cognition, edited by Cecilia Heyes and Ludwig Huber, 2000

Origination of Organismal Form: Beyond the Gene in Developmental and Evolutionary Biology, edited by Gerd B. Müller and Stuart A. Newman, 2003

Environment, Development, and Evolution: Toward a Synthesis, edited by Brian K. Hall, Roy D. Pearson, and Gerd B. Müller, 2004

Evolution of Communication Systems: A Comparative Approach, edited by D. Kimbrough Oller and Ulrike Griebel, 2004

Modularity: Understanding the Development and Evolution of Natural Complex Systems, edited by Werner Callebaut and Diego Rasskin-Gutman, 2005

Compositional Evolution: The Impact of Sex, Symbiosis, and Modularity on the Gradualist Framework of Evolution, by Richard A. Watson, 2006

Biological Emergencies: Evolution by Natural Experiment, by Robert G. B. Reid, 2007

Modeling Biology: Structures, Behaviors, Evolution, edited by Manfred D. Laubichler and Gerd B. Müller

Modeling Biology

Modeling Biology
Structures, Behaviors, Evolution

edited by Manfred D. Laubichler and Gerd B. Müller

The MIT Press
Cambridge, Massachusetts
London, England

For information about special quantity discounts, please email special_sales@mitpress.mit.edu

This book was set in Times New Roman on 3B2 by Asco Typesetters, Hong Kong.
Printed on recycled paper and bound in the United States of America.

Library of Congress Cataloging-in-Publication Data

Modeling biology : structures, behaviors, evolution / edited by Manfred D. Laubichler and Gerd B. Müller.
p. cm. — (The Vienna series in theoretical biology)
Includes bibliographical references and index.
ISBN-13: 978-0-262-12291-7 (hardcover : alk. paper)
1. Biology—Mathematical models. 2. Biology—Simulation methods. I. Laubichler, Manfred.
II. Müller, Gerd B.
QH323.5M623 2007
570.1′1—dc22 2006029825

10 9 8 7 6 5 4 3 2 1

Contents

Series Foreword

Biology is becoming the leading science in this century. As in all other sciences, progress in biology depends on interactions between empirical research, theory building, and modeling. But whereas the techniques and methods of descriptive and experimental biology have evolved dramatically in recent years, generating a flood of highly detailed empirical data, the integration of these results into useful theoretical frameworks has lagged behind. Driven largely by pragmatic and technical considerations, research in biology continues to be less guided by theory than seems indicated. By promoting the formulation and discussion of new theoretical concepts in the biosciences, this series intends to help fill the gaps in our understanding of some of the major open questions of biology, such as the origin and organization of organismal form, the relationship between development and evolution, and the biological bases of cognition and mind.

Theoretical biology has important roots in the experimental biology movement of early twentieth-century Vienna. Paul Weiss and Ludwig von Bertalanffy were among the first to use the term *theoretical biology* in a modern scientific context. In their understanding the subject was not limited to mathematical formalization, as is often the case today, but extended to the conceptual problems and foundations of biology. It is this commitment to a comprehensive, cross-disciplinary integration of theoretical concepts that the present series intends to emphasize. Today theoretical biology has genetic, developmental, and evolutionary components, the central connective themes in modern biology, but also includes relevant aspects of computational biology, semiotics, and cognition research, and extends to the naturalistic philosophy of sciences.

The Vienna Series grew out of theory-oriented workshops organized by the Konrad Lorenz Institute for Evolution and Cognition Research (KLI), an international center for advanced study closely associated with the University of Vienna (http://kli.ac.at). The KLI fosters research projects, workshops, archives, book projects, and the journal *Biological Theory*, all devoted to aspects of theoretical biology, with an emphasis on integrating the developmental, evolutionary, and cognitive sciences. The series editors welcome suggestions for book projects in these fields.

Gerd B. Müller, University of Vienna, KLI
Günter P. Wagner, Yale University, KLI
Werner Callebaut, Hasselt University, KLI

Preface

Throughout the biosciences the rapidly growing number of experimental results and the increasing complexity of the investigated phenomena pose an unprecedented challenge for the interpretation of the accumulated data and the integration of this knowledge into the theoretical framework of biology. Therefore, abstract and conceptual models have become an indispensable tool for the analysis and interpretation of empirical results, for the organization of data, for theoretical integration, and for the formulation of new hypotheses. At the same time, bottom-up modeling elucidates properties of natural biological systems that can subsequently be explored empirically. Therefore, modeling represents a major component of scientific endeavors. Any advancement in the biological sciences thus depends on different kinds of modeling, whether they are used tacitly or strategically, and computation has become central to all these procedures. The present volume investigates the many ways in which new modeling strategies in the biosciences assist and influence our understanding of biological phenomena.

This book grew out of a workshop titled "Biological Modeling: Structures, Behaviors, Evolution," which was organized by the KLI at Altenberg, Austria. The workshop brought together scientists from fields ranging across mathematics, physics, neurobiology, developmental biology, paleobotany, behavior, and theoretical biology. Three days of discussion reinforced the initial view that biological modeling will play an ever increasing role in the analysis and understanding of the properties and dynamics of biological systems, and therefore will be an important part of the resulting biological theories. Additional authors in fields of expertise that were not represented at the workshop kindly provided additional perspectives included in this volume.

The book is structured into parts according to the areas in which we feel the major advancements are taking place. We begin with the analysis of important conceptual issues regarding biological modeling in general, and then provide introductions to a number of specific approaches in the organismal domains of morphology, development, and behavior. We conclude with a part that addresses the modeling of these properties in the context of evolution, with a particular emphasis on the emerging field of EvoDevo. While in both the workshop and the volume we had to be selective, and even to a degree idiosyncratic, the contributions nevertheless amply demonstrate how modeling mediates in an increasingly critical fashion between theory and empirical research in a wide range of biological fields.

We thank the workshop participants and the additional contributors to the volume for their enthusiasm for theoretical biology that appears throughout these chapters. We also thank the staff of the KLI for their continued efforts in organizing high-quality workshops that promote a creative exchange of ideas in the conducive environment of the Lorenz mansion in Altenberg, Austria. And, as before, we thank the motivated people at the MIT Press for their continued support of the Vienna Series.

Manfred D. Laubichler
Gerd B. Müller

I INTRODUCTION

1 Models in Theoretical Biology

Manfred D. Laubichler and Gerd B. Müller

All branches of the biological sciences rely on models. This simple observation comes as no surprise to anyone who has either participated in or thought about biological research. Experimental studies investigate model systems or model organisms that are carefully selected for a particular research problem; these experimental data are assembled and incorporated into computational, diagrammatic, or conceptual models that are used to organize and analyze these results; and even abstract formal theories—those elusive targets of scientific research—come to life in the form of formal and predictive models. There are thus many different types of models within the biological sciences that fulfill a variety of important practical, conceptual, and epistemological functions. The literature on this topic is extensive; see, for instance, chapters 2 and 3 in this volume, and Morgan and Morrison (1999), de Chadarevian and Hopwood (2004); Hacking (1983), Giere (1988), Cartwright (1983), Wimsatt (1987), Krohs and Callebaut (2007), and Wolkenhauer and Ullah (2007).

Among the reasons why models are so important within the biological sciences are the challenges posed by the rapidly growing amount of experimental results and the increasing complexity of the phenomena under investigation. Conceptual abstractions and mathematical modeling have always been part of the interpretation and integration of these accumulated data. Today, due to the improvement of computational methods, models of biological phenomena have reached, in many ways, a new dimension. It is now possible to model and simulate complex biological structures and processes in hitherto unknown detail. At the same time, bottom-up models, such as those adapted from the physical sciences, elucidate the properties of natural biological systems. Therefore, as can be seen in all chapters of this volume, modeling and simulation have been the key to most major advances in the biological sciences in recent decades.

The complex dynamics of biological processes at all levels of organization represent one of the key issues of modeling biology. These dynamical processes can be best understood in terms of a set of relations between individual entities and their environment, which often includes other individual entities. For example, while a specific cell inside an organism interacts with its neighbor cells as well as with the extracellular medium, and specific organs interact with the rest of the body, the organism itself interacts with the surrounding environment. All these processes, taking place along several spatial and temporal scales, are essential for defining the reproduction and survival not only of specific individual organisms, but also of their species as a whole. And because the environment itself is affected by such interactions, important feedback loops that ultimately define the biological world are established. In addition, a better understanding of these interactions is essential for most human concerns, ranging from the predictability and treatment of diseases to the proper management of the environment.

The study of form, both organismal and molecular, is another important area in which modeling approaches have become central. Mainstream research during the last decades has focused largely on molecular, biochemical, and environmental issues, which has ultimately contributed to the consolidation of genetics, molecular biology, evolutionary biology, ecology, and animal behavior as well-established fields. At the same time, only limited attention has been paid to the important problem of biological form (i.e., the characterization and analysis of the geometrical and topological properties of biological entities). Together with the respective biochemical features, the morphological properties of any biological entity represent an important element in the dynamic interactions between individuals and their environment. The shape of a specific cell, for instance, constrains and is constrained by the morphology of the neighboring cells.

At a more macroscopic level, the density of branching structures, such as blood vessels or trees, determines the efficiency of material and energy exchanges with the environment. And at the organismal scale, the shape of the avian beak, for example, is critical for access to sources of food. Thus, generally speaking, the morphology of biological entities mediates in a critical fashion a large number of interactions between the individual and the environment. Recent breakthroughs in modeling biological forms (several discussed in this volume) have provided important insights into these crucial relationships between form and function.

Understanding the structural properties of biological forms is one important dimension, but it is not complete without considering the fact that all biological systems develop from less complex precursors. Development is thus another central aspect of modeling biological systems. The essential features of this temporal, and in a sense also historical, dimension of biological systems can be understood only through models, since any detailed reconstruction would face the Borges dilemma in biology—the map of the system would soon be as complex as the system itself, and therefore would be of no use (Laubichler and Pyne, 2006). Models that are based on the right kind of abstractions, both in the form of the relevant parts of systems as well as of their causal interactions, are thus the only way we can arrive at an understanding of development. These models combine the structural features of models of form with those interactive aspects that are characteristic of dynamical models. Like models of form, models of development also combine biological and physical properties of complex organismal systems.

Organisms not only develop, they also interact with each other and with their environment. In studying these forms of behaviors, researchers have always relied on models that enabled them to go beyond mere descriptions and to develop causal hypotheses about the triggers and purposes of observed behaviors. Behavior has been modeled in the context of physiology and even of mechanics—here the focus was on understanding the triggers and stimuli of behaviors, of their evolution—with an emphasis on the adaptive value of behavior, and of economic theory—that brought a perspective of strategies, gain, and "rational decision-making" to explanations of behavior. More recently a number of integrative approaches introduced robots into the study of behavior. This no longer represents merely a new conceptual approach or a new methodological tool; rather, these studies create a physical-technical model system with the goal to embody rather than to describe behavior.

Mathematical models have a long tradition in evolutionary biology. They are also one of the prime examples of successful conceptual abstractions within the biological sciences. These abstractions—from the concept of a population to that of a replicator—also have enabled a more generalized mathematical theory of evolving systems captured, for instance, by the replicator or the Price equations. The fundamental problem of modeling phenotypic evolution is how to combine these abstract dynamical models of population dynamics with concrete properties of organisms as developing systems that interact with each other and their environment. This will require a new class of models that integrate biological processes at different scales. In order to accomplish this goal, a new set of models covering the whole spectrum from experimental model systems and new kinds of model organisms to conceptual and mathematical models are currently being developed. The details of these resulting models differ, depending on whether they focus on morphology (part III), development (part IV), behavior (part V), or evolution (part VI), but they all span different scales and levels of complexity.

Recently philosophers and historians of science concerned with actual scientific practice and knowledge production have focused their attention on models and modeling strategies (e.g., Hacking, 1983; Giere, 1988; Morgan and Morrison, 1999). This literature has already yielded a more complex understanding of how knowledge has been generated in particular experimental contexts, and it has contributed to a more detailed and less schematic philosophical conception of the structure of scientific theories and explanations. This new focus on conceptual issues in the biological sciences, triggered to no small degree by the emphasis on models, is indeed a profound addition to theoretical biology at large. The chapters in part II of this volume discuss models and modeling strategies from the perspective of philosophy, history, and applied mathematics. These chapters also provide a critical evaluation of the theoretical literature on modeling.

Models in Theoretical Biology

Today theoretical approaches are gaining in prominence throughout the biological sciences, contributing to what can, without exaggeration, be called a renaissance of theoretical biology. Models and modeling strategies are an important part of this trend. Indeed, in many instances, developments in theory are a direct outgrowth of modeling efforts. However, the relationship between models and theories, and between modeling and theoretical biology, are complex and manifold. Following Laubichler et al. (2005), we distinguish at least four different, yet interrelated, areas of theoretical biology in the postgenomic age. Different types of modeling and modeling strategies define the distinctions between these areas.

The first dimension of theoretical biology reflects the need to analyze, organize, and manage large amounts of data. This is the domain of bioinformatics and of certain areas of computational biology, which deal with the substantial challenges of data analysis and representation. The problem here is not only that biology is currently experiencing a phase of exponential growth in the amount of data available; there are also different kinds of data that need to be connected (integrating sequence information and medical histories, for instance). Consequently, data mining is nowadays a desired skill for theoretical and experimental biologists alike. But many of the bioinformatics tools, which the average user simply wants to apply to analyze data, actually represent heuristic and theoretical models of the underlying complexities of biological systems.

Even though one can look at the computational and mathematical tools developed in bioinformatics as the microscope of the twenty-first century, allowing us to "see" new connections and structures (Laubichler et al., 2005), these nevertheless embody more conceptual and theory-driven dimensions of biological research and are in fact models that represent certain aspects of biological systems. A DNA chip and the computational tools that enable us to analyze molecular data are, in a way, a model of the transcriptional state of a cell or a tissue. These and other aspects of merging new data-rich fields with traditional modeling strategies are particularly relevant in the emerging field of systems biology (Krohs and Callebaut, 2007).

A second area of theoretical biology—for many, the primary aspect of theoretical biology—involves mathematical and computational model-building and analysis (see also Levins, 1984). Over the last decades, simulation and modeling approaches have increased dramatically. Today these routinely include standard systems of differential equations, agent-based models, and spatial models. All this has been made possible by the substantial increase in computational capacity. While model-building and simulation have become an integral part of most areas of biology, they also remain a core part of theoretical biology proper.

In our view, the main emphasis of *theoretical* biology is not so much on creating an accurate mathematical representation of a particular event, but rather on gaining insights of a more general nature by means of abstract representations of biological processes. These representations are, of course, a form of model of the phenomenon in question. Only from the perspective of such a model is it possible to inquire to what extent different phenomena can be understood as instances of similar underlying processes or dynamics. This is a primary goal of theoretical biology.

A focus on concept formation and conceptual analysis constitutes the third area of theoretical biology. While this is an important and sometimes overlooked part of biology, it is also related to many ongoing efforts within the philosophy of biology. Conceptual analysis is often closely linked to mathematical analysis, including several mathematical modeling strategies—even more so as many theoretical ideas in biology become expressed in formal terms. However, conceptual analysis is by no means restricted to mathematical models. As several examples in this volume demonstrate, it is precisely through the application of a *concept*, such as constraint or optimization, to a wide range of phenomena that new insights will be gained.

Darwin's own work, for instance, contains several examples of this kind. Probably the best-known one is the conceptual model provided by artificial selection, which he knew from firsthand experience and then employed it to infer processes in natural populations. We submit—and the chapters in this volume support this claim—that as more data and more powerful mathematical and computational tools become available, the importance of this sort of conceptual biology to aid in the discovery of fundamental similarities between different systems and processes is likely to increase.

The fourth area of theoretical biology involves theory integration. This function arises from the fact that formerly distinctly separate disciplines and research domains have begun to merge, both methodologically and conceptually. Theory integration requires different types of models, most prominently heuristic and conceptual models. The goal is to arrive at formal and theoretical models that capture the essential features of a wide range of biological phenomena. Theory integration is often connected to the idea of a "synthesis." One prominent example of this kind of integration is evolutionary developmental biology (EvoDevo), which combines several fields of investigation in a practical and theoretical sense, leading to new research strategies and agendas (Müller, 2005, 2007). As the discussion in chapter 16 of this volume shows, achieving such a synthesis depends in crucial ways on the availability of the right kinds of models, both conceptual and material. It is all the more important today because the life sciences are in a phase of rapid expansion and ever increasing specialization, while at the same time biological theories and concepts have become increasingly more widespread outside the traditional boundaries of the life sciences (from evolutionary psychology to artificial life).

In summary, in our conception, theoretical biology is both foundational and practical, mathematical and conceptual, and always in a close partnership with empirical research. Furthermore, all areas of theoretical biology involve different kinds of models and, indeed, represent different types of modeling strategies within biology. Theoretical biologists are engaged in all areas of biological research: managing and analyzing data, developing new strategies for representing and visualizing data, building mathematical models and developing simulations that are designed to capture essential and general aspects of biological processes, as well as fine-tuning these models to allow for precise predictions; formulating concepts that adequately represent the underlying biological phenomena and their associated mathematical representations; and, finally, contributing to the theoretical integration of the life sciences. This complexity and diversity of theoretical biology is matched by the diversity and complexity of the attempts to model biology.

Classes of Biological Modeling

Models have many different functions within biology and even, as we have seen, within theoretical biology. These functions are reflected in a variety of attempts to classify and analyze different types of models and modeling strategies (e.g., Hacking, 1983; Levins, 1984; Wimsatt, 1987; Giere, 1988; Morgan and Morrison, 1999). Each of these schemes is guided by a particular question or philosophical approach. Here we provide a different sort of classification, one that is derived from an analysis of the various and varied uses of models within theoretical biology as represented in this volume (table 1.1). Our classification is thus a preliminary summary of modeling strategies in biology derived from what practitioners in different fields emphasize as being central to their efforts. But we would argue that while it certainly is not complete or the only way of classifying models, table 1.1 is a good representation of bio-

Table 1.1
Classes of biological modeling

Material	Physical	objects, robots, model organisms	Chapters 3, 11, 13
	Computational	2-, 3-, 4-D virtual models and simulations	Chapters 6, 7, 9, 10, 14, 15
Theoretical	Diagrammatic	functional, organizational	Chapters 4, 5, 7, 8, 13, 15
	Mathematical	formal, analytical, predictive	Chapters 4, 5, 6, 8, 9, 10, 12, 14
Heuristic	Conceptual	interpretive, heuristic	Chapters 3, 4, 5, 11, 12, 14, 16

logical modeling in general. It is also ambiguous in the sense that our descriptions are more centers of emphasis than categorical distinctions, which is a more adequate reflection of the varied uses of models within biology.

We distinguish primarily between different reference frames and functional roles of models in biology. Models can be *material* (i.e., concrete physical models, such as model organisms and robots, or virtual reconstructions and simulations of concrete situations (such as the virtual 3- and 4-D reconstruction of a developing embryo or the simulation of the population dynamics of a given species). Models can be *theoretical*, which implies a higher degree of abstraction and/or generality, and includes diagrammatic models of the functional and organizational properties of organisms and organ systems (such as the nervous system or the vasculature, cells, and intracellular structures), as well as abstract representations of relational properties of biological systems (such as network models and graphical representations), and analytical and predictive models of dynamical systems (such as models of natural selection). And models can be *heuristic*, which tends to be a conceptual representation of a particular problem. These conceptual models are often the first descriptions of a research problem and stand at the beginning of investigative pathways (then they serve mainly as heuristic tools), but they also complete a research program when problems are really understood and can be summarized within a conceptual interpretative framework (then often including other types of models, such as predictive formal models, simulations, or representative models).

The function of models within biology is thus dynamic rather than static. At all stages of biological research, models play an important role, whether explicitly or implicitly. Often this means that we see a progression from heuristic to material and theoretical models of a particular problem. But just as often it means that one particular model is transformed in light of ongoing research, and different dimensions (heuristic, theoretical, etc.) gain in prominence. The success of research programs thus depends on the right kind of modeling strategies. In that sense almost all research in biology amounts to "modeling biology."

References

Cartwright N (1983) How the Laws of Physics Lie. New York: Oxford University Press.

De Chadarevian S, Hopwood N (eds) (2004) Models: The Third Dimension of Science. Stanford, Calif.: Stanford University Press.

Giere R (1988) Explaining Science: A Cognitive Approach. Chicago: University of Chicago Press.

Hacking I (1983) Representing and Intervening: Introductory Topics in the Philosophy of Science. Cambridge: Cambridge University Press.

Krohs U, Callebaut W (2007) Data without models merging with models without data. In: Systems Biology: Philosophical Foundations (Boogerd FC, Bruggeman FJ, Hofmeyer J-HS, Westerhoff HV, eds), 181–213. Amsterdam, Reed-Elsevier.

Laubichler MD, Hammerstein P, Hagen E (2005) The concept of strategies and John Maynard Smith's influence on theoretical biology. Biol Philos 20: 1041–1050.

Laubichler MD, Pyne L (2006) The Borges challenge in biology: Review of Kenneth M. Weiss and Anne V. Buchanan, *Genetics and the Logic of Evolution*. BioEssays 28: 768–769.

Levins R (1984) The strategy of model building in population biology. In: Conceptual Issues in Evolutionary Biology (Sober E, ed), 18–27. Cambridge, Mass.: MIT Press.

Morgan MS, Morrison M (eds) (1999) Models as Mediators: Perspectives on Natural and Social Science. Cambridge: Cambridge University Press.

Müller GB (2007) Six memos for EvoDevo. In: From Embryology to EvoDevo: A History of Embryology in the 20th Century (Laubichler MD, Maienschein J, eds), 499–524. Cambridge, Mass.: MIT Press.

Müller GB (2005) Evolutionary developmental biology. In: Handbook of Evolution, vol. 2 (Wuketits FM, Ayala FJ, eds), 87–115. San Diego: Wiley.

Wimsatt WC (1987) False models as means to truer theories. In: Neural Models in Biology (Nitecki MH, Hoffman A, eds), 23–55. New York: Oxford University Press.

Wolkenhauer O, Ullah M (2007) All models are wrong. In: Systems Biology: Philosophical Foundations (Boogerd FC, Bruggeman FJ, Hofmeyer J-HS, Westerhoff HV, eds). Amsterdam: Reed-Elsevier.

II CONCEPTUAL ISSUES

Modeling has always been central in biological research. This fact raises several fundamental questions about the role of models in scientific explanations, about the details of the historical development of the biological sciences, and about adequate methods for biological research. As we have seen in chapter 1, models occupy a variety of niches within the research enterprise of the biological sciences. Some of these are conceptual; some, material; some, theoretical and mathematical. The dependence on models also brings into focus the core epistemic values and assumptions of biological research. Researchers have more choice when selecting a model for a particular research problem than in making simple observations or measurements of the parameters of a biological process. Compared with the freedom associated with the selection of models, even experimental interventions are rather constrained by the specific details of the phenomenon in question. The selection of one particular type of model over another therefore reveals a good deal about the core values of the scientific community at a given time and place. Thus, emphasizing the role of models in the history of biological research contributes to a "philosophy of epistemological detail," which aims at a careful analysis of the scientific process.

The choice between different types of models also highlights one of the fundamental predicaments of the scientific representation of natural phenomena. As Richard Levins noticed, it is impossible to simultaneously maximize generality, tractability, and realism within a single manageable model. One consequence of Levins's analysis is that in many instances more than one model is needed in order to arrive at an intelligible theory. This characteristic feature of biological theories—that they are composites of multiple models, often of different types—raises important philosophical questions about the nature of explanation within biology.

Mathematical and computer-based models have become increasingly widespread throughout the life sciences. The possibility to implement iterative rules has enabled the simulation of a variety of complex natural processes. But as computations have become more complex, interpreting the results of these simulations has challenged some of the core assumptions about what it means to explain a biological phenomenon. A detailed mechanistic understanding is, in many cases, no longer possible. Rather, biologists increasingly rely on strategies similar to those of physicists, who have developed a sophisticated framework of statistical mechanics (as well as quantum mechanics) that abstracts from individual interactions and describes the relations between aggregate variables. Similarly, abstract mathematical methods, such as those developed in graph theory, set theory, or topology, also allow for general descriptions of a wide range of phenomena. These new techniques amount to a major conceptual shift in biological modeling.

Sabina Leonelli (chapter 2) provides a philosophical perspective on modeling strategies within biology. She takes the diversity of existing modeling practices as her starting point and argues that such a plurality is indeed necessary in order to accomplish the many different epistemic goals associated with the scientific enterprise. She takes issue with earlier normative claims in the philosophy of science that have attempted to restrict the function of models within the context of scientific explanation. In contrast, Leonelli advocates a systematic study of the many different ways in which scientists pursue theoretical knowledge, often, if not most of the time, employing one or more specific types of model. She argues that depending on the particular epistemic values of scientists, preference will be given to distinct modeling strategies.

Leonelli's own conception of scientific theories is based on the epistemic function of intelligibility. Within this framework she distinguishes between the different roles of theoretical and material models, and argues that both are necessary elements of intelligible scientific theories. Theoretical models, which are manipulated conceptually, are tools for increasing the explanatory power of theories, while material models, which can be largely independent from a specific theoretical account, are powerful tools not only for assessing empirical values of parameters but also for discovering new phenomena.

Hans-Jörg Rheinberger (chapter 3) introduces a "philosophy of epistemological detail" as an epistemological framework that more adequately captures actual scientific activities than do traditional philosophical accounts that focus on the logical analysis of theories and the validity of their claims. Taking the actual research practices of molecular biology as his starting point, Rheinberger observes that scientific progress is mostly a consequence of researchers transforming existing experimental systems, and their material and conceptual models, from within by reaching and finally overcoming their limits, rather than a theory-driven, logical, and planned pursuit. He calls for a detailed historical reconstruction of these processes, one that emphasizes the contingencies and patchwork nature of scientific activities.

Models are a central part of such a philosophy of epistemological detail. Experimental systems often combine several types of models, and they themselves can, in turn, be seen as material and heuristic models. Rheinberger also argues that concepts, if they are seen not as static linguistic terms but as a form of conceptual model, allow for a more adequate reconstruction of the historical development of science. Historical epistemology, as advocated by Rheinberger in his chapter, thus is also a way to understand the role of models in biology.

Luciano da Fontoura Costa (chapter 4) proposes that modeling strategies based on complex network analysis can become the basis of systems biology. Complex networks represent an elegant and economical way to describe, analyze, and model relations and interactions between parts, one of the fundamental properties of all biological systems. Costa gives a systematic introduction to the graph theoretical foundations of network analysis and also discusses the links between network analysis and statistical mechanics. These connections are particularly relevant because complex systems and networks share many properties with thermodynamic systems, such as the existence of phase transitions and percolations. In his presentation, Costa pays special attention to the small-world properties and the consequences of scale-free relationships in complex biological networks. He also emphasizes that the representational flexibility of these models greatly expands the range of applicability of network analysis within biology.

2 What Is in a Model? Combining Theoretical and Material Models to Develop Intelligible Theories

Sabina Leonelli

Both biologists and philosophers have widely acknowledged models to be crucial epistemological components of scientific reasoning and experimentation. This volume represents yet another instance of such recognition, while at the same time illustrating an important characteristic of modeling practices: their diversity. Modeling strategies in biology are extremely varied in both the form that they take (e.g., diagrammatic, computational, robotic models) and in the way they bridge between theories and data. This chapter contends that such variation not only is irreducible to an all-encompassing, unique account of what a model is or should be, but also that it is actually necessary to the pursuit of the several epistemic goals of interest to practicing biologists. This argument is intended as a contribution to a pluralistic account of modeling strategies, the development of which complements what I call "single-model approaches" to the subject. As an instance of the fruitfulness of a multimodel approach, I consider the differences between what I call *theoretical* and *material* models, and I argue that the complementary use of both types of models is crucial to achieving *intelligible* theories about biological phenomena.

As carefully spelled out by Schank and Koehnle (chapter 11 in this volume), research in biology involves the manipulation of several types of models, each of which plays a different role in the achievement of scientific understanding. How to assess the epistemological significance of such diversity? Perhaps unsurprisingly, many of the several philosophers who have been struggling with this question have privileged a focus on generality over attention to diversity. Especially between the 1960s and the 1990s, a good part of the flourishing philosophical debate on scientific models concentrated on establishing which features, if any, are *common to* all models. This choice is understandable, given the need for at least a minimal definition of model that would make it possible to distinguish it from other elements involved in scientific experimentation and reasoning (and thus clarify its distinctive epistemological role). And indeed, it generated a vast body of literature fruitfully analyzing the modalities by which models abstract, represent, illustrate, and/or explain the aspects of reality and/or the scientific theories that they embody.

However, this focus also took attention away from the glaring differences among models and among the types of theoretical knowledge produced and the epistemic goals achieved by their use. In much the same way, the importance, to practicing scientists, of using several types of models in combination did not receive much attention. The attempt to analyze the diversity among models turned into an effort to explain it away—either by arguing that only certain kinds of models should be characterized as such[1] or by elaborating a unifying description of the processes of abstraction, representation, and explanation characterizing modeling activities.[2]

I shall refer to the tendency to explain away, rather than value and analyze, the diversity among models as the "single-model approach." In the first section of this chapter, I spell out some reasons why this approach provides a necessary but insufficient basis for analyzing the epistemic utility of modeling practices within any given research context. I then argue for the development of a pluralistic account of modeling, which focuses not so much on a general definition of what models are as on an analysis of their epistemological functions. Such analysis draws explicit connections between the different features of models and their role in securing one or more of the several epistemic goals of potential interest to practicing scientists.

In order to investigate what is in *one* model, we need to recognize and explore how and why biologists choose specific *combinations of* them in order to pursue a desired research outcome.[3] As an illustration of this, the bulk of the chapter applies this pluralistic approach to the study of one prominent epistemic goal, the achievement of intelligible research outcomes. Specifically, I argue that an *intelligible theory* about a biological phenomenon can be developed only via the combined use of what I call "theoretical" and "material" models.

After exploring the epistemological differences between these two broad classes of models, I claim that they are irreducible to a single account of modeling practice. While theoretical models are powerful tools for increasing the explanatory power (and predictive value, when relevant) of a given theory, material models guarantee that the terms and symbols used within the theory secure and retain empirical content. Which is all very well, I conclude, since the need to combine these epistemological features accounts, at least in part, for biologists' need to adopt multimodeling approaches to the investigation of phenomena.

Toward a Pluralistic Approach: Refocusing on the Usefulness of Modeling

Empirical evidence overwhelmingly indicates that modeling strategies within biology are hugely diverse and, at least in practice, irreducible to each another. The contributions to this volume represent a good instance of this. We find here discussions of *mathematical models*, such as Rasskin-Gutman's genetic algorithms used to predict ciliary movements; *geometrical maps*, such as the one used by McGhee to study theoretical morphospace; two-, three-, or four-dimensional *computer simulations*, such as Bajaj's virus maps and Ascoli's study of the morphology of dendrites; *robotic models* such as Dautenhahn's human-sized robots and Schanks's robotic rat pups; and *model organisms*, both simulated and real, such as Palsson's slime molds. Each of these types of models is constructed—or even "discovered," as is arguably the case with model organisms—in different ways and for different purposes.

Further, none of those types of models is ever used on its own; all contributions to this volume underscore the need for a combination of different types of models in order to answer specific theoretical questions.[4] Consider Niklas's simulations of the adaptive walks characterizing the morphological evolution of ancient plants. The aim of his study is to reconstruct the evolutionary history of plant biomechanics as a demonstration of their increasing efficiency in facilitating the survival, growth, and reproduction of plants. To this purpose, he needs to integrate a variety of data sets generated by reference to data from the fossil record, by empirical observation of the physiology and morphology of currently existing plants, and by mathematical computation of geometrically possible morphologies and physiological constraints (which provide further clues about the physiology of ancient plants). Each of those data sets is organized, visualized, interpreted, and eventually integrated via different modeling strategies, including geometrical maps, propositional and symbolic descriptions, diagrams, equations, and three-dimensional simulations. Therefore, a major philosophical question raised by this case, as by so many others, is the following: Why such need for multiple representations and, consequently, for integration?

As the first step toward answering this question, I shall consider the views proposed by Morgan, Morrison, and their associates in their 1999 edited volume *Models as Mediators*.[5] Their analysis is congenial to my present concerns insofar as it focuses closely on the scientific practice of modeling and it takes interest in the overwhelming evidence for the diverse modeling practices used within any one research context. Making sense of the multiplicity of models and their uses is actually a major goal of the mediator's view, which is one of the reasons for the broad definition given of models themselves: anything used by practicing scientists to mediate between theory and phenomena can be called a model.

The notion of mediation is used to suggest that a model serves "both as a means to and as a source of knowledge" (Morgan and Morrison, 1999:35). The model functions as a *representative of* one or more phenomena as well as a *representative for* a given theory taken to apply to such phenomena (see also Morgan, 2003:230). In short, models constitute the meeting point between knowledge and reality, thus providing "the kind of information that allows us to intervene in the world" (Morgan and Morrison, 1999:23). By representation, Morgan, Morrison, et al. do not necessarily denote some kind of mirroring relation or structural (isomorphic) similarity between what is represented and the representation itself.[5] Rather, they imply "a kind of rendering—a partial representation that either abstracts from, or translates into another form, the real nature of the system or theory, or one that is capable of embodying only a portion of a system" (Morgan and Morrison, 1999:27).

This account is very permissive, encompassing a potentially enormous diversity of types of representations (including all the ones listed above). Indeed, I see this breadth of scope as one of its main strengths.[6] While useful in framing the analysis of modeling practices, however, the mediator approach does not venture into a systematic study of variability among models. This is arguably because it overlooks a crucial feature of the research context in which models are employed: the type of research outcome, or epistemic goal, envisaged by the investigators. Scientists make specific choices about what they wish to achieve within any given research project. Such choices inform the whole research process, including the selection and use of theoretical assumptions, descriptions of phenomena, and modeling strategies. Further, extended philosophical reflection on this matter has made it clear that (1) there are several possible outcomes to be obtained via scientific research and (2) some of those outcomes are incompatible with each other.

Point (1) is exemplified by the suggestion that both theoretical and nontheoretical types of knowledge constitute typical targets for biological research. Investigation often aims at improving control over phenomena by developing new ways of intervening in the world.[7] The goal of this type of knowledge can be referred to as "knowing how" (Polanyi, 1998:56). In this context, the relevant research outcomes are, for example, the development of procedures, protocols, and instrumentation that allow scientists to modify the entities and processes of interest (where the usefulness of specific modifications might depend largely on the agenda of funding agencies interested in the applications of such knowledge—as is often the case, for example, with contemporary research on functional genomics).[8]

On the other hand, a lot of biological knowledge aims at "knowing what," that is, at acquiring knowledge of (what we regard as) facts, processes, explanations, concepts—in other words, knowledge about nature itself. This is what I would call theoretical knowledge, a broad characterization of "theory" that encompasses a wide variety of expressions. We find biologists pursuing theoretical knowledge in the form of, among others:

- Propositionally or diagrammatically expressed *explanations* providing mechanistic, structural, functional, causal, phylogenetic, and/or narrative information (as illustrated by Machamer et al., 2000; Salmon, 1998)
- Ensembles of propositions that are *intelligible* to a given audience (this idea will be expanded upon in the second part of this chapter)

• Lawlike generalizations (where "lawlike" includes accounts as varied as Dretske's [1977] universally valid generalizations and Cartwright's [1999] nomological machines)

• Mathematical formulas

• Networks of concepts organized into some type of part-whole relationships (as instantiated in gene ontologies[9])

• Collections of models and their robust consequences, as coordinated by a theoretical perspective (as advocated by Levins, 1984; Griesemer, 2000).

These different ways to conceive of and express theoretical knowledge are all alive within scientific research, some of them more popular in specific disciplines than others.[10] In fact, especially within the biological sciences, scientists are not necessarily divided as to which of these types of theoretical knowledge is the most informative or correct. More often than not, it is acknowledged that these types of knowledge, as well as knowledge aimed at intervention, can all have a *relative significance* for the understanding of a given phenomenon (Beatty, 1997:S433). Most important for my purposes, each of these forms of theory constitutes a different type of research outcome and thus might require, in order to be achieved, a different set of modeling practices.[11] Hence the necessity of a framework in which the study of the diversity and integration of modeling practices is combined with a focus on the research outcome that each model system is supposed to yield.

Now, what about pluralism? Nothing in my considerations so far implies that we should not strive for a universally applicable account of how biologists model phenomena. The crucial step away from single-model approaches follows from an elaboration of observation (2) above, according to which the knowledge contained within these various theoretical outcomes is largely nonoverlapping. Each way of theorizing implies a different perspective, a different way of carving nature at its joints. An obligatory passage point in dealing with this point remains Richard Levins's work on the multiplicity of epistemic virtues that any one model can possess. Levins characterizes such virtues as features that a model is expected to maximize in order to accomplish its mediating function in the most successful way. These features range from the requirement of generality (denoting the range of phenomena to which the model can be applied) to the ideas of tractability (the ease with which it can be computationally or experimentally manipulated) and realism (the degree of empirical accuracy with which it represents relevant elements of the phenomenon under investigation).

Levins's important observation is that no single model can posses all virtues at once. Emphasis on one of them is possible only at the expense of some other. For instance, all concessions to the realism of a model unavoidably detract from its tractability or generality. Trade-offs among the epistemological merits of models are therefore unavoidable; scientists need to employ a "multi-modelling approach" involving the integration of several different models, each of which maximizes at least some of those dimensions.[12] This allows biologists to optimize their overall modeling strategy at any given moment by constantly shifting and calibrating priorities from one set of requirements to another, depending on the changing demands of their research context (in turn dictated, at least in part, by the pursuit of the above-mentioned research outcomes). As Levins concludes, "It is desirable, of course, to work with manageable models which maximise generality, realism and precision toward the overlapping but not identical goals of understanding, predicting and modifying nature. But this cannot be done" (1984:19).

Coming back to the present discussion, this implies that the types of theoretical outcomes listed above differ in, among other things, the epistemic requirements they impose upon research practices. A cursory second look through the list confirms this intuition. While the development of a mechanical explanation of a particular process might sacrifice generality to realism (to use Levins's own selection of criteria), the pursuit of a mathematical formalization of the same process, or cluster of processes, will in most cases privilege generality over realism. Similarly, attempts to build models for interdisciplinary research will favor tractability over realism and precision.

I thus conclude that the initial choice of an epistemic goal, as well as the heuristic role that such a choice plays throughout the process of investigation, crucially influence both the way in which models are manipulated and the assumptions about what models are supposed to represent. Given the multiplicity of epistemic goals of potential interest to scientists, a clear analytic distinction between the notions of model, theory, and phenomenon can emerge only from an assessment of the modalities and the extent to which the use of models facilitates the achievement of a specific type of outcome.[13] Such an assessment provides insights into the epistemological utility of models. I also think that a renewed focus on the utility of models can help us make sense of their variability, as well as of the need manifested by practicing scientists to use more than one model in order to tackle a specific query.

In her analysis of the multiplicity of modeling strategies characterizing the history of experimental biology, Evelyn Fox Keller points to similar conclusions:

> To be sure, a model is expected to bear some resemblance to that which is being modeled, but in science as in art, the degree of resemblance is generally understood to be a matter of perspective. The more critical question is whether it is a "good" model, and in both science and art the measure of how good a model is varies notoriously. If it is possible at all to make any generalisation, one might say that it is here, in the criteria brought to the measure of "goodness", that models in science and art most clearly depart. For the value of a scientific model is judged, first and foremost, by its *utility*... There is a lot of hedging here, but—once one recognises the enormous variability in the meaning of scientifically productive—*necessarily so* (2001:47; my emphasis)

My intention here is to pursue this suggestion in a way that Keller does not explicitly propose, that is, by articulating a pluralistic framework for reflecting upon the epistemological role of models in science. Such a framework could complement single-model approaches in two ways:

a. By taking explicitly into account the relation between modeling practice and its goals within its research context

b. By building a taxonomy of types of models based on their differing ability to facilitate the achievement of one or more epistemic goals.

Recommendation (b) is particularly relevant, given that philosophers have hitherto paid very little heed to it. The development of utility-based taxonomies of models would provide us with a general analytic framework for the study of models, without encountering the problems afflicting single-model approaches. Moreover, it would throw further light not only on the diversity among models but also on the reasons for scientists to use more than one type of model within any one research setting. Once we acknowledge that we can build a typology of models based on their differing epistemic functions (rather than on their structural properties and/or the properties of the theory/phenomenon that they represent), it becomes easier to see how the choice of a specific combination of models reinforces investigators' commitment to a specific research outcome (as emphasized by Griesemer[14]), and allows them to pursue more than one epistemic goal within the same research context (as exemplified by Levins).

An in-depth study of how various combinations of modeling dimensions might facilitate or prevent the development of different theoretical outcomes is far beyond the scope of this chapter. What I have tried to point out within this section is the feasibility—indeed, the necessity—of such study as part of a pluralistic account of scientific modeling. In the next section, I provide a sample of such an approach that focuses on a specific (and arguably popular) epistemic goal: the struggle to develop *intelligible* theories about biological phenomena. I shall examine a way to distinguish among types of models that is based on their relevance to securing this goal.

Modeling to Develop Intelligible Theories

Let me start this second part of my discussion by qualifying what I mean by intelligible theories and how I think they can be developed. I rely on the general framework provided by Henk de Regt,[15] which broadly defines intelligibility as the epistemic feature of theories that guarantees their relevance to a scientific understanding of the world. De Regt emphasizes the strong heuristic role played by understanding in science: the vast majority of scientists strive to increase our scientific understanding of natural phenomena. Next, de Regt stipulates that in order to obtain such understanding, scientists need to develop intelligible theories. This position is summarized by his Criterion for the Understanding of Phenomena (CUP): "A *phenomenon P is understood iff a theory T of P exists that is* *intelligible* *(and meets the usual logical and empirical requirements)* (de Regt and Dieks, 2005).

Within this framework, de Regt elaborates what I regard as his most significant claim. He argues that intelligibility is a context-dependent feature. It depends on the content and applicability of the theory to which it applies (that is, on its epistemic virtues) *as much as* it depends on the skills, background, and commitments of the scientist employing that theory. Hence, not all scientific theories are intelligible to everyone at every given time; researchers need to acquire specific skills and background knowledge in order to use a theory to gain understanding of a given phenomenon. From a philosophical perspective, this implies that an analysis of scientific understanding needs to focus on the context in which scientific theories are produced and applied: information about *who* understands, *what* is understood, and *how* it is understood becomes crucial to the assessment of the epistemic value of a theory.[16]

My present discussion upholds the idea that developing an understanding of phenomena is of major interest to practicing scientists, as well as the intuition that both the epistemic virtues of theories and the specific skills of scientists are relevant to its investigation. Indeed, within this section I abide by the CUP by focusing strictly on what I call theoretical knowledge, or knowledge aimed at "knowing what." Before further elaboration, however, I should note two important differences between my intuitions and de Regt's account of the role of theory.

The first is a matter of definition: the scope of my notion of theoretical knowledge is potentially much broader than what the definition of De Regt and Dieks allows for, or so it may seem since they do not provide a clear specification of what they mean by "usual logical and empirical requirements." The second difference concerns the status of theory as a necessary requirement for understanding. My remarks on the importance of "knowing how" in biology make this position very difficult to maintain. Theories are certainly an important component in the acquisition of understanding, but they need not be a necessary one. Knowledge about how to modify and control specific aspects of scientific phenomena also adds to the scientific understanding of those phenomena. My present concern with intelligible theories as an important epistemic goal in biology therefore represents just a subset of the many ways in which scientists understand the world.

Given these qualifications, let us now zoom in on "knowing what" claims and examine more closely what it means for a biological theory to be intelligible. I want to propose that in order to provide understanding about a phenomenon, a theory needs to satisfy two main requirements.[17] First of all, it must possess empirical content. This means that it should be possible to trace an explicit relation between at least some terms within the theory and some aspects of the phenomenon to which it applies. The second requirement for an intelligible theory is that it should have explanatory power. The theory should trace a narrative that clarifies why focus on its terms accounts for a specific feature of the phenomenon—or, as is often the case in biology, for its very existence. This narrative can take the form of any available type of explanation, conceptualization, or perspective; what matters for it to be explanatory is that it specifies how the relationships among its components account for the behavior, structure, or existence of the entities or processes to which the theory applies.

Conceptual and Material Manipulation

So how can a theory that has both empirical content and explanatory power be obtained? An obvious answer, in the light of Morgan and Morrison's account, is that, like all theories, such a theory is developed by using models. The focus of this section will therefore be on the kinds of modeling practices that can help to secure a potentially intelligible theory. For a start, let us turn briefly to Morgan and Morrison in order to examine another relevant component of their view. This is the recognition that modeling is, first and foremost, a human activity. This activity is directed at the construction and constant modification of objects that can mediate between theories and phenomena. These objects can be actual entities, such as mechanical or scale models, or mental representations, such as mathematical formulas, diagrams, or even thought experiments (where diagrams and equations also can, and often do, take the form of actual objects whenever they are reproduced on paper, computer screens, or blackboards).

In both of these cases, using models to learn something about what they represent always implies their manipulation: "Models are not passive instruments, they must be put to work, used or manipulated" (Morgan and Morrison, 1999:32; see case studies in the same volume for an illustration of this point). In my view, the notion of manipulation includes all conceivable ways of "putting [the model] to work" (Morgan and Morrison, 1999:33). I interpret the claim that models are not "passive instruments" as indicating that both the building of a model and its use within a research context involve constant attempts to modify some of its features. In short, all manipulation involves modification—and such endless bickering and transforming of the model exemplifies scientists' active engagement with it.

Remarkably, this engagement requires expertise, encompassing adequate training as well as the experience acquired via iteratively handling the appropriate instruments, procedures, and material. In other words, the successful manipulation of models requires both background knowledge and skills (as implied by de Regt's analysis). Taking this into account, I shall now distinguish between two ways of manipulating models, each of which plays a different, yet indispensable, role in fulfilling the above-mentioned conditions for the intelligibility of theories. A hint to this view was presented in the above remarks on manipulation when I suggested a difference between manipulating a model via physical interaction and manipulating it conceptually. Indeed, I see the degree to which a model can be manipulated materially—that is, by constant appeal to information provided by sense perception (including touch, smell, sight, hearing, and sometimes even taste), rather than to reasoning—as a prominent source for variation among the features of models, as well as their epistemological utility.

In particular, I want to argue that subjecting models to extensive material manipulation is usually done at the expense of conceptual manipulation. As I explain below, this is not because the two types of manipulation are incompatible. Rather, it is because models that are manipulated materially seem to have different epistemological functions with respect to models manipulated mostly conceptually—so that a shift, for instance, from material to conceptual manipulation implies a shift of epistemological focus in the use of a model. Elaboration of this claim will bring me to assert that models subjected to extensive material manipulation are especially conducive to the fulfillment of the first condition for the intelligibility of theories: securing the empirical content of the theory developed via the model. By contrast, models developed via largely conceptual manipulation help to satisfy the second condition: to enhance the explanatory power of the theory.

Before turning to a comparative analysis of these two types of manipulation and their effects on the epistemological usefulness of models, I should clarify that I take the distinction between material and conceptual to be descriptive of actual modeling practices in a very broad and yet a very limited sense. It seems intuitively correct to attempt an analytic distinction between material and conceptual manipulation. For instance, we can easily point to cases in which the material manipulation of a model is almost absent, yet that model conveys a lot of knowledge and can serve as a tool for discovery. Consider the process of thinking about a mathematical equation. Most often, this type of model is manipulated conceptually as a mental representation, in which case no physical movement on the part of the investigators is necessary in order to use and modify it as required. This case is very different from the one of a scale model, such as models of ships used in hydraulic engineering or models of riverbeds used by physical geographers, ecologists, and geologists, whose adequate use requires skillful physical interaction with three-dimensional objects.

Yet, it would be wrong to think that equations are not manipulated materially, too, or that use of scale models does not involve conceptual manipulation alongside the material. More often than not, manipulating an equation involves scribbling it down and deleting or adding terms to it; and in some cases these physical actions may yield precious epistemological insight (for instance, when playing around with symbols on a piece of paper unexpectedly leads to an illuminating new formula). Equally often, the material manipulation of a scale model is guided by the conceptual manipulation of some of its features. Given that the mixing of conceptual and material manipulation unavoidably characterizes the handling of a model, my distinction between the two types of manipulation should be read more as an analytic tool than as a descriptive statement.

What I want to stress are two basic intuitions: that just as there are differences in the degree to which a model is materially manipulated, and that there are cases in which such material manipulation is largely predominant over conceptual manipulation, so there are cases to which the inverse applies. The distinction is not meant to be rigorous, but rather to provide ground for a comparison between the epistemological role played by "theoretical models" (largely manipulated conceptually) and "material models" (largely manipulated materially) in the development of intelligible theories.

Theoretical Models

Let us examine more closely the cases in which model manipulation is largely conceptual. In those cases, the main goals of the manipulation of the model are the testing, elaboration, and/or illustration of a given theory about the phenomenon that is modeled. The goal of manipulation is, in other words, to uncover ways in which a model can be *representative for* a given theory. The choice of the parameters used within the model is thus characteristically informed by a well-defined hypothesis about the theoretical outcome that the model is supposed to illustrate, test, and/or elaborate. This is because we start from a theoretically informed "prepared description" of the phenomena under scrutiny.

The term "prepared description" was introduced by Cartwright, who defines it as "presenting the phenomenon in a way that will bring it to the theory" (1983:133). Importantly, she also argues that "the check on correctness at this stage is not how well the facts known outside the theory are represented in the theory, but only how successful the ultimate mathematical treatment will be" (Cartwright, 1983:133). Thus, the properties of the phenomena that are abstracted in order to be represented within the model are properties that are either more causally relevant or more general (less context-dependent) than others; both generality and causal relevance are assessed in the light of a given theory.

I refer to models that are thus conceptually constructed as "theoretical models," in order to indicate how strongly their use is related to the theory that they represent. The use of theoretical models is increasingly widespread among biologists. Take mathematical models, whose crucial role in the establishment of the Modern Synthesis in the 1920s and 1930s (Mayr and Provine, 1980), resulting in the birth of a discipline relying on statistical methods of analysis (population genetics), was only a prelude to their growing application across almost all biological disciplines. Even more evident is the pervasive use of simulations and algorithms to visualize empirical data, not to mention the push toward formalization and away from the laboratory brought about by the increasing use of bioinformatics to store, organize, and integrate data.

These models are especially useful for elaborating explanations or confirming predictions stemming from given hypotheses (they are what Cartwright calls *interpretative* models (1999:181). They are also fundamental to the integration of biological knowledge concerning specific phenomena (for instance, bringing together insights from physiology, molecular biology, functional genomics, and cell biology in order to understand root development in plants). However, because of their strict reliance on theory, theoretical models are not the best epistemic tools when the goal of their manipulation is to improve the empirical content of a theory (to fulfill condition [1] for the intelligibility of theories). Theoretical models give few indications as to which feature of the phenomenon under scrutiny should be considered relevant to the development of explanatory knowledge about that phenomenon. Further, a theoretical model does not help with testing the empirical (descriptive) accuracy of the relation it stipulates between theoretical terms and aspects of the phenomenon.[18]

Material Models

When thinking about the type of description needed in order to model a phenomenon, Cartwright proposes also to focus on "unprepared descriptions." These are descriptions that (1) "contain any information we think relevant, in whatever form we have available" and (2) "are chosen solely on the grounds of being empirically adequate" (1983:133). I consider the notion of unprepared description as a fruitful recognition, contra more traditional accounts of modeling that insist on referring to theoretical physics as a "role model" for all other sciences, that within many experimental sciences the testing of theories need not be the starting point of investigation. What we witness in experimental biology is the use of several types of models that are not built by relying on a given theory. Their use is unavoidably guided by background knowledge and a commitment to the investigation of specific conceptual issues. Yet, the background knowledge needed to formulate a question should not be confused with the knowledge produced by trying to answer it. In preparing a description, scientists rely on an already formed hypothesis about how to answer a given theoretical question. The manipulation of unprepared descriptions, by contrast, does not require the choice of a specific interpretation to start with. That is to say, the manipulation of models sometimes requires no more than a general interest in exploring one or more aspects of the phenomena that they are taken to represent.[19]

In fact—and here I part company with Cartwright's 1983 account and move into her revised 1999 framework—there are cases in which the unprepared description of a model is constituted by diagrams, objects, or even samples of the phenomenon. These are cases where the model is a two- or three-dimensional object that is taken to be *representative of* a set of phenomena, in the sense of being used to explore which properties of the phenomena could turn out to be relevant to a theoretical account of it. Thus the model provides the epistemic access to phenomena that is necessary in the first place, in order to infer the kind of unprepared description that Cartwright is taking about.[20] Epistemic access is granted first and foremost by material manipulation, since the amount of conceptual manipulation necessary to handle these models is minimal. The theoretical framework for which they are representative does not need to be specified in order for scientists to use them, since it will eventually be developed via their physical manipulation.

In other words, contrary to my characterization of theoretical models, these "material models" belong to a "proto-explanatory context" (Ankeny, 2001) where scientists not only have not agreed on a theoretical explanation of the phenomena under investigation, but also have not yet settled on which properties of those phenomena could be relevant to the explanation (a decision that is crucial to the building first of an unprepared, then of a prepared, description, thus enabling the shift from largely material to largely conceptual manipulation, and back.

Material models are tangible objects such as scale models,[21] samples, robots, or, most emblematically, model organisms,[22] with which scientists interact through their sense perception. As put by Hacking in a different context, they are tools "for doing, rather than thinking" (1983); the action implied by "doing" here is the material modification of characteristics of the models. A scale model or a model organism might seem far too complex an object to play a representational role, yet interacting with it—feeling, choosing, discarding, and comparing its characteristics (whether implicitly or explicitly)—often leads scientists to consider alternative explanatory frameworks, precisely because of the largely underdetermined and dynamic nature of their features (on this point, see also Magnani, 2001; Polanyi, 1998). This is possible at least in part because material models are both *constructs* and *samples* of the integrated wholes of which they are representative. As Griesemer makes clear, "Material models are able to serve certain sorts of theoretical functions more easily than abstract formal ones in virtue of their material link to the phenomena under scientific investigation.... They are robust to some changes of theoretical perspective because they are literally embodiments of phenomena" (1990:80).

It is thus the degree to which material models embody phenomena that makes them, in my view, take a substantially different epistemological role with respect to theoretical models as characterized above. Acknowledgments of the tight connection between learning and the material manipulation of objects have a long history and are increasingly accepted within current cognitive science.[23] The philosophy of modeling is also, albeit more slowly, turning to the role of physical action and sensory perception in scientific reasoning.[24] This move arguably results from a double recognition already hinted at within this chapter: that of scientific epistemology as often geared toward intervention and control, rather than "pure knowledge"; and that of experimental reasoning as yielding not only new ways of seeing, thinking of, and talking about the world, but also new ways of interacting with it.

An important implication of my characterization of material models is that they do not require the same type of abstraction processes that characterize theoretical models. Here the selection of parameters to be used in the model results from trial-and-error processes of standardization and experimentation, rather than from the exclusion of features deemed irrelevant to the research context on the basis of a given theoretical hypothesis. As John Bonner famously put it, when discussing his discovery of chemotaxis as obtained by experiments on slime molds: "I had not carefully designed an experiment that would prove diffusion; I had managed it by accident. That and all the other observations I had made told me that the slime molds were in charge, not I. They would let me know their secrets on their terms, not mine" (2002:77–78). The trial-and-error selection of features to be standardized, and thus incorporated within a model, is due both to what is possible to maintain in practice and to what is "felt" by scientists, via epistemic action, to be a promising representation of a particular class of phenomena (where what is "promising" is not necessarily determined on the basis of an explicit theoretical framework, but typically depends on some vague intuitions on which properties of a phenomenon might turn out to be theoretically relevant).

Conclusion

On the basis of the above characterizations of the differences in epistemological functions characterizing theoretical and material models, I infer the following conclusions. Material models, by virtue of their large degree of independence from a theoretical account and their strong link to the phenomena of which they are representative, are especially useful in assessing the empirical value of parameters to be used in a theory. In other words, materially manipulated models are instrumental in the fulfillment of the first condition for the achievement of intelligible theories: clarity about the empirical content of the theory eventually derived from their manipulation. In order to secure our chosen epistemic goal, however, another type of model is needed to provide explanatory power to the theory. It is not sufficient to know the properties of the phenomenon that it should refer to; we also want to know what the establishment of a relation between properties of the phenomenon and theoretical terms adds to our knowledge of the phenomenon. This is the function of theoretically manipulated models. No matter how blurred the distinction between these two types of models is, I believe that at least one of each type is needed in order to successfully conduct research leading to the development of an intelligible theory about a biological phenomenon.

As an example of how this works in practice, consider the current attempt, within large communities of biologists working on the same model organism, to gather all available data about the best-known model organisms into large, highly standardized digital databases. The goal of this type of project is to bring together the vast amount of data on a specific organism collected within several biological fields, so as to integrate those data and thus enhance our overall understanding of the biology of that organism. Examples of such enterprises are resources such as the Generic Model Organism Database (GMOD), bringing together data from different organisms but thought to have relevance for many life-forms; The Arabidopsis Information Resource (TAIR), storing data from research on the plant *Arabidopsis thaliana*; FlyBase, the database for *Drosophila*; and BioCyc, assembling information over metabolic cycles in the main model organisms.[25]

These resources construct digital models of various aspects of the biology of these organisms, including their molecular structure, genomics, metabolism, and developmental processes. The visualizations of data proposed by these databases are based on a variety of abstract concepts and theories about the phenomena to which the data refer. Only through reference to theories and concepts can the vast amount of data available on specific organisms be assembled and visualized; further, the concepts used also enhance the integrative power of those models, which bring together data collected within different fields, thus enhancing interdisciplinary dialogue and information exchange within model organism communities. The resulting visualizations of data can thus be viewed as theoretical models: their construction is based on a well-prepared description of the phenomena to which they apply; its explanatory power comes from its reliance on specific theories concerning those phenomena.

Scientists leading the resources that produce such theoretical models insist that the increasing sophistication of these images could, at some point in the future, render much of the current experiments on real organisms obsolete. Namely, data visualizations in TAIR or GMOD could be used as "discovery tools," thus fulfilling the exploratory function that I assigned to laboratory organisms earlier: theoretical models could replace material models as tools for the elaboration of intelligible theories about organismal biology.

In light of my considerations above, I find this expectation entirely misguided. There is no way in which the conceptual manipulation of a virtual model displayed by one of those databases could replace the material manipulation of an actual organism in the laboratory, because it is thanks to the latter research activity that the virtual model acquires empirical content. Scientists working on databases seem to think that reference to empirical data will suffice to confer empirical content to a model. However, by the time the data have been organized and displayed according to the theoretical frameworks adopted within the databases, it becomes very difficult for a biologist to establish the relation between the resulting digital model and the phenomenon (process, component, function of the biology of the organism) to which it refers. Virtual displays of data are thus certainly priceless in their contribution to our understanding of organismal biology; they constitute theoretical models of great explanatory power, insofar as they use abstract concepts in order to classify, organize, and relate the empirical data gathered from a given organism to the theoretical knowledge available on that organism. However, they cannot be used to formulate intelligible theories about the phenomena that they depict, unless the biologists referring to them do not attribute empirical content to that knowledge by manipulating actual organisms.

Research on actual model organisms is indeed about assessing what organisms are made of, how organismal components interact with each other, and which process gives rise to which other. This type of knowledge is needed to enrich the theoretical models provided within databases. The images therein are so abstract that they cannot be brought to bear on the phenomna that they depict unless biologists can complement them with experimental research on actual organisms.

My analysis of the complementary epistemic utility of theoretical and material models constitutes, in my view, a convincing argument for maintaining an analytic distinction between conceptual and material manipulation of models, despite its above-mentioned fragility when considering how tightly these two dimensions are intertwined in scientific practice. Another important motivation is the need to counter a growing tendency among biologists of all trades: the tendency to value conceptual manipulation of models over and above their material manipulation. As I hope to have shown, the usefulness of splendid research tools such as computer simulations and mathematical models to achieve an improved understanding of the natural world is due as much to their own characteristics as to their complementation by material models. The largely happy marriage of empirical accuracy with theoretical tractability is, after all, one of the characteristic strengths of the biological sciences.

Favoring predictive accuracy over empirical accuracy would amount to threatening the intelligibility of the knowledge produced. The combined use of what I have called theoretical and material models—and thus the constant recourse to both conceptual and material manipulation in modeling—is the only possible guarantee for the acquisition of intelligible theories allowing one to explain and predict the behavior of the system under scrutiny, while at the same time maintaining and constantly redefining the descriptive value of the parameters that are used.

As stated earlier, this argument is also meant as a step toward the development of a pluralistic account of modeling strategies. Such a philosophical account would focus on the epistemological significance of differences among models rather than emphasizing what unifies them (again quoting Levins: "understanding is not achieved by generality alone, but by a relation between the general and the particular"; 1984:26). My discussion of the distinct features of theoretical and material models, I hope, illustrates that a renewed attention to the pluralism of models may lead to the formulation of research questions that differ considerably from the ones cherished within "single-model approaches," and yet are in dire need of philosophical attention. Among those questions are, for instance: What makes the combination of different models so valuable? How are relevant types of models chosen and integrated within any specific research context? How do such choices relate to the expected outcomes of research, and what impact do they actually have on such outcomes when implemented?

As testified by the contributions to this volume, these questions are very close to the ones posed by biologists themselves when confronting and discussing their modeling strategies. The development of a pluralistic account of modeling practices might thus provide fertile ground for an increase in collaboration between philosophers and scientists, which might in turn go a long way toward unraveling the multifaceted contributions of models to scientific theorizing.

Acknowledgments

I thank all participants in the KLI workshop "Modeling in Biology" (July 2004), in particular Werner Callebaut and Gerd Müller, for their valuable feedback and generous hospitality. Interaction with the research group, Knowledge, Practice, and Normativity, at the Free University of Amsterdam, in particular Henk de Regt and Hans Radder, fundamentally shaped and improved my ideas. I would also like to thank Rasmus Winther and James Griesemer for illuminating discussions on these topics.

Notes

1. For instance, see Patrick Suppes's restriction of the definition of model to set-theoretical constructs (2002) and Daniela Bailer-Jones's view of models as mental representations (2003).

2. Again, two illustrations among many possible examples: Hesse's (1966) early work on model-based analogical reasoning as heuristic to the development of theories and Cartwright's (1983) account of how features of representationally powerful models are abstracted from phenomena and idealized from theories (which she modified by focusing on the distinctions among models in her later work).

3. Note that most philosophers hitherto busy with "single-model approaches" are trying to develop a more pluralistic approach. Their work directly addresses the piecemeal nature of modeling practices. For

instance, Cartwright's (1999) framework is unique in its attempt to deal with different types of models across the natural and the social sciences, while Suarez (1999) carefully considers the pluralism among notions of theory idealized within models. Bailer-Jones published both on the different ideas that practicing scientists hold on what models are (2002) and on the use of "submodels" in order to obtain an overall model of a phenomenon (2000). Arguably, however, these philosophical accounts focus on how data from different models can be made to overlap into a coherent story, rather than looking at why and how models provide different types of information.

4. Historians of science contributed extensive evidence for the importance of combining different models for the development of scientific research. For instance, some controversial cases of theory choice characterizing the history of physics and biology are being usefully reconstructed and understood as depending on changes in the range and interpretation of empirical evidence as much as on creative transformations in the number and sophistication of the modeling practices favored within each of the rival theories (see, for instance, Keller, 2000, 2002; Boumans, 2001; and Chang and Leonelli, 2005). Equally indicative are the analyses of the evolution of research systems presented by Kohler (1994), Rheinberger (1997), Griesemer (2006), and Galison (1997), among others.

5. This view, characteristic of the semantic account of models, found a strong proponent in Suppes (1969, 2002), among others. For forceful criticisms of what Giere refers to as the "instantial view" on models, see Giere (1999) and Suarez (1999).

6. This feature also makes this notion of representation rather slippery, for how are we to draw clear distinctions between a representation and "the real nature of the system or theory" that it represents? Schank and Koehnle's analysis of modeling within behavioral biology (chapter 11 in this volume) can be used as a forceful illustration of this problem. In their view, most elements used by practicing biologists as tools for investigation—*including* theories, data, and material samples—should be thought of as models, that is, as representations conveying information about the real nature of what they represent in a limited and provisional way. This argument comes from the recognition that within any research context, the stipulation of what the phenomenon and theory in question are, depends on the tools used to investigate them just as much as the choice and handling of those tools are based on what scientists endorse as theoretical hypotheses and descriptions of reality.

How, then, should we draw boundaries between the notion of "mediators" and the notions of theory and/or phenomena that they are supposed to mediate? The risk of a slippery slope, leading to the denial of the possibility to distinguish between the model and what it models, represents a very serious problem for Morgan and Morrison: it runs counter to their basic intuition, which I share and value, that an analysis of the representational value of models necessarily includes making explicit what it is that models represent at any given moment of the research process. Morgan and Morrison try to avoid the problem by a sophisticated discussion of what they mean by "theory" and "phenomenon," a detailed assessment of which eludes the purposes of this chapter (see Morgan and Morrison, 1999; Cartwright, 1983, 1999). I thank James Griesemer for discussions on this point.

7. As Claude Bernard famously put it, in new ways of acting, rather than seeing, talking, or thinking; see also Longino (2002:124) on this point.

8. Another important set of considerations underscoring the importance of experimental protocols for achieving scientific understanding is provided by Hans Radder (2002), who argues that the achievement of methods to replicate experimental results is a necessary step toward the development of theories.

9. See Smith (2004) and Leonelli (2007b) on the increasing use of gene ontologies as a framework for data organization and storage.

10. See Winther (2003, 2006) for a discussion of the differences between styles of theorization adopted within what he calls "formal" and "compositional" biology.

11. I exclude here the idea that the biological sciences do not produce any kind of theory (e.g., Cooper, 1996). This claim is based on restricting the notion of theory to indicate subsumption under a universal law, as in the "received view" on scientific explanation (e.g., Hempel, 1965). Case-based analyses within both the life and the physical sciences indicate that this latter notion of theory is far too restrictive to account for all the types of generalizable knowledge acquired via scientific practice. Rather than interpreting this finding as a denial of the relevance of the notion of theory to biological knowledge, I see it as calling for its reformulation.

12. For details on this view, see Levins (1984), Puccia and Levins (1985), and the elaborations of Levins's argument presented by Wimsatt (1981, 1987), Griesemer (2000, 2006), Schank and Koehnle (chapter 11 in this volume), and Van der Steen (1995:26ff).

13. A similar argument is offered by Bailer-Jones (2000).

14. On the notion of commitment and its variety, see also Polanyi (1998:320).

15. See de Regt and Dieks (2005) and de Regt (2004).

16. For the purposes of this chapter, I will scrutinize only the role of models in the acquisition of understanding, thus leaving aside considerations about the individual and/or community to which theories are intelligible. For some insight into this difficult topic, see Longino (2002), Solomon (2001), and Leonelli (2007b).

17. I do not believe that these two criteria represent sufficient conditions for a theory to be deemed intelligible. I do, however, take them to specify two necessary conditions for this. Several scientific theories do not conform to one or the other of these two requirements. Think of string theory as a glaring case of a highly explanatory theory whose empirical content cannot be specified; and of early versions of classification systems (such as the Linnaean taxonomy) as highly accurate in their empirical content, but arguably containing little or no narrative to make sense of how the categories that they use to individuate relevant properties of specimens (the theoretical terms in this case) may account for the structure, function, and/or existence of such properties. In both cases, intelligibility does not figure among the many virtues of these theories; the former contributes an explanation without specifying the phenomena to which it applies, while the latter provides a description without clarifying its explanatory relevance.

18. I am upholding the empiricist view that it is not sufficient to rely on convention or whim in stipulating and maintaining such relation in order to secure the empirical content of a theory.

19. For further reflections on models as exploratory tools within an experimental context, see Radder (2003) and references therein.

20. Cartwright recognized this in her most recent work and thus distinguished these "representative models" from the interpretive ones (1999:180).

21. On scale models, see Giere (1999).

22. For arguments on model organisms as models, see Ankeny (2001) and Leonelli (2007a).

23. Indicative of this trend is the recent focus of fields such as artificial intelligence on the idea of "embodied cognition"—see, for instance, Anderson (2003) and references therein.

24. See, for instance, Morgan (2003) and Magnani (2001).

25. For further information about these projects, see the relevant Web pages. GMOD: http://www.gmod.org/home; TAIR: www.arabidopsis.org; FlyBase: http://flybase.bio.indiana.edu; BioCyc: http://biocyc.org/metacyc.

References

Anderson ML (2003) Embodied cognition: A field guide. Art Intell 149: 91–130.

Ankeny RA (2001) Model organisms as models: Understanding the "lingua franca" of the Human Genome Project. Phil Sci 68(3): S251–S261.

Bailer-Jones D (2000) Modelling extended extragalactic radio sources. Stud Hist Phil Mod Phys 31(1): 49–74.

Bailer-Jones D (2002) Scientists' thoughts on scientific models. Perspect Sci 10: 275–301.

Bailer-Jones D (2003) When scientific models represent. Intl Stud Phil Sci 17: 59–74.

Beatty J (1997) Why do biologists argue like they do? Phil Sci 64: S432–S443.

Bonner JT (2002) Lives of a Biologist. Cambridge, Mass.: Harvard University Press.

Boumans M (2001) "Ceteris paribus" conditions: Materiality and the application of economic theories. J Econ Methodol 8(1): 11–26.

Cartwright N (1983) How the Laws of Physics Lie. New York: Oxford University Press.

Cartwright N (1999) The Dappled World. Cambridge: Cambridge University Press.

Chang H, Leonelli S (2005) Infrared metaphysics: Theory-choice and the ontology of radiation (part 2). Stud Hist Phil Sci A36(4): 687–706.

Cooper G (1996) Theoretical modeling and biological laws. Phil Sci 63: S28–S35.

De Regt H (2004) Discussion note: Making sense of understanding. Phil Sci 71: 98–109.

De Regt H, Dieks D (2005) A contextual approach to scientific understanding. Synthese 144: 137–160.

Dretske FI (1977) Laws of nature. Phil Sci 44: 248–268.

Galison P (1997) Image and Logic: A Material Culture of Microphysics. Chicago: University of Chicago Press.

Giere R (1999) Using models to represent reality. In: Model-Based Reasoning in Scientific Discovery (Magnani L, Nersessian NJ, Thagard P, eds), 41–57. New York: Kluwer Academic/Plenum.

Griesemer JR (1990) Material models in biology. PSA 2: 79–93.

Griesemer JR (2000) Development, culture and the units of inheritance. Phil Sci 67: S348–S368.

Griesemer JR (2006) Genetics from an evolutionary process perspective. In: Genes in Development (Neumann-Held EM, Rehmann-Sutter C, eds), 343–375. Durham, N.C.: Duke University Press.

Hacking I (1983) Representing and Intervening. Cambridge: Cambridge University Press.

Hempel C (1965) Aspects of Scientific Explanation. New York: Free Press.

Hesse M (1966) Models and Analogies in Science. Notre Dame, Ind.: University of Notre Dame Press.

Keller EF (2000) The Century of the Gene. Cambridge, Mass.: Harvard University Press.

Keller EF (2002) Making Sense of Life. Cambridge, Mass.: Harvard University Press.

Kohler RE (1994) Lords of the Fly: *Drosophila* Genetics and the Experimental Life. Chicago: University of Chicago Press.

Leonelli S (2007a) Cultivando hierba, produciendo conocimiento. Una historia epistemológica de *Arabidopsis thaliana*. In: Variedad sin Limites. Las Representaciones en la ciencia (Suarez E, ed). Mexico City: Universidad Autónoma de México Editorial Limusa.

Leonelli S (2007b) Weed for thought. Using *Arabidopsis thaliana* to understand plant biology. Ph.D. dissertation, Vrije Universiteit Amsterdam.

Levins R (1984). The strategy of model building in population biology. In: Conceptual Issues in Evolutionary Biology (Sober E, ed), 18–27. Cambridge, Mass.: MIT Press.

Longino H (2002) The Fate of Knowledge. Princeton, N.J.: Princeton University Press.

Machamer P, Darden L, Craver C (2000) Thinking about mechanisms. Phil Sci 67: 1–25.

Magnani L (2001) Abduction, Reason, and Science: Processes of Discovery and Explanation. New York: Kluwer Academic/Plenum.

Mayr E, Provine W (eds) (1980) The Evolutionary Synthesis. Cambridge, Mass.: Harvard University Press.

Morgan M (2003) Experiments without material intervention. In: The Philosophy of Scientific Experimentation (Radder H, ed), 216–235. Pittsburgh, Pa.: University of Pittsburgh Press.

Morgan MS and Morrison M (eds) (1999) Models as Mediators. Cambridge: Cambridge University Press.

Morrison M (2002) Models and statistics: Pearson and Fisher on Mendelian populations. Brit J Phil Sci 53: 39–68.

Polanyi M (1998 [1958]) Personal Knowledge. Towards a Post-Critical Philosophy. London: Routledge.

Puccia CJ, Levins R (1985) Qualitative models. In: Qualitative Modeling of Complex Systems: An Introduction to Loop Analysis and Time Averaging (Puccia CJ, Levins R, eds), 1–11. Cambridge, Mass.: Harvard University Press.

Radder H (1996) In and About the World: Philosophical Studies of Science and Technology. Albany: State University of New York Press.

Radder H (2002) How concepts both structure the world and abstract from it. Rev Metaphys 55: 581–613.

Radder H (2003) Technology and theory in experimental science. In: The Philosophy of Scientific Experimentation (Radder H, ed), 152–173. Pittsburgh, Pa.: University of Pittsburgh Press.

Rheinberger H-J (1997) Toward a History of Epistemic Things. Stanford, Calif.: Stanford University Press.

Rose S (1997) Lifelines. Oxford: Oxford University Press.

Salmon WC (1998) Causality and Explanation. Oxford: Oxford University Press.

Smith B (2004) Beyond concepts: Ontology as a reality representation. In: Formal Ontology and Information Systems (Varzi A, Vieu L, eds), 73–84. Amsterdam: IOS Press.

Solomon M (2001) Social Empiricism. Cambridge, Mass.: MIT Press.

Star SL, Griesemer JR (1989) Institutional ecology, "translations" and boundary objects: Amateurs and professionals in Berkeley's Museum of Vertebrate Zoology, 1907–39. Soc Stud Sci 19: 387–420.

Suarez E (2001) Satellite-DNA: A case-study for the evolution of experimental techniques. Stud Hist Phil Biolog Biomed Sci 32(1): 31–57.

Suarez M (1999) The Role of Models in the Application of Scientific Theories: Epistemological Implications. In: Models as Mediators (Morgan M and Morrison M, eds), 168–195. Cambridge, UK: Cambridge University Press.

Suppes P (1969) A comparison of the meaning and uses of models in mathematics and the empirical sciences. In: Studies in the Methodology and Foundations of Science. Selected Papers from 1951 to 1969 (Suppes P, ed), 10–23. Dordrecht: Reidel.

Suppes P (2002) Representation and Invariance of Scientific Structures. Stanford, Calif.: CSLI Publication.

Trout JD (2002) Scientific understanding and the sense of understanding. Phil Science 69(2): 212–233.

Van der Steen WJ (1995) Facts, Values and Methodology: A New Approach to Ethics. Amsterdam and Atlanta: Rodopi Press.

Wimsatt WC (1981) Robustness, reliability and overdetermination. In: Scientific Enquiry and the Social Sciences (Brewer M, Collins B, eds), 124–163. San Francisco: Jossey-Bass.

Wimsatt WC (1987) False models as means to truer theories. In: Neutral Models in Biology (Niteki M, Hoffman A, eds), 23–55. London: Oxford University Press.

Winther RG (2003) Formal and compositional biology as two kinds of biological theorizing. Ph.D. dissertation, University of Indiana, Bloomington.

Winther RG (2006) Parts and Theories in Compositional Biology. Biol Philos 21(4): 471–499.

3 Experimental Model Systems: An Epistemological Aperçu from the Perspective of Molecular Biology

Hans-Jörg Rheinberger

The following remarks on model systems in biology are those of a molecular biologist-turned-historical epistemologist, not a theoretical biologist or a philosopher of biology. I understand historical epistemology to be a reflection of the practical as well as the theoretical, of the material as well as the conceptual, of the scientific as well as the cultural conditions of knowledge acquisition in their historical entrenchment. I came to practice molecular biology in the late 1970s, and I have experienced it as a largely *empirical* culture relying heavily on lab work systems. My epistemological remarks inevitably carry the mark of this enculturation.

An anecdote might help to set the stage. In the late 1980s, I visited one of Berkeley's molecular biologists, Gunther Stent, who was spending some time as a fellow at the Wissenschaftskolleg in Berlin. I knew him in his capacity as a member of the advisory board of the Max Planck Institute for Molecular Genetics in Dahlem, where I was then working. We had lunch, and he asked me what I was doing. I told him that besides doing experiments on protein biosynthesis, I was working on the history of evolutionary theory, in particular on the concept of selection. "But 'selection' is no longer a concept," he stated. "Selection is a process that can be produced and manipulated in the laboratory." During the long and heated debate that followed, he stuck to his claim that science, in the end, is not an activity that leads to entities such as concepts or even theories; rather, it is an activity that ultimately resolves concepts into something for which there is a very appropriate old German word: a *Sachverhalt*, a relation between things. For Stent, theory was something that bridged gaps. If the gaps disappeared, so did the need for the bridges. In this discussion, Stent certainly played the role of a devil's advocate, but his statement intrigued me, and it has continued to intrigue me ever since. To put it in other terms: The claim is that the apotheosis, and ultimately the fulfillment, of a theory is its dissolution. Let me call this Stent's paradox.

If it is not a conceptual framework in the first place, how is molecular biology to be addressed? We can look at it as a network of more or less articulated experimental systems, or model systems, evolving around scientific objects that enter into more or less specified relations to each other. We must look seriously at what molecular biologists—and this holds for other areas of biology as well—*do* when they occupy themselves with the objects of their analytic and synthetic efforts in the context of such systems. I will use the notions of experimental system, model system, and experimental model system synonymously, as scientists do when they talk about their work. The term "model" is here taken in its generic meaning as an experimental setup that appears to be particularly apt for the analysis of a particular phenomenon and that usually involves a corresponding model organism. Every valuable experimental system is a model system in this sense.

There can be no doubt that the objects of molecular biological research are physicochemical things. In its deliberate search for the physicochemical basis of life, molecular biology has frequently been criticized and accused of driving reductionism to its extremes. But it is precisely in following such an epistemically reductive modeling strategy that molecular biology has been confronted with an obstacle from within: that these physicochemical objects exhibit quite singular and historically unique features that are, as products of evolution, at times not fully understandable by their physicochemical nature alone.

Let me mention just one fundamental example: At least as far as can be said today, there appears to be no possibility to deduce the specific letters of the genetic code, that is, to assign a particular nucleic acid triplet to a particular amino acid solely on the basis of stereochemical principles. Obviously, we have to do here with historical contingencies that are accessible only through an analysis of the empirical detail, and not deducible from first principles. Although this might be a matter of debate for some, Michael Polanyi very early on concluded from this fact that the objects of biology in this sense are categorically distinct from the objects of the other sciences (Polanyi, 1969).

Several years ago, Richard Burian remarked, and I think rightly so, from the point of view that I adopt here, that historical contingency is probably the deepest reason for the overwhelming importance of empirical research in virtually all disciplines of modern biology (Burian, 1995). Of molecular biology in particular, he has even claimed that it looks more like a battery of techniques than a general theory. If it is not a mere battery of techniques, an articulated ensemble of model systems will certainly be the view to take here. Finding appropriate model systems is the way to carry on productive research in this situation.

A short look at the history of elucidating the genetic code between 1953 and 1963 is indeed revealing in this respect. If we follow Lily Kay's historical analysis (Kay, 2000), all efforts of theoretical physicists after the establishment of the structure of the double helix to deduce the code from a priori considerations failed: George Gamow's, Francis Crick's, Henry Quastler's, and those of many others. It was only in 1961 that a biochemical model system of in vitro protein synthesis, originally set up for reasons unrelated to what was to become molecular biology, entered the scene of experimental molecular genetics. To the disappointment of all those who had so desperately been racking their brains over coding rules, the code was solved on the basis of an empirical assay. None of those who finally managed to do the job—Heinrich Matthaei, Marshall Nirenberg, and all the others thereafter—had been involved in the painstaking theoretical speculations that preceded this experimental feat, and all those involved in the early speculations were left behind.

On the level of conceptualization, things do not look very different. Recently, I conducted a historical analysis on the introduction of the information concept in the scientific papers of François Jacob and Jacques Monod, molecular biologists of the first generation at the Pasteur Institute (Rheinberger, 2006). The analysis has led to the conclusion that their papers are not very different with respect to the introduction of the language of information from those of their fellow biochemists Mahlon Hoagland and Paul Zamecnik at Massachusetts General Hospital in Boston, which I had looked at earlier (Rheinberger, 1997). They show an obviously quite general feature of the emerging discourse of molecular genetics: that there is no sharply defined conceptual break. Instead, different layers of notions were successively added to form a genuinely hybrid discourse.

In the case of Jacob and Monod, the initial structural vocabulary of specificity and determination was first supplemented by a performative vocabulary of prescription, execution, and control that had been taken over from the study of phage lysogeny, on the one hand, and the induction of lactose metabolism in *E. coli* on the other. To this already hybrid discourse were added the language of information, message, and code; the language of communication and signaling; and finally the language of reading and writing a text. All these idiomatic layers carried along with them quite different conceptualizations. The different idioms did not replace each other in clearly separated steps, and they were not used as clearly distinct model conceptions. They coexisted. They became superimposed upon each other and supplemented each other as different registers of one single experimental model system which was itself a hybrid system combining the study of sugar metabolism and viral gene transfer in the bacterium *E. coli*.

These registers allowed for very subtle experimental modulations of the system. The language in which an experimental science expresses its results is differentiated and changes in tight connection with the technical means and the media of realizing and developing the experimental systems on which the science relies and on whose outcomes the further avenues of research depend. The notion of information in particular—so pervasive in later years—only slowly became part and parcel of that patchy discursive network. And it is not by chance that it never took on the technical, purely quantitative meaning bestowed on it by information theory. "Information" instead came to occupy the place of preceding versions of biological specificity. It took on the meaning of a representation of the structural repository of biological function.

"The living world is one of complexity, the result of innumerable interactions among organisms, cells, molecules. In analyzing a problem, the biologist is constrained to focus on a fragment of reality, on a piece of the universe which he arbitrarily isolates to define certain of its parameters. In biology, any study thus begins with the choice of a 'system'" (Jacob, 1988:234). What transpires through these words, taken from the autobiography of Jacob, is the conviction that because of the very conditions of possibility of biological experience and knowledge acquisition, the fragmentation of reality, the cutting of the universe of the living into pieces is a prerequisite for the constitution of biological knowledge. Jacob held that although there are generalizations in biology that can go farther or less far at times, there are very few, if any, theories (Jacob, 1974). We might well claim, with the historian of biology Georges Canguilhem, that for most cases in the history of biology, *concepts* rather than theories have been the organizing centers. And if we follow Stent in addition, the successful ones among them interact and become amalgamated with particular scientific objects that can be manipulated in experimental systems. Canguilhem has concluded from his studies that the history of biology is best written as a history of concepts. We could go on and say that in biology the modular structure of its experimental practice with its experimental model systems is reflected in the theoretical realm by conceptual modularity.

Consequently, what we need on the epistemological level, is, as I have argued in my book *Toward a History of Epistemic Things*, in following the French epistemologist Gaston Bachelard, a "philosophy of the epistemological detail" (Bachelard, 1966:12). Philosophers have often seen such fragmentation as a deplorable, albeit possibly inescapable, limitation of empirical knowledge. We can view it, however, quite differently. I would rather contend that this imagined limitation is a condition of the possibility for the production of scientific knowledge. Science, in the authentic sense of the word, is a way of gaining *unprecedented*, that is, *new*, knowledge. For this to happen, conditions must be created that, in contrast to deductive systems, allow for epistemic events that *cannot* be anticipated.

Now, experimental systems are precisely the arrangements that allow for the creation of cognitive, spatiotemporal singularities, and thus allow for unprecedented epistemic events. The attentive study of experimental model systems in molecular biology and the networks they constitute leads to the conclusion that these systems are, according to the words of Jacob once again, "machines for making the future" (Jacob, 1988:13). They are instances of iteration, of differential reproduction in the realm of empirical knowledge acquisition. A history of molecular biology, I think, will have much to contribute to such an epistemology of iteration, exactly because of the eminently patchwork structure of its practices entrenched in experimental systems. In order to assess this broader epistemic network, we need to understand, in the words of the Belgian philosopher Isabelle Stengers, the material as well as the conceptual "operations of propagation" and "operations of passage" that shape the network into a more or less cohesive, but highly mobile, field of knowledge production (Stengers, 1987).

Elsewhere I have shown in detail that powerful experimental systems must be located—quite literally—at the cutting edge of the fractionation process mentioned above (Rheinberger, 1997). They operate most productively at the fluctuating, unstable border between the trivial and the complex. For a particular experimental system, the broader network represents complexity as a kind of "epistemic horizon." Within a particular experimental system, the tendency to experimental reduction of complexity prevails. There are no general rules which guarantee that these reductive moves proceed in directions that will be accepted in the long run. In the end, it is the network of experimental model systems that decides about the epistemic value of its elements. If, as we may take for granted, ontic complexity *has* to be reduced in order to make experimental research possible and experimental systems work, the complexity of the living is epistemically retained and constantly reconstituted in the rich contexture of an experimental landscape, in which new connections and disconnections between model systems can happen at any time, and where the eruption of one "volcanic system" can change the whole landscape, through passage and through propagation.

With concepts such as conjuncture, hybridization, and ramification, I have tried to capture some relations within ensembles of experimental systems and their intricate interactions. The concepts express a vision of a structure of articulated experimental networks of objects and practices whose coherence, just as in the case of individual experimental systems, is patched and whose meaning is laterally constituted by horizontal concatenation. The cohesion of the systems reaches as far as epistemic entities can circulate throughout the network. Conjunctures, hybridizations, and ramifications describe the dynamics of reorientation, fusion, and proliferation of particular experimental systems in terms of shifts, links, and descent. These processes translate the microdynamics of localized and situated experimental model systems into the macrodynamics of broader fields of experimentation.

With this, we come very near to an idea that Stuart Kauffman developed in his book *At Home in the Universe* and that he calls the patch procedure.

> The basic idea of the patch procedure is simple: take a hard, conflict-laden task in which many parts interact, and divide it into a quilt of non-overlapping patches. Try to optimize within each patch. As this occurs, the couplings between parts in two patches across patch boundaries will mean that finding a "good" solution in one patch will change the problem to be solved by the part in the adjacent patches. Since changes in each patch will alter the problems confronted by the neighboring patches, and the adaptive moves by those patches in turn will alter the problem faced by yet other patches, the system is just like our model coevolving ecosystems.... We are about to see that if the entire conflict-laden task is broken into the properly

chosen patches, the coevolving system lies at a phase transition between order and chaos and rapidly finds very good solutions. Patches, in short, may be a fundamental process we have evolved in our social systems, and perhaps elsewhere, to solve very hard problems. (Kauffman, 1995:252–253)

In its application to model systems, we may call this the patchwork view of research. It leads us to look for a logic that is exhausted neither by the rationality of individual actors nor by the social conventions of a broader discipline. It lies somewhere between the extremes of Cartesian egos and Kuhnian paradigms. Patchwork epistemology is the quest for conceptual tools that may help to clarify a little better the historical trajectories of the sciences.

In his encompassing conceptual history of biology, *The Growth of Biological Thought* (1982), Ernst Mayr devoted the last chapter to molecular biology. While praising its technical breakthroughs, he considered it of minor impact with regard to the great theoretical challenges of biology which he thought had basically been solved—solved, of course, by the synthesis of evolutionary theory and classical genetics. Mayr was right and he was not. He was right in that molecular biology started as a technical, physicochemical conquest. It was not a ready-made notion of genetic information that guided molecular biology's efforts from the beginning. Rather, the new concept of biological specificity was shaped gradually and diffused only slowly—as I have tried to show—into the partial systems of the new biology. And for quite a while, molecular biology appeared to live happily with a gene concept, for instance, that was in no way hurting the claims of classical genetics. By pinning the gene down to DNA, molecular biology simply appeared to add a material, molecular dimension to it, and to open a new field of experimentation: bacterial and phage genetics.

But Mayr was also wrong, because the situation has profoundly changed along the way. And yet, the quite radical changes that our views of biological specificity and of genes and genome structures have undergone since the 1950s are by no means the result of deliberately alternative approaches called onto the scene in order to counteract reductionist genetics and molecular biology. On the contrary, local experimental sophistication with experimental models, reaching down to the molecular level as a kind of obligatory passage point, has exploded a rather coarse and simplistic gene concept from deep within molecular biology, leading to quite unprecedented vistas on the structure and regulation of the genome. A battery of mechanisms and entities has been identified—often by exploiting the tools of gene technology—that constitute what could be called a system of hereditary "respiration" (Rheinberger, 2000).

A few years ago, at a conference on "Rethinking the Enlightenment," the historian of science Yehuda Elkana asked the rhetorical question whether the time was ripe for challenging and daring alternatives in the sciences, or whether we needed to do just more of the same. My equally rhetorical answer to this perennial question is that we need to do more of the same *precisely* in order to arrive at unprecedented alternatives. The only *fruitful* and lasting way to reach beyond the confines of a system, in particular of any model system, is to try hard to explore its limits and finally to transform or even founder it from within.

At the end of *At Home in the Universe*, Kauffman ponders: "I wonder if we really understand very much of what we are creating." And he continues: "All we can do is be locally wise, even though our own best efforts will ultimately create the conditions that lead to our transformations to utterly unforeseeable ways of being" (Kauffman, 1995:298, 303). Local wisdom is what characterizes the practices of an endeavor that, since the Enlightenment and contrary to all appearance and experience, has never ceased to depict itself as an allegedly global undertaking: the making of "Science." I contend that the order of the day for epistemology is learning to understand how local wisdoms, embedded in research attractors such as experimental systems, get connected to knowledge patchworks, to ever new and changing configurations of niches of knowing. Fragmentation—"cantonization," in the language of Bachelard (1949)—far from being deficient, deleterious, and deplorable, then must appear as one of the basic conditions of unprecedented development.

In *Epigenetic Inheritance and Evolution*, Eva Jablonka has made a strong argument for taking seriously the different epigenetic transmission systems that organisms appear to have acquired in the course of evolution (Jablonka and Lamb, 1995). She could not have written this book had card-carrying molecular biologists not been puzzled for many years by observations they eventually came to call chromatin-marking strategies, the most prominent of which are based on patterns of DNA methylation and on DNA binding proteins. These findings did not come about by following a Lamarckian research strategy in a quest for tracing the inheritance of acquired characters. They came about, once again, by doing "more of the same," but, at the same time and as an essential epistemological increment, by being attentive to what would prove to be both resilient and refractory to the standard view taken for granted at the time.

Since the 1980s we have witnessed a similar scenario in the realm of developmental biology as it has been forced through the bottleneck of molecular genetics. The work of Christiane Nüsslein-Volhard, Ed Lewis, Eric Wieschaus, Walter Gehring, and others on the homeobox system and on other regulatory master switches in early development has brought long-elusive morphological patterns into the realm of molecular interpretation. Developmental biology had been widely set aside by classical transmission genetics since the 1920s, and by early molecular biology as well. Yet, here again, it was neither an alternative organismic approach nor a simple reevaluation of earlier developmental theories that paved the way to new vistas on "homology" and to what now is called the "zootype." It was, rather, the intelligent exploitation of a set of new techniques belonging to genetic engineering that helped to open the box again.

With a lengthy quote from John Maynard Smith and Eörs Szathmáry's *Major Transitions in Evolution*, I come back to my initial considerations:

> There are some profound consequences of this definition of animals. The zootype is based on functions that are informational: it is easy to imagine that different systems could have evolved in different phyla. Thus the organization of the zootype is gratuitous, and indicates common ancestry, rather than convergent adaptation or developmental constraints. . . . This observation is a rather strong blow at the essentialist or structuralist approach to body plans: the definition of animals rests on genetic ancestry, rather than on first principles of form. . . .
>
> How can we expect to gain further insight? The best hope lies in molecular genetics. Molecular data are the main reason for accepting the monophyletic origin of metazoans. The homeobox story . . . tells us that the common ancestor of the chordates and the arthropods probably had a differentiated head, middle, and tail. The regulatory genes responsible for the differentiation were already present. Since their DNA-binding region, the homeobox, is also present in the gene-determining mating type in yeast, it is likely that the common ancestor of these genes already existed in their protist ancestor. No-one could have foreseen these observations. We can reasonably hope that future discoveries will be equally illuminating. (Maynard Smith and Szathmáry, 1995:224, 254).

Understanding the experimental dynamics that lies behind these transformations of our views on development and evolution is what a "philosophy of the epistemological detail," in the sense of Bachelard, has to address. This does not, of course, amount to theoretical biology, nor is it philosophy of biology from an ontological perspective. If anything, these considerations are meant to contribute elements to a theory, or philosophy, *about* the constitution of biology as a science. Within this *epistemological* context, I have argued, experimental model systems play a major role. If I may draw a conclusion, it is the following: Historical epistemology, too, must become quasi-molecular and experimental if it wants to live up to the scientific practices it tries to analyze and understand. Well-selected case studies are, then, *its* model systems. Such an endeavor has begun, but there is certainly still a long way to go in this direction.

References

Bachelard G (1949) Le Rationalisme appliqué. Paris: Presses Universitaires de France.

Bachelard G (1966 [1940]) La Philosophie du non. 4th ed. Paris: Presses Universitaires de France.

Burian R (1995) What is this science we have made? Some epistemological issues regarding molecular biology. Unpublished manuscript.

Jacob F (1974) Le Modèle linguistique en biologie. Critique 322: 197–205.

Jacob F (1988) The Statue Within: An Autobiography. New York: Basic Books.

Jablonka E, Lamb MJ (1995) Epigenetic Inheritance and Evolution: The Lamarckian Dimension. Oxford: Oxford University Press.

Kauffman S (1995) At Home in the Universe. Oxford: Oxford University Press.

Kay LE (2000) Who Wrote the Book of Life? A History of the Genetic Code. Stanford, Calif.: Stanford University Press.

Maynard Smith J, Szathmáry E (1995) The Major Transitions in Evolution. New York: Freeman.

Mayr E (1982) The Growth of Biological Thought: Diversity, Evolution, and Inheritance. Cambridge, Mass.: Belknap Press of Harvard University Press.

Polanyi M (1969) Life's irreducible structure. In: Knowing and Being. Essays by Michael Polanyi (Grene M, ed), 225–239. Chicago: University of Chicago Press.

Rheinberger H-J (1997) Toward a History of Epistemic Things. Synthesizing Proteins in the Test Tube. Stanford, Calif.: Stanford University Press.

Rheinberger H-J (2000) Gene concepts: Fragments from the perspective of molecular biology. In: The Concept of the Gene in Development and Evolution (Beurton P, Falk R, Rheinberger H-J, eds), 219–239. Cambridge: Cambridge University Press.

Rheinberger H-J (2006) Notions of regulation, information, and language in the writings of François Jacob. Biol Theory 1, 261–267.

Stengers I (1987) La Propagation des concepts. In: D'une Science à l'autre. Des Concepts nomades (Stengers I, ed), 9–26. Paris: Seuil.

4 The Role of Complex Networks in Biological Modeling and Systems Biology

Luciano da Fontoura Costa

Truth in science can be defined as the working hypothesis best suited to open the way to the next better one.
—Konrad Lorenz

There are no definite truths in science. We live in a continuously changing universe where one can never step twice in the same river. At the same time that changes allow, through natural evolution, the appearance of more complex species, they also imply continuing challenges at the individual level. One of the most effective mechanisms adopted by individuals in order to cope with the pressures of survival and reproduction is *imitation* of the behavior of other individuals, a subject which was treated with great insight by Konrad Lorenz. By imitating, an individual can bypass the efforts originally required to learn a given procedure, thus saving time and effort. Because errors often take place during imitation, variations of the imitated procedures are obtained which may, eventually, lead to more effective performance. In this way, behavioral imitation mimics genetic evolution, characterized by imitations and mutations at the molecular level.

One possible way to understand human intelligence is as an imitation process which thrives particularly on effective creativity and experiments, rather than only on procedural errors, for improvements. A direct by-product of survival pressures and curiosity, science can be best understood as the effort to obtain particularly good imitations—that is, models—of parts of the natural world which not only can explain observed processes, but also can help to predict them. It may constitute the leading edge of evolution. Modeling and imitation are closely related because both try to reproduce a dynamic phenomenon. While a goose may imitate a procedure performed by other geese as a way to obtain food, science is totally oriented toward imitating natural phenomena in order to obtain predictions. Both procedures involve observation, identification of the important elements to be considered, and simulation to achieve the objectives.

Like imitation, most (if not all) natural processes can be understood as involving interactions between constituent parts. For instance, the Earth is attracted by the sun, and vice versa. One animal species feeds on other species. Animals from the same species interact in order to form bonds. Organs in our bodies interact in order to maintain our lives and perpetuate our species. By interaction, proteins provide the basis for most cellular processes. Basic particles interact among themselves in order to define matter. We live in a changing world pervaded by connections and associations to which biology is no exception.

Because of their suitability for representing biological structures and systems undergoing development, complex networks are poised to provide one of the main pillars of systems biology, integrating not only the components in each subsystem but also the subsystems themselves. Such prospects provide the main motivation for this chapter. It starts by reviewing the basic concepts from graph theory and statistical physics, the two areas most immediately associated with complex networks, and then presents the main ideas and methods in complex network theory, and by discussing and illustrating the application of such concepts, tools, and methods to modeling biology.

Basic Concepts

This section presents the basic concepts underlying complex networks. For simplicity's sake, such concepts are covered separately in the following sections with respect to traditional graph theory and statistical physics, and then from the integrated perspective of complex networks.

Graph Theory

A Historical Glimpse The beginnings of graph theory are usually identified with Leonhard Euler's solution of the Königsberg problem: the possibility of covering all river-delimited terrains of that town by crossing each of the seven existing bridges only once. By representing the terrains and bridges as nodes and edges between points, Euler devised the first graph and, by considering the number of connections of each node, concluded that there was no solution to the Königsberg problem. However, in case we are less strict about graph theory as a formalized area, its rudiments can be traced back to the very beginnings of humanity, since most visual representations are actually kinds of graphs. For instance, each depiction of animal hunting on the walls of ancient caves can be understood as a graph where the main elements of the chase are represented as nodes, while the actions of sighting, pursuing, and animal hitting would correspond to edges of action. Such graphic representations would therefore correspond to models of important activities at the time,

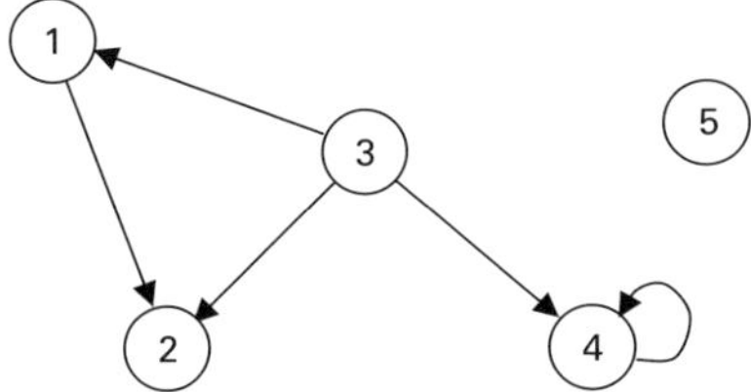

Figure 4.1
A simple directed graph composed of five nodes and five directed edges, one of which is a loop. The outdegrees and indegrees of the nodes in this graph are given in table 4.1.

represented as graphs. Perhaps as old as such chase scenes, maps are even more directly related to graphs. Actually, maps *are* graphs of very particular importance.

Since Euler, graph theory has become the subject of intense formal interest from mathematicians who, through the centuries, have developed a massive framework of related knowledge. However, until Flory (1942), Rappoport (1957), and Erdös and Rényi (1959), graph theory was mostly concerned with static graphs. As implied in the name, static graphs are structures whose nodes and edges never change with time. A major contribution of the above-mentioned researchers was to develop the theory of random graphs, where edges are added randomly (with uniform probability) among the existing nodes. Unfortunately, very few natural systems follow such a model, a fact which recently was fully addressed only through complex network theory.

Elementary Concepts in Graph Theory A graph is a diagram composed of *nodes* and of *edges* which can be established between nodes, as illustrated in figure 4.1. A graph can be completely represented in terms of its *adjacency matrix*. The adjacency matrix, A, of the graph in figure 4.1 is

$$A = \begin{bmatrix} 0 & 0 & 1 & 0 & 0 \\ 1 & 0 & 1 & 0 & 0 \\ 0 & 0 & 0 & 0 & 0 \\ 0 & 0 & 1 & 1 & 0 \\ 0 & 0 & 0 & 0 & 0 \end{bmatrix}$$

The element $A(i, j)$ of this matrix (the value at row i and column j) reflects the connectivity from node j to i, assuming either null (absence of connection) or unit (presence of connection) values. For instance, the fact that node 1 connects to 2 implies $A(2, 1) = 1$. The number of edges leaving a node is known as the *outdegree* of that node, and the number of edges arriving at a node is the *indegree* of that node. The outdegrees and indegrees of the nodes in the graph in figure 4.1 are given in table 4.1. Note that the outdegree of a node i can be obtained by adding the adjacency matrix entries along its i-th column. Similarly, the number of edges arriving at a node

Table 4.1
The outdegrees and indegrees of the nodes of the graph in figure 4.1

Node	Outdegree	Indegree
1	1	1
2	0	2
3	3	0
4	1	2
5	0	0

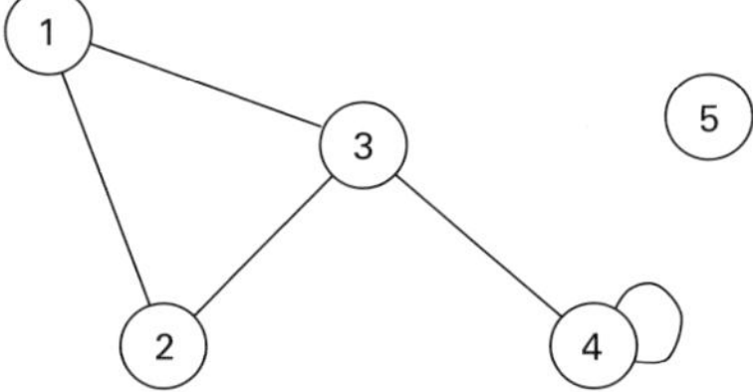

Figure 4.2
A simple graph composed of five nodes and five undirected edges, one of which is a loop. Note that this graph can be obtained from that in figure 4.1 by replacing each original directed edge with an undirected edge. The degrees of the nodes in this graph are given in table 4.2.

corresponds to the *indegree* of that node, which can be obtained by adding along the respective matrix row.

A graph (as in figure 4.1) which involves directed edges is called a *digraph*; the term *graph* is typically reserved for undirected structures such as that in figure 4.2. The adjacency matrix, A, for that graph is

$$A = \begin{bmatrix} 0 & 1 & 1 & 0 & 0 \\ 1 & 0 & 1 & 0 & 0 \\ 1 & 1 & 0 & 0 & 0 \\ 0 & 0 & 1 & 1 & 0 \\ 0 & 0 & 0 & 1 & 0 \end{bmatrix}$$

As in all undirected graphs, the adjacency matrix of the graph in figure 4.2 is *symmetric*; that is, $A(i,j) = A(j,i)$ for any i and j. The degree of a node i of a graph is defined as the number of edges attached to that node, and can be obtained by adding the values along the i-th row (or column) of the adjacency matrix. The degrees of the nodes of the graph in figure 4.2 are shown in table 4.2.

Given a graph, it is possible to define a series of *subgraphs*. This is achieved by selecting a subset of nodes and a subset of edges between those nodes. Figure 4.3 illustrates some of the subgraphs which can be obtained from the graph in figure 4.2.

Table 4.2
The degrees of the nodes of the graph in figure 4.2

Node	Degree
1	2
2	2
3	3
4	2
5	0

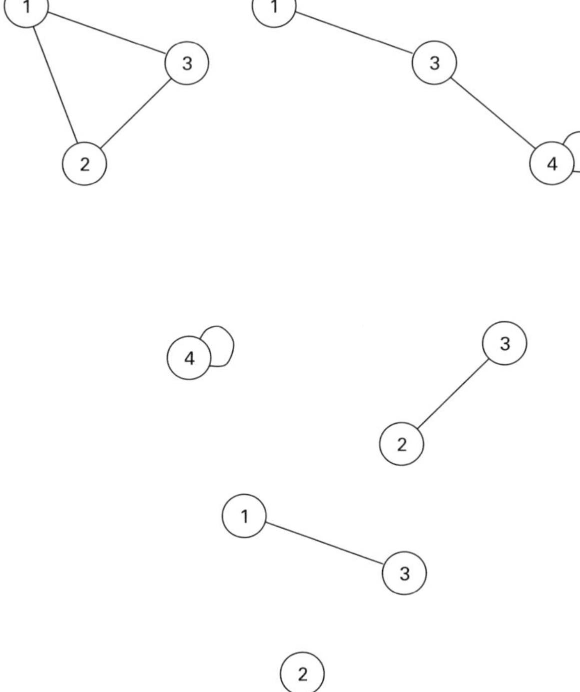

Figure 4.3
Five of the possible subgraphs which can be obtained from the graph in figure 4.2.

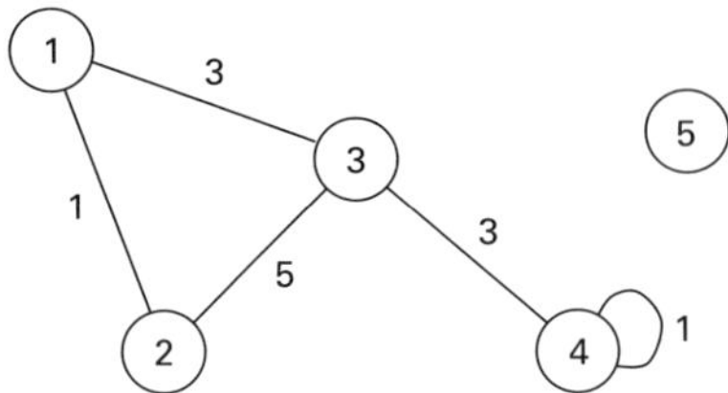

Figure 4.4
A simple weighted graph, characterized by the weights assigned to each edge.

Graphs and digraphs can be generalized in many ways. For instance, it is possible to assign weights to each edge, thus obtaining a *weighted graph*. Figure 4.4 illustrates this type of graph. If each node has a defined position in space, represented in terms of its Euclidean coordinates (x, y), the graph is said to be *geographical*.

An important property of graphs regards the *accessibility* of nodes. For instance, given two nodes i and j, it is interesting to know if it is possible to reach node j by starting at i and moving along the graph edges. If this is possible, it is said that there exists a *path* between those two nodes. In a weighted graph, the length of a path can be defined either as the number of edges along the path or as the sum of the weights of those edges. A cycle is a closed path; a loop is a special cycle where a node is connected to itself (e.g., the loop of node 4 in figure 4.4).

Given a graph, a particularly important type of subgraph which can be obtained is its *connected components*. Such a subgraph is characterized by the fact that any of its nodes can be reached while moving from any other of its nodes.

Many measurements other than the node degree can be obtained from a graph or digraph, including the *clustering coefficient* (indicating the connectivity among the neighbors of a node), the *number of cycles*, the *number of connected components*, and so on. A survey of measurements can be found in Costa et al. (2005a).

Graph Representation of Biological Systems Graphs can be used to represent an almost unlimited number of systems and processes. Examples of biological applications of graphs include, but are by no means limited to, the following:

1. *Protein-protein interaction networks.* Each protein is represented by a node, and the possibility of docking between any two proteins is indicated by an edge. The edges are undirected.

2. *Predator/prey networks.* Each animal species is represented as a node, and the fact that one species preys on another is represented by a directed edge, yielding a digraph representation.

3. *Metabolic networks.* Biochemical molecules are represented as nodes, and the transformation from one molecule into another is represented by a directed edge. It

is possible to express the effect of modulating agents (e.g., catalysts) in terms of weights associated with the edges, yielding a weighted digraph.

4. *Phylogenetic networks.* By representing each species as a node and the derivation of a species from another by directed edges, it is possible to obtain phylogenetic trees, which are specific cases of graphs.

5. *Imitation networks.* Each animal species is represented as a node, and the fact that one species imitates another is indicated by a directed edge.

6. *Disease transmission networks.* Each individual is represented as a node, and the act of transmitting the disease is indicated by a directed edge. Note that, provided the spatial positions of the individuals are known, it is possible to obtain a dynamic geographical network in which the positions of the nodes are allowed to vary as the individuals move in their environment.

7. *Social interaction networks.* Each individual is represented as a node, and social relationships are expressed as edges.

8. *Neuronal structures.* Graphs provide a natural means for representing neuronal systems at both the microscopic and the macroscopic levels. At the micro level, each neuron can be represented by a node, and synaptic connections between neuronal cells are indicated by directed edges. At the same time, connectivity between cortical areas can be naturally obtained by expressing each area as a node and connections between regions as directed edges (e.g., Costa and Sporns, 2005).

9. *Channel/flow networks.* The channel network underlying structures such as bones can be effectively represented by a graph whose edges stand for the channels, and each node for the confluence of two or more channels. The average width of a channel can be considered as the weight of the respective edge, indicating the flow capabilities of the network (Costa et al., 2006; Costa et al., 2005b).

There is virtually no limit to the potential of networks in representing biological structures and systems; moreover, such a potential can be further enhanced by combining distinct networks. For instance, it is possible to associate social networks and disease transmission networks in order to obtain a more complete model of the spread of disease. However, the full potential of such representations can be harnessed only by considering network representations where the activity at the nodes and the connectivity are allowed to vary with time. Such possibilities are covered in more detail in the remainder of this section.

Statistical Physics

One of the best-established areas of physics, statistical physics (or statistical mechanics), is aimed at establishing the relationship between the microscopic and macroscopic aspects of physical systems. For instance, the state of a gas inside a bottle can be described microscopically (up to quantum limits) in terms of the position and velocity of all the involved particles—appearing in a very large number, impossible to be stored in any modern computer—or by macroscopic measurements of temperature, volume, and pressure, which are *thermodynamic variables*. A great part of statistical physics has been developed by considering systems at thermodynamic equilibrium, allowing the proper representation of the state of the system in terms of stable averages of the thermodynamic variables.

The beginnings of statistical physics are associated with Daniel Bernoulli's developments aimed at relating the macroscopic variables of temperature and pressure to the microscopic measurement of the speed of the molecules. After its formalization by J. Willard Gibbs, statistical mechanics received contributions from many great physicists. One of the current challenges in this important area concerns its extension to systems out of equilibrium. Good textbooks on statistical physics include Mandl (1988) and Landau and Binder (2002). Although very important as an overall characterization of the dynamic evolution of a thermodynamic system, statistical physics's equilibrium configuration is very particular and does not correspond to many of the situations typically met in nature and biology. Formally speaking, a system is at equilibrium when the probabilities of occupancy of all its states are stabilized and no longer change with time. Alternatively, at the level of each individual state, the probability of leaving that state becomes equal to the probability of entering that same state.

The time taken by a system to enter equilibrium can vary substantially. In biology, however, most agents are operating far from equilibrium while replying to stimuli and perturbations from the environment. As expected, nonequilibrium statistical physics is much more challenging, since very often the dynamic equations are not known and must be predicted in some way. At the same time, many such investigations have to rely on intensive computations required for simulations.

Despite the large and diverse framework of concepts and methods developed in statistical physics, here we concentrate our attention on the concepts of *phase transition* and *percolation*. Informally speaking, a phase transition is characterized by the existence of a point along a dynamic process where the physical properties of the system of interest are observed to undergo an abrupt modification. For instance, the hardness of water increases suddenly as it transforms from its liquid to its solid state. Interestingly, the underlying dynamics of such physical phase transitions can be captured completely in analogous geometrical systems, without the need of any physical quantity: a phase transition is observed as the connectivity of a random graph increases steadily.

Informally speaking, percolations are phase transitions characterized by the sudden change of the connectivity of the constituent elements of the structure under analysis. For instance, the progressive removal of isolating obstacles along a nearly

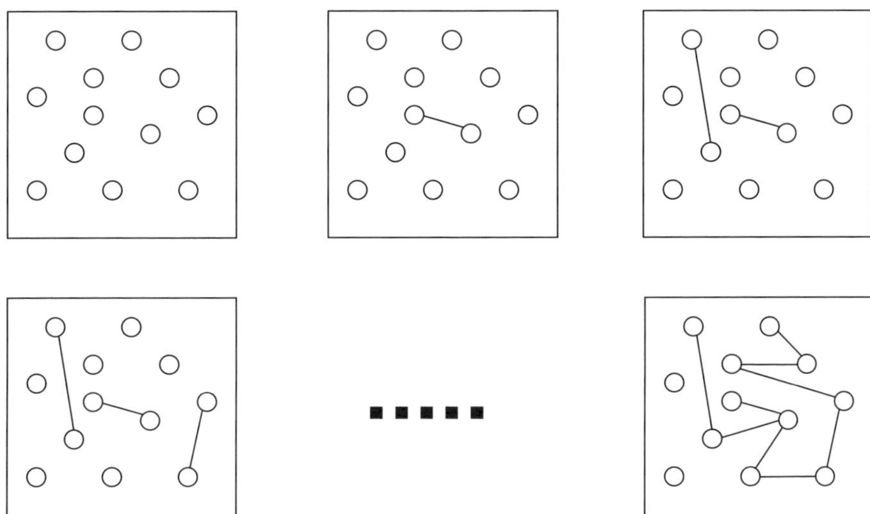

Figure 4.5
The topology of a network can be allowed to change by adding new edges between any two nodes chosen with uniform probability. This process ultimately leads to percolation, characterized by the appearance of a giant connected component.

porous channel uniting two reservoirs will eventually result in the establishment of a flow between those reservoirs. The percolation occurs at the point where the physical connection is thereby established. Because graphs are mathematical abstractions optimized for the representation of connectivity, they are inherently suitable for representing and modeling structures undergoing percolation.

Figure 4.5 illustrates a procedure to obtain an evolving random graph. It starts with a set of nodes, and pairs of nodes are successively chosen at random (with uniform probability) and connected through an edge. An immediate effect of such a growing strategy is that the average node degree will increase steadily while nodes are connected in pairs, and then in trees and cycles. When applied to a large graph containing many more nodes than figure 4.5, this process of increasing the number of edges leads to a point where a giant cluster appears in the graph (i.e., a connected component containing a large portion of the nodes in the graph). This *critical point* corresponds to unit average node degree. It can also be verified that after percolation, the number of nodes not connected to the giant cluster decreases exponentially. Such a phase transition is also called *percolation*, characterized by the sudden increase of connectivity among the involved elements.

Although ubiquitous in physics and material science, phase transitions are also essential for biology. For instance, the appearance of life, by corresponding to a sudden change of the property of matter, can be understood—though not yet explained—as a phase transition. Similarly, the emergence of consciousness can be considered a phase transition in which somebody enters a state of self-awareness and separation from the environment. Several other biological phenomena involve phase transitions, such as the onset of epidemics and the appearance of patterns during epigenesis. Interestingly, the accurate definition of phase transitions and percolations in biology is a tricky issue because the definition of the key concepts (e.g., life, consciousness, etc.) remain illusive.

One of the few exceptions to this situation is the appearance or extinction of a given species as a consequence of dynamic changes in environment and other species. Possibly many of the investigations in systems biology will eventually turn around the chicken-and-egg problem of defining more sophisticated biological properties and explaining them in terms of phase transitions. Actually, it may even turn out that the concept of phase transition, in relation to objectively quantified basic biological elements (e.g., molecules and reactions), will be used to define those properties.

Complex Networks

Complex networks can be understood as beginning with the random graph investigations of Flory (1942), Rapoport (1957), and Erdös and Rényi (1959). Note that, for historical reasons, we henceforth use the term *graph* rather than *network*. Although they are not identical, according to graph theory, recent common usage has made these two terms synonyms.

Random Networks One of the most important properties of random networks, introduced and discussed in the two previous sections, is that their connectivity can be reasonably characterized in terms of their average node degree. In other words, since this measurement is found not to vary too much along the several nodes of a random network, its average can be used as a good descriptor of its overall connectivity. In mathematical terms, this implies that the statistical distribution of the node degree has relatively small dispersion and is centered at the average. Indeed, the probability distribution of the node degree in random networks can be verified to be well fitted by a Gaussian, which implies a severe limitation to nodes with degrees that are too high or too low. Because of the statistical uniformity of the connections (i.e., nodes are chosen indiscriminately for connections), random networks define a good statistical reference against which other networks can be compared. At the same time, random networks are characterized by a relatively simple and regular connectivity which, as commented above, can be well characterized in terms of the node degree average alone.

Although the investigations on random networks led to a series of important theoretical results, especially those related to the dynamic evolution of the properties of the networks, unfortunately they proved not to be particularly useful for representing and modeling natural phenomena. The problem is that such phenomena and structures are not obtained by connecting pairs of nodes indiscriminately, but by taking into account properties of such nodes.

Small-World Networks Following random networks, the next important step along the path of the history of complex networks can be identified as corresponding to the identification and characterization of small world network models. It has been said for a long time (see Barabási, 2003) that human beings are related through just a few intermediate contacts; hence the name *small world*. For instance, in a given society, it is very likely that given any two people **A** and **B**, there will be a path of acquaintances from **A** to **B** involving just a few individuals (e.g., **A** knows **C**, who knows **D**, who knows **E**, who knows **B**). Such a striking property of networks was investigated in systematic fashion by Stanley Milgram and Mark Granovetter during the 1960s, and later by several other researchers, including Duncan Watts and Steven Strogatz (Watts and Strogatz, 1998). Figure 4.6 illustrates one of the possible means

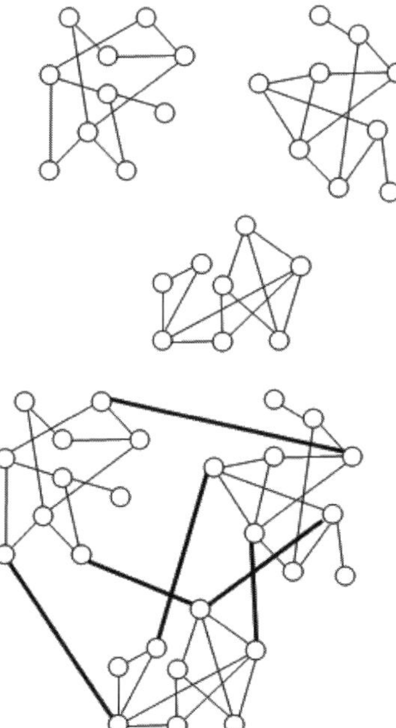

Figure 4.6
Start with some densely connected, but isolated, networks (a). Add some links (thicker lines) between nodes in different networks (b). The resulting network will exhibit the small-world property, characterized by the fact that there are relatively few links along the shortest path of most of pairs of nodes.

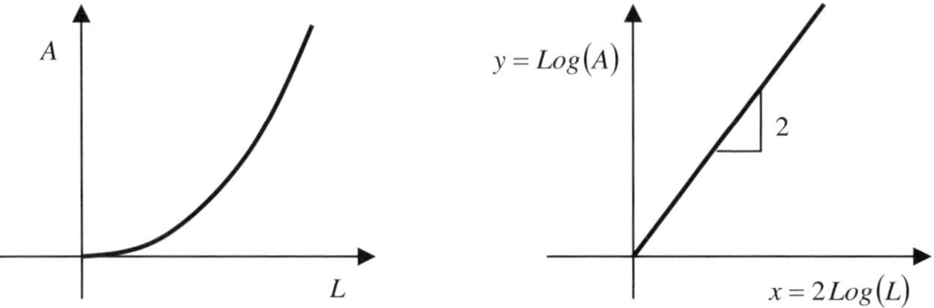

Figure 4.7
The relationship between the side of a square and its area shown in normal coordinates (a) and log-log coordinates (b). The fact that no special feature (e.g., a peak) can be found along the straight line relationship in (b) as one moves along the x axis reflects the scale-free relationship between A and L.

to obtain small world networks. One starts with some densely connected networks (three in this example) and adds some links between such networks. The resulting network will exhibit the small-world property.

While almost no real-world structure follows a uniformly random scheme of connections, most of them present the small world property. Therefore, while the value of random networks was impaired by the lack of adherence, small world networks proved to be too ubiquitous. For instance, even random networks are verified to exhibit the small world property.

Scale-Free Networks Let L be the length of the side of a square. The area of such a square therefore is given as $A = L^2$. After taking the logarithm at both sides of this equality, we have

$$Log(A) = Log(L^2) = 2\ Log(L).$$

If we now make $y = Log(A)$ and $x = 2\ Log(L)$, the graph of y versus x will yield a straight line with slope equal to 2, as illustrated in figure 4.7. Two entities (such as A and L) which are related through a *power law* (such as $A = L^2$) reveal a *scale-free* relationship between those two measurements. Note, however, that power laws have been more traditionally associated with negative relationships such as $Y = cX^{-b}$, where c is a constant and b is a positive real value.

The term *scale-free* is used to reflect the fact that there is no characteristic change of behavior of the relationship between Y and X as one moves along the horizontal axis (the scale). For instance, the rate of increase of the area A of the square of side L is always the same, independent of the value of L (the derivative of a linear relationship is constant). There is no value of L for which the variation of the area exhibits a peculiar behavior. This is not the case of the node degrees in a random network, since it can be shown that they can be well fitted by a Gaussian distribution which,

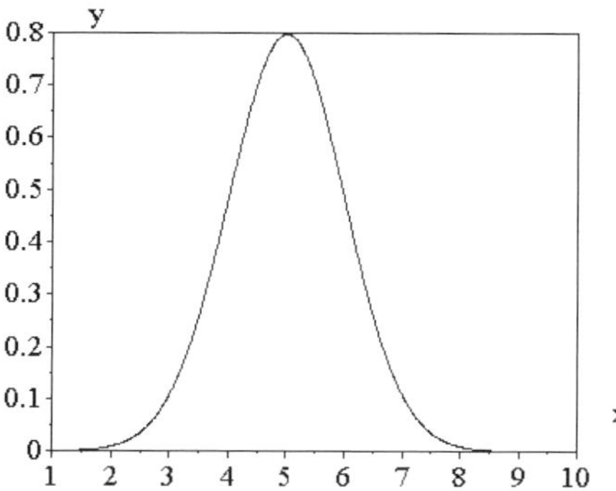

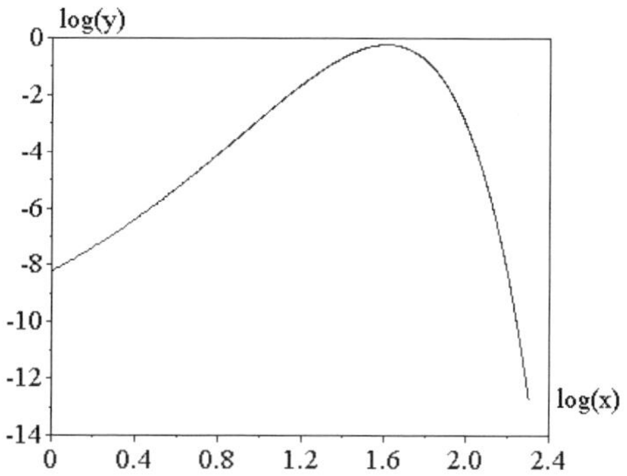

Figure 4.8
A Gaussian distribution of the values of x (a) and its representation in log-log axes (b). Since the rate of variation of $log(y)$ in terms of $log(x)$—the derivative of the curve in (b)—is not constant, there is *no* scale-free relationship. The peak occurring near $log(x) = 1.6$ in the log-log representation can be understood as the characteristic scale of the distribution.

when mapped into log-log axes, will not lead to a straight line. Indeed, a curve characterized by a peak will be obtained, as illustrated in figure 4.8.

A great part of the current interest in complex networks stems from the fact that several structures and phenomena in nature have been found to be characterized by node degree distributions which are well approximated by a power law or scale-free relationship. Although power laws had been observed for several natural and human-related measurements, such as the number of inhabitants of towns and cities (the so-called Zipf or Pareto distributions; e.g., Newman, 2004), the finding that important systems such as the Internet presented node degrees close to a scale-free relationship had major implications for the popularization of complex networks. Good overviews on complex networks can be found in Albert and Barabási (2002), Dorogovtsev and Mendes (2002), and Newman (2003), and a survey of measurements can be found in Costa et al. (2005a).

Figure 4.9 shows an example of a power law distribution: $y = x^{-3}$, plotted in standard and log-log axes. Note the lack of characteristic scales, as revealed by the constant derivative of the curve in (b). No part of the curve can be found which exhibits different variation; it is always constant.

In addition to defining a uniform behavior along the observation scales, distributions characterized by a scale-free relationship imply that nodes with high degree tend to be more likely in those distributions than in distributions such as the Gaussian. Actually, because the node degree values are intensely concentrated around the average degree, there is very little chance of observing many nodes with particularly large degrees. This is generally not the case with scale-free distributions, which

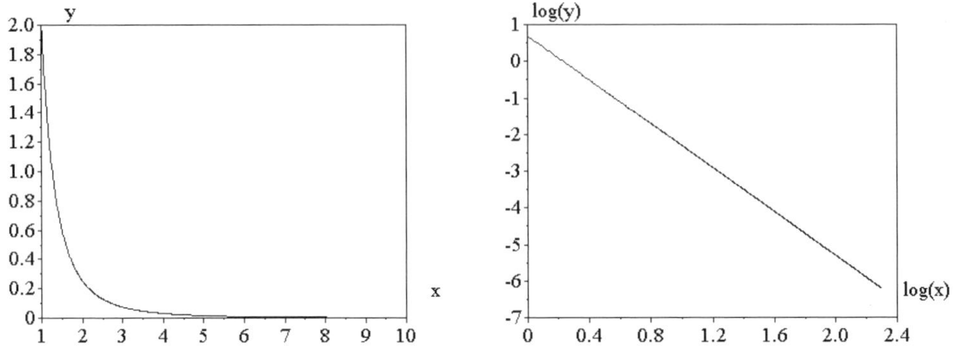

Figure 4.9
The power law relationship $y = x^{-3}$ shown in standard (a) and log-log (b) axes.

implies the existence of a considerable number of nodes with high degree, the so-called *hubs*.

Hubs are important because they concentrate connections, thus contributing substantially to the overall properties of the networks. For instance, given a generic node i in a scale-free network, there is a good probability that it will be connected to a hub. Since hubs are connected to several other nodes, the hub will provide a bypass, involving only two edges, to many other nodes. Consequently, hubs become critical to several processes performed on the network. For instance, in protein-protein interaction networks, where each protein is represented as a node and dockings are expressed as edges, the knockout of a hub protein may imply the collapse of several fundamental reactions, and eventually the death of the organism.

Geographical Networks Many biological networks, including protein-protein interaction systems, are characterized by presenting nodes which do not have a specific position in space. Indeed, while a specific protein will have a well-defined position inside a cell or organism, the nodes in protein-protein interaction networks do not stand for a specific protein molecule, but for *all* instances of that protein in the organism of interest. If the nodes can have their position clearly specified, we obtain a *geographical network*. Examples of geographical networks include the railway connections between cities and the networks of neurons in the nervous system. One particularly interesting aspect of geographical networks is the spatial adjacencies implied by the distribution of nodes along a space (typically 2-D or 3-D). For instance, the afferent synapses of a neuronal cell are likely to connect with neighboring cells. Similarly, animals in the wild tend to interact with those which live in neighboring regions. Interestingly, such a tendency to interconnect between neighbors, allied to a relatively uniform distribution of individuals, tends to yield near-regular networks (graphs whose nodes have a characteristic degree).

Communities One issue which has received growing interest in complex networks research is the identification of *communities* (e.g., Girvan and Newman, 2001). Informally, a community is characterized by presenting nodes which connect more intensely with one another than with the rest of the network. For instance, the network in figure 4.6b contains three clearly defined communities.

Interestingly, because of random fluctuations, communities can be found even in random networks, which, as discussed previously, are characterized by uniform connectivity. The identification of communities is an interesting task because it provides valuable insights about the organization and relationship between nodes. Finding communities in biological networks is particularly relevant. For example, the nodes belonging to communities identified in protein-protein interaction networks are likely to be part of specific metabolic cycles and other types of reactions. Similarly, communities in neuronal networks may correspond to modules of neuronal processing.

Dynamics in Networks While several efforts in complex networks research have been aimed at obtaining meaningful and informative measurements of the connectivity of the networks and at studying the evolution of the connectivity according to several growth schemes, an increasing number of works have focused on the investigation of dynamic processes taking place *on* complex networks. For instance, one may be interested in modeling and studying the evolution of neuronal activity in a network of cells interconnected in terms of random or scale-free networks (e.g., Stauffer et al., 2003; Costa and Stauffer, 2003). Another interesting subject has been the investigation of the synchronism between networks of oscillators. One particularly challenging possibility for investigation regards the study of complex networks whose connectivity changes as a consequence of the dynamics of the nodes, such as Hebbian reinforcement of connections between neuronal cells.

Some Examples of Complex Networks' Applications to Biology

This section describes three examples of applications of complex networks to biological systems. Although such developments have been reported continuously in the literature, the following examples more closely reflect recent developments by the author. Additional information about the application of complex networks to biology can be obtained from Barabási and Oltvai (2004) and from the *archive* http://arxiv.org/archive/ or a public repository or articles.

Relating Protein Connectivity and Essentiality

The web of interaction between proteins provides one of the most fundamental scaffoldings for sustaining life because through such interactions the all-important structure of proteins can be changed. Recall that proteins are extremely important not

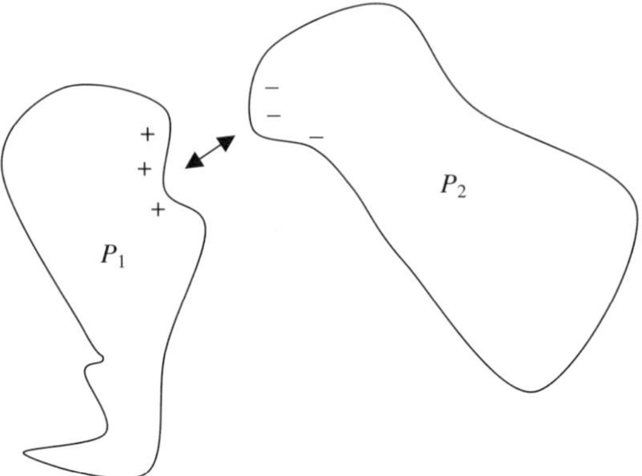

Figure 4.10
The interaction (docking) between two proteins P_1 and P_2 requires geometrical and force field compatibility at the region of contact (arrow). In this case, the concave shape at the connecting place in protein P_1 is matched by a convex region in P_2. In this case, affinity between electric fields of force is also observed.

only because of their structural role (e.g., in defining the cytoskeleton's mechanical properties) but also as regulators (catalysts) of many biochemical reactions. Typically, the interaction between two proteins occurs as a consequence of the compatibility between geometry and force fields at the site of connection (or docking). In other words, at such a place, the geometry of one of the proteins has to allow the coupling (e.g., a concave part docks at a convex site). At the same time, the forces at these sites have to imply attraction between the proteins (e.g., positive against negatively charged fields at each protein). The physical interaction (docking) between two proteins is symbolically illustrated in figure 4.10.

As a consequence of continuing technological and basic progresses, a growing amount of data on protein-protein interaction has been obtained and become publicly available through Internet repositories (e.g., http://cyvision.if.sc.usp.br/~francisco/networks). Increasing amounts of information have also been obtained about the special role of some proteins, such as those called *essential* or *lethal*, which play a decisive role in maintaining the life of the cell or individual. Such networks have repeatedly been verified (e.g., Jeong et al., 2001) to exhibit scale-free distributions of node degrees.

In one of the first remarkable applications of complex network concepts to protein-protein interaction networks, Jeong and Barabási (Jeong et al., 2001) verified that essential proteins in *S. cerevisae* tended to exhibit many connections, thereby establishing an interesting putative relationship between essentiality and hubs. Since then, other related works have been reported. In Wuchty (2001, 2002), networks between protein domains have been obtained and analyzed. Two principal methods have been considered: (a) single-domain proteins and (b) domains which appear at least once in the same protein are connected. It has been shown that the degree of the domain follows a power law. Another investigation (Ng et al., 2003) considered weights assigned to the edges and interactions in protein complexes.

Neuromorphic Complex Networks

Neuroscience provides one of the most challenging and exciting areas (see chapters 6 and 10 in this volume) in which to develop and test concepts from systems biology. As described above, given a nervous structure, it is possible to obtain a complex network whose nodes correspond to the neurons, and its directed edges, to synaptic connections between neurons. Thus, the system of synaptic connectivity between the neuronal cells is completely captured by the complex network representation. A large number of investigations on both topological and functional properties of neuronal networks can be obtained by considering the complex network representing the connectivity of a given neuronal subsystem.

From a purely topological perspective, it is possible to use the average degree of the nodes in the network to characterize the overall connectivity of the neuronal structure, in the sense that the higher this value, the more connected the neurons are on average. In turn, the clustering coefficient provides an indication of how much the neurons in the neighborhood of each neuron are interconnected. Additional information about the connectivity of the neuronal structure under analysis can be obtained in terms of the distribution of shortest paths. While the neurons distributed spatially along a small cortical area tend to establish connections in their more immediate neighborhood, intra- and intercortical axonal projections contribute to reducing the average shortest path between pairs of neurons, eventually yielding small world properties. Another interesting possibility stems from the fact that neuronal cells have well-defined spatial positions, allowing the derivation of geographical complex network representations, from which additional insights about the architecture of the neuronal structure can be inferred. For instance, it is possible to try to correlate the degree of a node in a specific position with the fractal dimension defined by the distribution of its most immediate neighbors.

While many other possibilities regarding the topological characterization of complex neuronal networks can be explored, one issue deserving special attention concerns the investigation of the relationship between the shape/structure of the neuronal cells and structures and their respective functionality (e.g., Costa, 2005a). This important issue has been somewhat overlooked because neuronal structures are usually observed and analyzed after their development. A direct consequence of this approach is to associate the dynamic behavior of the neuronal cells to the static pattern of synapses. However, there are at least three reasons why such a perspective is incomplete: (1) the neuronal connectivity is not static, but changes continuously with the addition and deletion of new synapses, consequences of the change of the neuronal morphology; (2) the current connectivity pattern is a consequence of a long developmental process in which the cells, by changing their shapes, seek to establish proper synapses as a consequence of their biochemical preprogramming and environmental stimuli; and (3) the activity of an individual neuronal cell depends not only on its efferent synapses but also on other factors, such as the local concentration of ions and other chemicals. Valuable insights about all these effects and processes can be obtained by considering complex networks as representations of neuronal structures.

As part of a continuing research line initiated in 1992, the author has investigated several aspects of the relationship between neuronal shape and function (Costa, 2005a). Initially, efforts were made to identify and propose measurements of the individual neuronal cell morphology, which were particularly meaningful with respect to several related questions, such as the identification of the potential connectivity of a given neuron. Efforts were also aimed at the use of representative sets of measurements in order to computationally synthesize neuronal cells whose morphological properties reflected those of a specific class of real neurons.

Programs capable of synthesis of neuronal cells have been developed (e.g., Coelho and Costa, 2002) which produce morphologically realistic neuronal cells by sampling, through the Monte Carlo methodology, statistical models describing the geometrical properties of the real neurons. Such programs have also been used to obtain morphologically realistic neuronal networks. After distributing the cell somata according to some statistical criterion (e.g., uniform distribution or normal distribution around a center of reference), the program is used to grow, step-by-step, the neuronal shape of each cell according to given statistical models. Finally, dynamic activity can be obtained from such a system by associating specific functional models of neurons to each of the cells in the growth structure. The considered functional models can range from coarse approximations such as Hopfield to models as complete as Hodgkin-Huxley and transmission cable, passing through the standard Rosenblatt neuronal model.

Although such investigations have been carried out since 1992, the recent consideration of the concepts and paradigms of complex networks has allowed a natural and powerful integration of such research activities. For instance, the concept of percolation in graphs has been applied to characterize the potential of morphologically realistic neurons to make connections (Costa and Manoel, 2003). More specifically, model neurons are incorporated into the system while the connectivity between them is monitored. Once the system has percolated, the density of neurons at that point is recorded and used as a quantifier of connectivity. A more recent study (Costa and Coelho, 2005) considered a fixed quantity of neurons which were allowed to grow according to specific statistical models until percolation was attained. Such an investigation made it clear that specific morphological features of the classes of cells play a decisive role in defining the overall connectivity pattern of the emerging network.

Investigations have also been conducted regarding the relationship between the shape of the neurons and their behavior. A recent related investigation about the relationship between neuronal shape and memory recall in Hopfield networks (Rodrigues et al., 2005) targeted the quantification of the importance of hubs, showing that although neurons with many connections (hub neurons) are particularly important for the network dynamics in recovering patterns, the removal of the same number of edges at random from the network may have a more intense effect.

More recently (Costa and Sporns, 2005), the set of hierarchical measurements proposed in Costa (2004) and Costa and Silva (2004) was applied in order to characterize the architecture of connections between the main cortical areas in the cat and the macaque. Because of the extensive information provided by the hierarchical measurements, which reflect a wider neighborhood than would be possible by using the traditional measurements of node degree and clustering coefficient, a particularly complete characterization of the connectivity of each cortical area was obtained, leading to the identification of well-defined, distinct properties between areas such as dorsal and ventral.

Characterizing Cortical Bone Structure

Although they involve a high percentage of dead matter, the interior of the bones of mammals present a very intricate network of voids and channels. Cortical bones, normally found along the middle part of elongated bones, are particularly important because of their mechanical properties and highly complex network of channels, which can be divided into two main groups: Havers and Volkmann. As proposed by Costa et al. (2006), it is possible to capture the essential architecture of such channel systems by a complex network whose nodes represent the confluence of channels, while the edges correspond to the channels. Such a complex network representation of the cortical structure of channels clearly expresses the topology of that network, allowing investigations of several biological and medically relevant issues (see figure 4.11).

For instance, unlike what would be expected in the vascular system, we identified several cycles present in the channel system (Costa et al., 2005b). The existence of such channels has been related to the need to provide some degree of redundancy and bypasses along the channel system, in order to cope with obstructions. This hypothesis was further investigated by performing edge attack on the network (Costa et al., 2005b); that is, successively selecting, in random fashion, an edge from the network and removing it. The resilience of the channel network resulted in a figure higher than that of a tree with the same number of nodes and edges, and similar

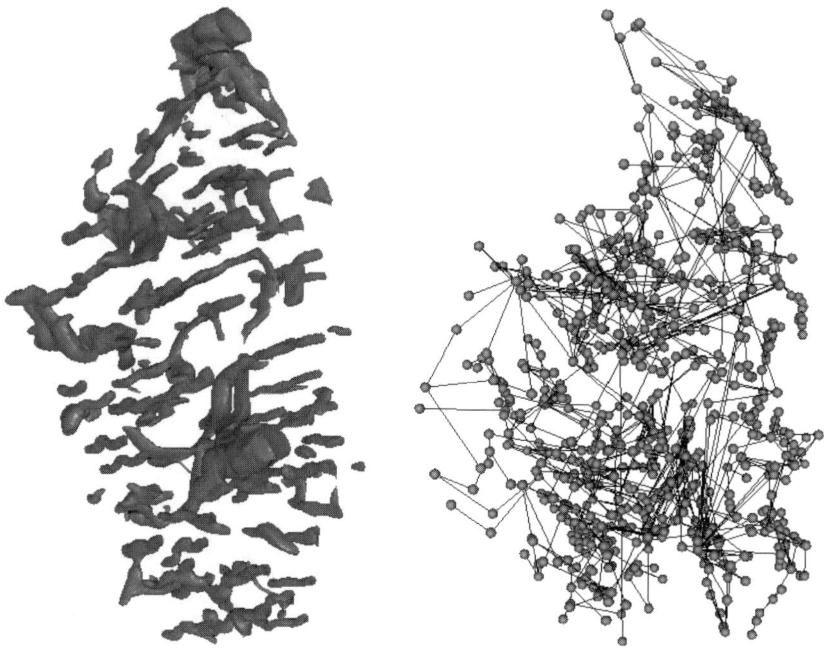

Figure 4.11
Three-dimensional reconstruction of part of the considered cortical bone channel system (a) and the complex network obtained from the whole reconstruction (b). (Histological preparation by Marcelo Beletti; reconstruction and graph extraction by Matheus P. Viana.)

to that of a regular network. By using the method described in Costa (2004), it was also possible to identify the longest hierarchical backbone of the network, and the edges remaining after removal of such a backbone were found to belong to well-defined spatial communities (subnetworks of connected nodes which are near one another).

The node degree distribution of the bone network was characterized by a small world nature, corroborating the effectiveness of the irrigation capabilities of the system. In addition, the log-log node distribution of this complex network came close to being scale-free, thus indicating the presence of hubs corresponding to particularly critical channel confluences.

Concluding Remarks and Future Perspectives

After revising the main basic concepts in complex networks theory, including several measurements of connectivity and the important issues of phase transition and percolation, I have discussed how complex networks can be particularly useful for representing and modeling biological structures and systems. Examples have been presented of the important issues of protein connectivity/essentiality, bridging the gap between neuronal shape and function, and the analyses of complex geometrical biological structures, such as the system of channels underlying cortical bone structure.

Although many investigations and breakthroughs in physics and the biological sciences have been founded by reductionism and specialization, the many dimensions and relationships involved in biological systems demand more integrative approaches capable of representing the myriad links between biological entities over time and space. By uniting powerful concepts and methods from the areas of graph theory and statistical physics, complex networks are one of the most suitable frameworks for representing and modeling natural and man-made systems.

Moreover, the integration of different networks provides an effective means for incorporating the different systems pervading biology, such as protein-protein interaction, metabolic networks, evolution of species, prey-predator relationships, and so on. In addition, given their inherent integrative nature, complex networks are a natural choice for addressing *systems biology* (Kriete and Eils, 2006). As is often the case with newly created terms, it is particularly difficult to provide a clear-cut definition of systems biology. Generally speaking, this area represents a reaction to reductionism, which was necessary in order to try to integrate and accommodate the virtually unending complexity of life and natural sciences. Because of their representational flexibility and ability to naturally integrate the connectivity and functionality of the analyzed systems, while the topology and activity of the network are allowed to follow an intertwined dynamics, complex networks are poised to provide one of the main means not only for achieving effective systems biology approaches but also for general scientific investigation. Such developments will involve three stages, possibly implemented simultaneously:

1. *Representation* of the several biological entities in terms of possibly hierarchical networks. Basic elements (e.g., proteins, genes, etc.) can be represented as nodes, with the related processes being represented as a series of hierarchically organized subnetworks (i.e., networks of subnetworks).

2. *Characterization* of such network representations in terms of statistical analysis of several measurements of network connectivity (e.g., node degree, clustering coefficient, shortest paths, etc.). The characterization of such systems is useful not just in itself but also for its valuable clues for development of respective models.

3. *Modeling and simulation* of the biological systems/subsystems represented in terms of biological networks. Such models and simulations should by no means be restricted to static networks, but also should encompass complex network representations which are allowed to have their connectivity changed over time, possibly as a

consequence of the history of their dynamic activity. Refinement of such models can be obtained through the confrontation of the obtained results with the natural data, as in the classical scientific method.

In this way, the contribution of complex networks to unifying biological investigation can be understood as providing the means for systematizing and enhancing the scientific method, so that ever more successful emulations of nature can be obtained and used for predictions and for driving new questions and truth in science.

Acknowledgments

I am grateful to Fundação de Amparo à Pesquisa do Estado de São Paulo—FAPESP (proc. 99/12765-2), the Conselho Nacional de Desenvolvimento Científico e Tecnológico—CNPq (proc. 308231/03-1); and the Human Frontier Science Program (RGP39/2002) for financial support.

References

Albert R, Barabási AL (2002) Statistical mechanics of complex networks. Rev Mod Phys 74: 47–97.

Barabási AL (2003) Linked. New York: Plume.

Barabási AL, Oltvai Z (2004) Network biology: Understanding the cell's functional organization. Nat Rev Gen 5: 101–113.

Coelho RC, Costa LdaF (2002) Realistic neuromorphic models and their application to neural reorganization simulations. Neurocomputing 48: 555–571.

Costa LdaF (2004) The hierarchical backbone of complex networks. Phys Rev Lett 93(9): 1–4.

Costa LdaF (2005a) Morphological complex networks: Can individual morphology determine the general connectivity and dynamics of networks? q-bio.MN/0503041.

Costa LdaF (2005b) Socioeconomic development and stability: A complex network blueprint. Physics/0505008.

Costa LdaF, Coelho RC (2005) Growth-driven percolations: The dynamics of community formation in neuronal systems. Eur Phys J 47: 571–581.

Costa LdaF, Manoel ETM (2003) A percolation approach to neural morphometry and connectivity. Neuroinformatics 1(1): 65–80.

Costa LdaF, Rocha LEC (2006) A generalized approach to complex networks. Eur Phys J B50(1–2): 237–242. Also cond-mat/0408076.

Costa LdaF, Rodrigues FA, Travieso G, Villas-Boas PR (2007a) Characterization of complex networks: A survey of measurements. Adv Phys 56(1): 167–242.

Costa LdaF, Silva FN (2004) Hierarchical characterization of complex networks. cond-mat/0412761.

Costa LdaF, Stauffer D (2003) Associative recall in non-randomly diluted neuronal networks. Physica A330(1–2): 37–45. Also cond-mat/0302040.

Costa LdaF, Sporns O (2005) Hierarchical features of large-scale cortical connectivity. Eur Phys J B48(4): 567–574.

Costa LdaF, Viana MP, Beletti ME (2005b) Hierarchy, fractality, small-world and resilience of Haversian bone structure: A Complex network study. q-bio.TO/0506019.

Costa LdaF, Viana MP, Beletti ME (2006) The complex channel networks of bone structure. Appl Phys Lett 88: 033903.

Dorogovtsev SN, Mendes JFF (2002) Evolution of networks. Adv Phys 51: 1079–1187. Also cond-mat/0106144.

Erdös P, Rényi A (1959) On the evolution of random graphs. Pub Math 6: 290–297.

Erez T, Hohnisch M, Solomon S (2004) Statistical economics on Solomon networks. cond-mat/0406369.

Flory PJ (1942) Molecular size distribution in three-dimensional polymers: I, II, III. J Am Chem Soc 63: 3083–3100.

Girvan M, Newman MEJ (2001) Community structure in social and biological networks. Proc Natl Acad Sci USA 99: 7821–7826.

Jeong H, Mason SP, Barabási AL, Oltvai ZN (2001) Lethality and centrality in protein networks. Nature 411: 41. Also cond-mat/0105306.

Kriete A, Eils R (eds) (2006) Computational Systems Biology. San Diego: Academic Press.

Landau DP, Binder K (2002) A guide to Monte Carlo simulations in statistical physics. Cambridge: Cambridge University Press.

Mandl F (1988) Statistical Physics. New York: Wiley.

Newman M (2004) Power laws, Pareto distributions and Zipf's law. cond-mat/0412004.

Newman MEJ (2003) The structure and function of complex networks. SIAM Rev 45: 167–256. Also cond-mat/0303516.

Ng SK, Zhang Z, Tan SH (2003) Integrative approach for computationally inferring protein domain interactions. Bioinformatics 19: 923–929.

Rapoport A (1957) Contribution to the theory of random and biased nets. Bull Math Bioph 19: 257–277.

Rodrigues EP, Barbosa MS, Costa LdaF (2005) On the importance of hubs in Hopfield complex neuronal networks under attack. cond-mat/0507677.

Stauffer D, Aharony A, Costa LdaF, Adler J (2003) Efficient Hopfield pattern recognition on a scale-free neural network. Eur Phys J 32(3): 395–400. Also cond-mat/0212601.

Watts DJ, Strogatz SH (1998) Collective dynamics of small-world networks. Nature 393: 440–442.

Wuchty S (2001) Scale-free behavior in protein domain. Mol Biol Evol 9: 1694–1702.

Wuchty S (2002) Interaction and domain networks of yeast. Proteomics 2: 1715–1723.

III FORM

Morphology, structural analysis, and other scientific approaches to the study of natural forms have always employed a variety of modeling methods ranging from the physical models of early anatomists and chemists to today's computer-based reconstructions, visualizations, and analyses. These disciplines thus look back at a long tradition of modeling in biology, a tradition that also includes many examples of successful interactions between modeling, empirical research, and educational endeavors, such as the wax models in anatomy and embryology, or the now famous model-building strategy that led to the discovery of the double helix. Over the last few centuries structural models have been developed at many different levels of complexity, a fact that is reflected in the three chapters of this section, which include models of organismal (McGhee), cellular (Samsonovich and Ascoli), and molecular morphology (Bajaj).

The modeling approaches discussed in this section share many elements, strategies, and goals, despite the differences in their objects of analysis. All distinguish, explicitly or implicitly, between the following dimensions of modeling in biology: (1) conceptual models, used to map and delineate problems and explore their essential characteristics; (2) complex but computationally tractable models, which incorporate as many aspects of particular phenomena as are feasible within the technical constraints of the modeling platform; (3) analytical models, which reduce the complexity of phenomena in order to discover their essential features and parameters, thus allowing for their reconstruction within the confines of the model; and (4) biophysical models, which attempt to explain the properties of the system in reference to underlying biophysical laws. These modeling strategies thus involve both the reconstruction of phenomena (including their visualization) and their explanation through the application of organizational principles, which often allows for a reduction of complexity.

George McGhee (chapter 5) discusses the modeling strategies and analytic techniques of theoretical morphology and their application to the analysis of form and constraints. Insofar as evolutionary biology investigates patterns and processes of life on Earth, theoretical morphology provides an important analytical model for this endeavor. As McGhee explains, the analysis of possible forms and the decomposition of the theoretical morphospace—an all-inclusive N-dimensional space that mathematically describes all possible forms—into sets of possible and impossible variants enables us to distinguish the effects of different types of constraints.

McGhee provides a set-theoretical framework for the decomposition of the generalized morphospace into different, mutually exclusive subsets, such as geometrically possible and impossible, functionally possible and impossible, phylogentically possible and impossible, and developmentally possible and impossible forms. He shows how constraint boundaries can be mapped as a consequence of these decompositions. Within the modeling framework of theoretical morphology, different types of constraints are defined as the boundaries between the possible and the impossible. However, since the exact locations of these boundaries in many instances have to be determined empirically, modeling in theoretical morphology guides and stimulates empirical research.

Alexei Samsonovich and Giorgio Ascoli (chapter 6) present computational models of dendritic morphology. The high degree of structural complexity and the natural variability of neuronal cells make a morphological description of their variable properties almost impossible. To address this problem, Samsonovich and Ascoli developed computational algorithms that are capable of generating neuronal morphologies with the same shapes found in nature. Their algorithms are based on parameters derived from empirical measurements and employ stochastic methods (Markov models) to simulate neuronal shapes. Samsonovich and Ascoli demonstrate that an elegant hidden Markov algorithm, where the hidden variables correspond to two measurable quantities (the expected number of terminal tips of a dendritic subtree and the path distance from the soma of the neuron), is able to accurately describe dendritic morphologies of two classes of pyramidal cells in rats, with all functionally relevant statistical details.

In this case computational modeling is employed as an analytical tool that (1) allows the generation of neuronal morphologies based on simple and elegant models, thus reducing the descriptive complexities to a manageable size; (2) reveals hidden variables that in turn suggest possible mechanistic causes of the development of these forms; and (3) allows for the detection of different mechanistic causes in the development of different cell types. The latter is based on a comparison with a different set of cells that could be simulated only with a different algorithm, one based on different hidden variables. Models of dendritic morphology are thus powerful tools of parsimonious description and also facilitate insights into the underlying biological processes that generate morphological complexity.

Chandrajit Bajaj (chapter 7) describes recent developments in computational modeling and visualization of viral ultrastructures. Viruses occupy a special niche in biological modeling; as the smallest nano-organisms, lacking most of the molecular and physiological machinery characteristic of cells, they closely resemble other self-assembling molecular structures. Several of the methods described by Bajaj are also employed in structural biology and chemistry, and, on the other hand, viral infection dynamics are among the most widely modeled population dynamic processes. In his chapter Bajaj focuses on modeling viral structures and the geometrical patterns and molecular interactions that generate these structures.

These models are, in part, motivated by the need for adequate visual representations of viral structures, since direct observations of these nanoparticles—at resolutions ranging from less than 4Å to 15Å—are not easily interpreted. Rather, empirical observations and measurements need to be transformed into visual representations with the help of geometric algorithms. Such an endeavor is possible because most viruses are highly symmetrical. The resulting models thus also reveal organizational principles on the molecular scale, not unlike the modeling strategies in theoretical morphology discussed by McGhee.

5 Modeling the Spectrum of Existent, Nonexistent, and Impossible Biological Form: A Research Program

George R. McGhee

The Theoretical Spectrum of Biological Form

The discipline of theoretical morphology is fundamentally concerned with the analysis of evolutionary constraint (McGhee, 2001). Theoretical morphology seeks to simulate the range of forms that biological entities can take and to discover why certain forms exist in nature while other forms, although geometrically possible, do not (Raup and Michelson, 1965; McGhee, 1999; Eble, 2000; Maclaurin, 2003).

Much confusion exists at present concerning the concept of evolutionary constraint and the various types of constraint, although the concept itself is not new (see the reviews of Maynard Smith et al., 1985; Antonovics and van Tienderen, 1991; McKitrick, 1993; Schwenk, 1995; and Blomberg and Garland, 2002). For example, one group of potential constraints has been variously called developmental, architectural, fabricational, constructional, ontogenetic, and morphogenetic. These have been contrasted with, or sometimes equated to, phylogenetic constraints, historical constraints, canalization constraints, and phylogenetic inertia. The resulting confusion in terminology has prompted one set of reviewers to propose, tongue-in-cheek, the term "ontoecogenophyloconstraints" to cover all concepts (Antonovics and van Tienderen, 1991).

In this chapter a theoretical morphospace analysis of the concept of evolutionary constraint will be used, and the types of constraints considered will be explicitly defined by the use of Venn diagrams and set theory notation. Any given biological form, f, may be described by a set of measurements taken from that form. Each type of measurement, such as length, width, or height, can be considered as a form dimension. The total set of possible dimensions of form can be used to construct a hyperdimensional space of possible form (figure 5.1). All coordinate combinations within this space represent a universal set of form, U. Some of the coordinate combinations within the total hyperspace of form represent the set of geometrically possible forms: $GPF = \{f \mid f = \text{geometrically possible forms}\}$. Other coordinate

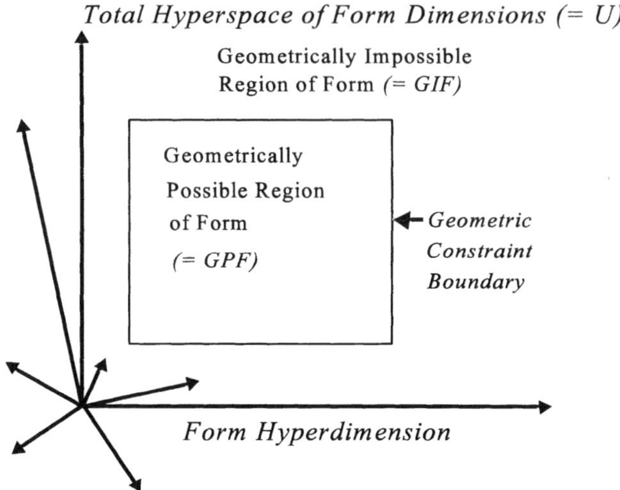

Figure 5.1
Geometric constraint as the boundary between the set of geometrically impossible forms (*GIF*) and the set of geometrically possible forms (*GPF*) in the total hyperspace of form dimensions (the universal set of form, *U*).

combinations within the total hyperspace represent the set of geometrically impossible forms: $GIF = \{f \mid f = \text{geometrically impossible forms}\}$.

The regions of impossible and possible forms within the total hyperspace form do not overlap one another; that is, a given set of coordinates within the hyperspace cannot simultaneously represent a possible geometry and an impossible geometry. The sets of form *GIF* and *GPF* are thus complements of one another:

$$GIF \cup GPF = \{f \mid f \in GIF \text{ or } f \in GPF\} = U,$$

$$GIF \cap GPF = \{f \mid f \in GIF \text{ and } f \in GPF\} = 0.$$

The boundary between the form sets *GIF* and *GPF* is here designated as the *geometric constraint boundary* (figure 5.1).

Within the region of geometrically possible forms there are two subregions: a region of possible forms that are functional in nature, and a region of nonfunctional forms (figure 5.2). The totality of these forms represents the two form sets $FPF = \{f \mid f = \text{functional possible forms}\}$ and $NPF = \{f \mid f = \text{nonfunctional possible forms}\}$. The word "nonfunctional" is used here in the sense of lethal function; that is, possession of nonfunctional form does not allow the organism to survive in nature.

The regions of functional and nonfunctional form within the region of geometrically possible form do not overlap one another; that is, a given set of coordinates within the geometrically possible region cannot simultaneously represent a form

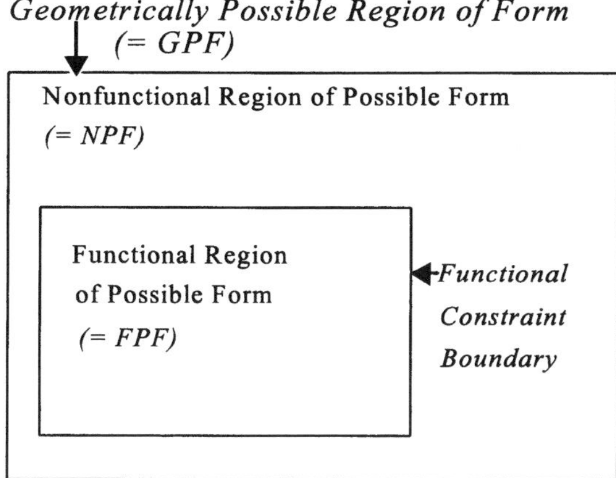

Figure 5.2
Functional constraint as the boundary between the set of nonfunctional possible form (*NPF*) and the set of functional possible form (*FPF*) within the set of geometrically possible form (*GPF*).

Table 5.1
Theoretical morphospace characterization of evolutionary constraints

I. Extrinsic Constraints (imposed by the laws of physics and geometry)
1. Geometric Constraint: the boundary between form sets *GPF* and *GIF*.
2. Functional Constraint: the boundary between form sets *FPF* and *NPF*.
II. Intrinsic Constraints (imposed by the biology of a specific organism)
1. Phylogenetic Constraint for taxon *x*: the boundary between form sets *PPFx* and *PIFx*.
2. Developmental Constraint for taxon *x*: the boundary between form sets *DPFx* and *DIFx*.

that is both functional and nonfunctional. Thus, the sets of form *FPF* and *NPF* are complements of one another and are contained within the form set *GPF*:

$$FPF \cup NPF = \{f \mid f \in FPF \text{ or } f \in NPF\} = GPF,$$

$$FPF \cap NPF = \{f \mid f \in FPF \text{ and } f \in NPF\} = 0.$$

The boundary between the form sets *NPF* and *FPF* is here designated as the *functional constraint boundary* (figure 5.2).

The concepts of geometric constraint and functional constraint are here considered to belong to the class of *extrinsic constraint*, where extrinsic constraints are those imposed by the laws of physics and geometry (table 5.1). Extrinsic constraints exist whether any actual biological form encounters them or not. A separate class of constraint is that of *intrinsic constraint*, where intrinsic constraints are those imposed by the biology of a specific organism. Intrinsic constraints do not exist in the absence of an actual organism.

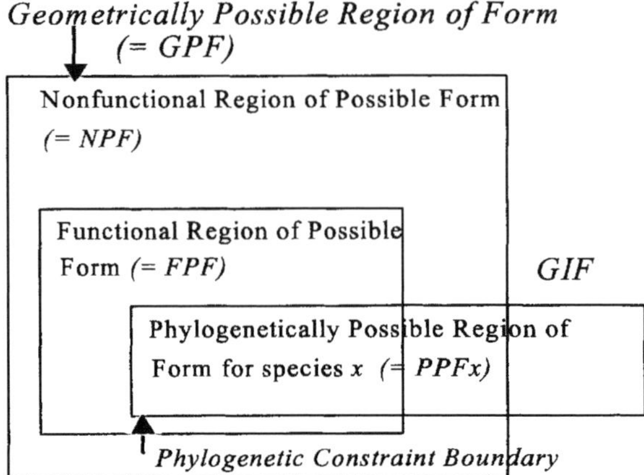

Figure 5.3
Phylogenetic constraint for species *x* as the boundary of the set of forms that are possible, given the genetic coding present in species *x* (*PPFx*). Note that this set has boundaries that hypothetically may cross the boundaries of both functional and nonfunctional possible form (*FPF* and *NPF*), and the boundaries of both geometrically possible and impossible form (*GPF* and *GIF*).

At least two conceptually different types of intrinsic constraint exist: phylogenetic constraint and developmental constraint (table 5.1). In figure 5.3, *phylogenetic constraint for species x* is illustrated as the boundary of the set of forms that are possible, given the genetic coding for form that is present in species *x*, and these forms represent the form set $PPFx = \{f \mid f = \text{phylogenetically possible forms for species } x\}$. Note that this form set has boundaries that hypothetically may cross the boundaries of both functional and nonfunctional possible form (*FPF* and *NPF*), and the boundaries of both geometrically possible and impossible form (*GPF* and *GIF*). That is, species *x* may possess the genetic coding to produce functional form, but it may also possess the genetic coding for nonfunctional form. Hypothetically, species *x* may also possess the genetic coding that could potentially produce coordinate combinations in the total hyperspace of form dimensions that are geometrically impossible (figure 5.3).

We can formally consider these three possibilities as set intersections:

$$PPFx \cap FPF = \{f \mid f \in PPFx \text{ and } f \in FPF\} = \text{potential existent form for species } x,$$

$$PPFx \cap NPF = \{f \mid f \in PPFx \text{ and } f \in NPF\} = \text{lethal form for species } x,$$

$$PPFx \cap GIF = \{f \mid f \in PPFx \text{ and } f \in GIF\} = \text{developmentally impossible form.}$$

Only the set intersection $PPFx \cap FPF$ represents genetic coding combinations that will produce functional form in nature. Lethal mutations lie in the set intersection

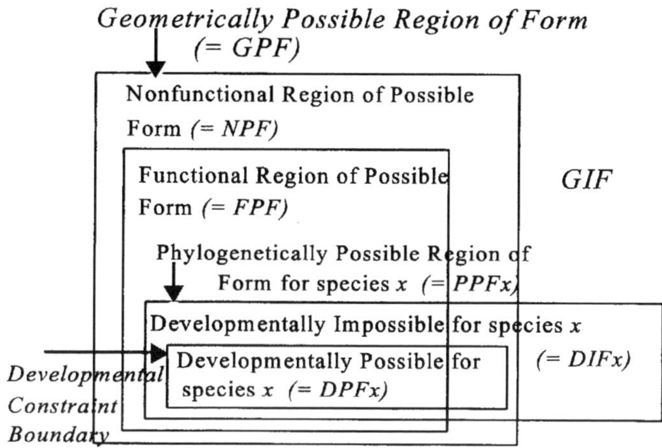

Figure 5.4
Developmental constraint for species x as the boundary of the set of forms that can be successfully developed by species x ($DPFx$) and the set of forms that cannot be developed by species x ($DIFx$) within the set of forms that are phylogenetically possible for species x ($PPFx$).

$PPFx \cap NPF$. The hypothetical possible intersection $PPFx \cap GIF$ (i.e., genetic coding for forms that are impossible to develop) is of interest because it leads us to the second intrinsic constraint, that of development.

In figure 5.4, developmental constraint for species x is illustrated as the boundary between the two sets $DPFx = \{f \mid f =$ developmentally possible forms for species $x\}$ and $DIFx = \{f \mid f =$ developmentally impossible forms for species $x\}$. The regions of developmentally possible and developmentally impossible form within the region of phylogenetically possible form for species x do not overlap one another; that is, a given set of genetic codings cannot simultaneously produce a form that can and cannot be developed. Thus $DPFx$ and $DIFx$ are complements of one another, and are contained within $PPFx$:

$$DPFx \cup DIFx = \{f \mid f \in DPFx \text{ or } f \in DIFx\} = PPFx,$$

$$DPFx \cap DIFx = \{f \mid f \in DPFx \text{ and } f \in DIFx\} = 0.$$

Note that the set of developmentally impossible forms for species x, $DIFx$, could hypothetically include forms that are geometrically impossible ($PPFx \cap GIF$), forms that are geometrically possible but nonfunctional ($PPFx \cap NPF$), and forms that are both geometrically possible and functional ($PPFx \cap FPF$), but cannot be developed (figure 5.4). Developmentally possible form must also be phylogenetically possible and geometrically possible, but can be either functional ($DPFx \cap FPF$) or lethal ($DPFx \cap NPF$).

For the programmer, the distinction between phylogenetic constraint and developmental constraint in morphogenesis may be illustrated by two analogous programming questions: (1) phylogenetic constraint: Is the source code (i.e., the genetic coding) for the morphogenetic program available? (2) developmental constraint: Will the program run, or will it crash? That is, if the code is present, we can potentially simulate the form; otherwise, we cannot. But, as every programmer knows, possessing the code specifically written to produce a form does not necessarily mean that the program will run on a particular computer or that the form can be simulated.

The theoretical morphospace characterization of the evolutionary constraint concept used in this chapter is summarized in table 5.1. The extrinsic constraint boundaries of geometric and functional constraint in form hyperspace are absolute, and do not vary with time. That is, the laws of geometry and the physics of swimming are the same today as they were 100 million years ago. This is not true of intrinsic constraint boundaries. Both phylogenetic constraint boundaries and developmental constraint boundaries are intrinsic to the biology of specific organisms, and as organisms evolve over time, their intrinsic constraint boundaries may evolve as well.

Figure 5.5 summarizes the total effect of evolutionary constraint for the development of form by species x within the total hyperspace of form dimensions. The subset of *actual existent form for species* $x \subset DPFx \subset PPFx \subset FPF \subset GPF =$

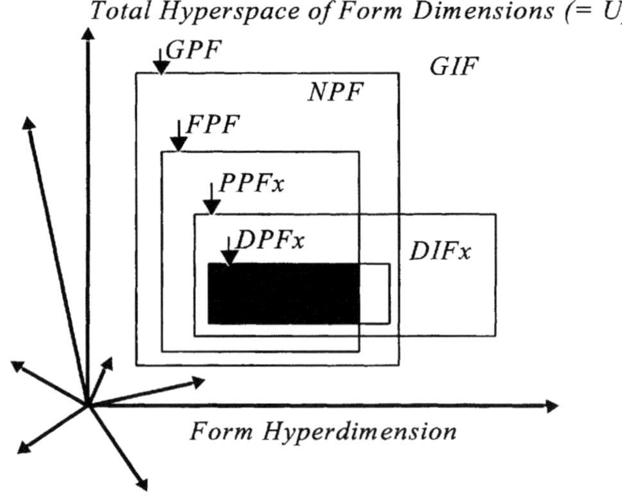

Figure 5.5
Existent form (black rectangle) as the subset of developmentally possible form for species x ($DPFx$), phylogenetically possible form for species x ($PPFx$), functional possible form (FPF), and geometrically possible form (GPF).

$\{f \mid f \in GPF, f \in FPF, f \in PPFx, f \in DPFx\}$. Forms outside the region of set intersections $DPFx \cap PPFx \cap FPF \cap GPF$ are not possible for species x, but they may be possible for species y, which has a different set of phylogenetically possible forms $PPFy$. That is,

$$DPFx \cap PPFx \cap FPF \cap GPF \neq DPFy \cap PPFy \cap FPF \cap GPF.$$

Thus intrinsic constraints will vary from taxon to taxon. Extrinsic constraints do not, however; and all existent forms must be in the form set $FPF \cap GPF$.

This is not to say that the phenomenon of intrinsic constraint must mean that each separate taxon occupies its own unique region of the total hyperspace of form dimensions. Each taxon has its own unique evolutionary history; thus, it is highly unlikely that the set of forms evolved by two taxa will be exactly alike. However, the phenomenon of evolutionary convergence in form, which is extremely common in nature, means that some forms function particularly well in particular environments, and that these forms are repeatedly evolved in many different taxa. That is, although the set of forms $DPFx \neq DPFy$, the phenomenon of evolutionary convergence in form means that for many separate taxa:

$$DPFx \cap DPFy = \{f \mid f \in DPFx \text{ and } f \in DPFy\} \neq 0.$$

Modeling Biological Form in Theoretical Morphospace

The analytic techniques of theoretical morphology allow us to explore the boundaries of geometric, functional, phylogenetic, and developmental constraints within the total hyperspace of form dimensions by the construction of theoretical morphospaces (figure 5.6). Theoretical morphospaces are n-dimensional geometric hyperspaces produced by systematically varying the parameter values of a geometric model of form (McGhee, 1999). Theoretical morphospaces are produced without any measurement data from real organic form. They not only are independent of existent morphology; they also can be used to produce nonexistent morphology. Such morphospaces exist in the absence of any measurement data (as opposed to empirical morphospaces, which are data-dependent; see the discussion in McGhee, 1999:22–26). Thus, theoretical morphospaces are topological spaces (for a discussion of pre-topological spaces in the analysis of biological form, see Stadler et al., 2001).

The dimensionality of a theoretical morphospace will always be less than the total hyperspace of form dimensions, U (figure 5.6); indeed, one of the goals of theoretical morphospace analysis is to create a morphospace with the lowest dimensionality possible while still capturing the essential aspects of the biological form under consideration (McGhee, 1999:18–22). However, the boundaries of the theoretical morphospace can hypothetically cross all of the evolutionary constraint boundaries (figure

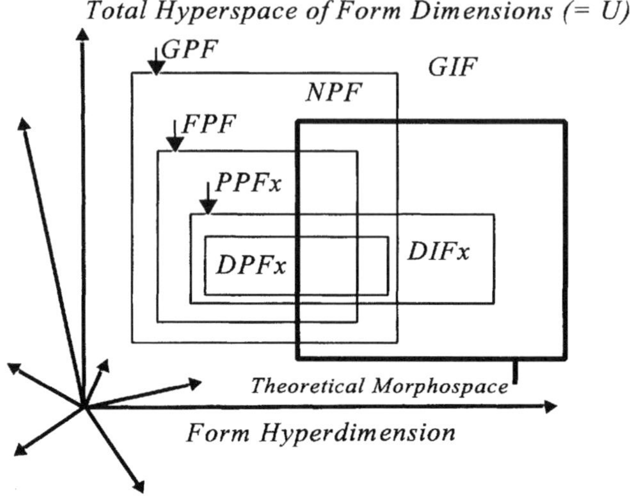

Figure 5.6
Boundaries of a hypothetical theoretical morphospace within the total hyperspace of form dimensions. Note that although the area of the theoretical morphospace (i.e., the dimensionality of the morphospace) is less than that of the total hyperspace of form, the boundaries of the theoretical morphospace may cross all the boundaries of form constraint within the total hyperspace.

5.6); that is, we potentially can actually map these boundaries within the theoretical morphospace for a given set of actual organisms.

Mapping Geometric Constraint Boundaries

Figure 5.7 illustrates a two-dimensional slice through a six-dimensional theoretical morphospace of potential brachiopod form. Brachiopods are filter-feeding marine invertebrates with two valves that are articulated together to form a bivalved shell. These shells can be simulated by using a simple logarithmic spiral model of growth (McGhee, 1999:209–225). In figure 5.7, the two dimensions show differential valve convexities in the shell, as measured by the tangent angle α in the logarithmic spiral model (the other four dimensions of the morphospace measure the position of the commissure, the angular orientation of the commissure, and the shape of the commissure for the two valves in the shell). Shells with large positive values of α in both valves are globose, subspherical in shape, whereas shells with small positive values of α in both valves have flat, compressed shapes (figure 5.7). Valves with negative values of α are concave rather than convex; negative values of α in the dorsal valve produce concavo-convex shells (upper left region in the morphospace), whereas negative values of α in the ventral valve produce convexi-concave shells (lower right region in the morphospace).

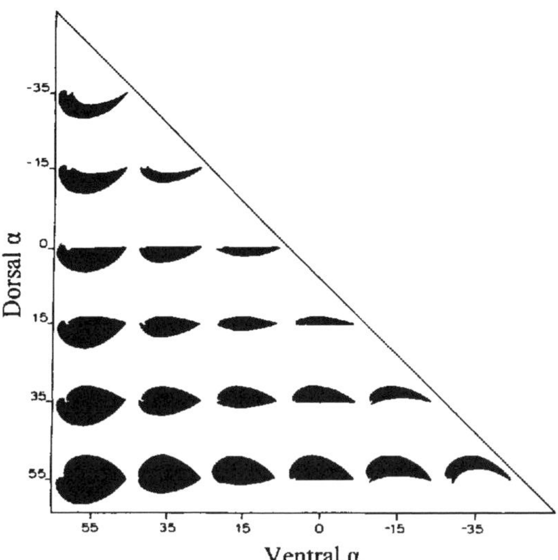

Figure 5.7
A two-dimensional slice through a six-dimensional theoretical morphospace of potential brachiopod form. (Modified from McGhee, 1999.)

All of the simulated shells shown in figure 5.7 are geometrically possible for brachiopods to evolve. However, if we venture beyond certain values of α in the two valves of the shell, we enter into geometrically impossible regions (figure 5.8). Two different types of geometric constraint emerge in these impossible form regions. First, valves with α values of 70° have curvatures that have increased to the point that the whorls overlap each other (see the simulation with dorsal $\alpha = 70°$ and ventral $\alpha = 80°$ in figure 5.8). It is impossible to articulate two valves together to form a bivalved shell if whorl overlap is present in either valve—the whorls would have to interpenetrate one another. That is, portions of the two separate valves would have to occupy the same space at the same time, which is geometrically impossible.

Second, it is impossible to form a bivalved shell if α is negative in both valves. Both such valves would have convex inner surfaces and concave outer surfaces (see the simulation with dorsal $\alpha = -70°$ and ventral $\alpha = -70°$ in figure 5.8). In essence, the region of negative α coordinates in both valves within the morphospace produces brachiopod shells that have been turned "inside out."

The triangular line surrounding the shaded region labeled "potential brachiopod morphospace" in figure 5.8 is an actual geometric constraint boundary, a real demonstration of a boundary between the set of forms *GIF* and *GPF* in figure 5.1. Thus, one who uses the analytic techniques of theoretical morphology can actually map geometric constraint boundaries, rather than simply thinking of such boundaries in a hypothetical and heuristic sense.

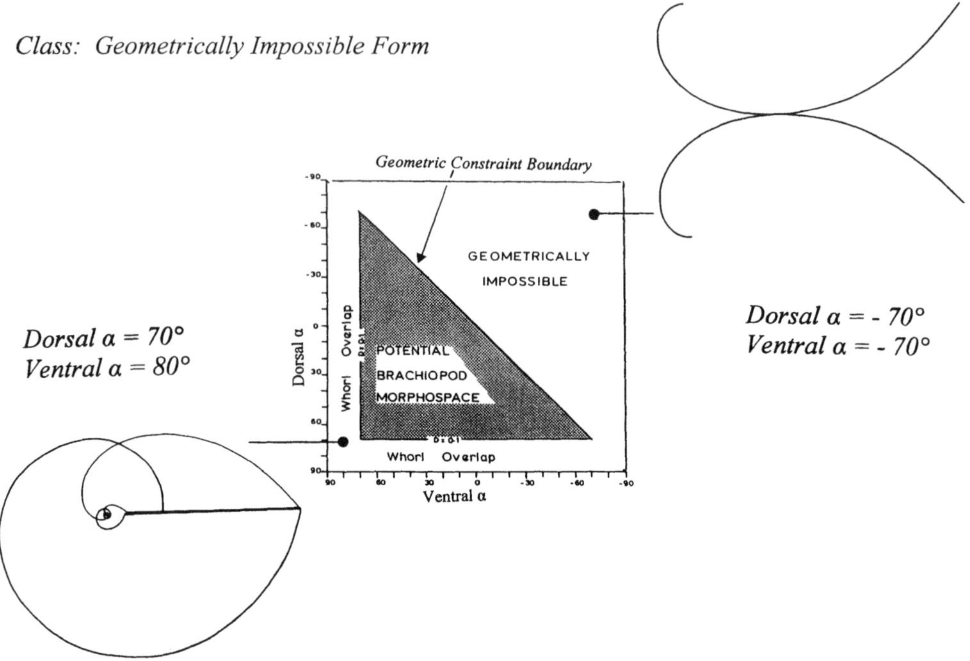

Figure 5.8
Illustration of a geometric constraint boundary within the theoretical morphospace of potential brachiopod form (modified from McGhee, 1999). Note that the morphospace coordinates dorsal $\alpha = 70°$ ventral $\alpha = 80°$, and dorsal $\alpha = -70°$ ventral $\alpha = -70°$ produce geometrically impossible forms.

Creating Nonexistent Form and Mapping Functional Constraint Boundaries

A two-dimensional slice through a three-dimensional theoretical morphospace of helical colony form in the Bryozoa is illustrated in figure 5.9, based on McKinney and Raup (1982), McGhee and McKinney (2000, 2002), and McKinney and McGhee (2003). Bryozoa are colonial marine invertebrates that produce intricate colony geometries composed of hundreds of small filter-feeding zooids. Shown within the morphospace are the boundary polygons and plots of data taken from measurements of 208 actual helical bryozoan colonies, representing four genera that have convergently evolved helical forms. Four simulations of helical colony form are also shown, and the coordinate combinations that will produce those forms are indicated within the morphospace.

The simulation in the lower left region of the morphospace shows a form actually found in the small helices of the bryozoan *Bugula plumosa*; the simulation in the bottom center region shows a form actually found in *Retiflustra cornea*; the simulation above this one shows the form most commonly found in many species of *Archimedes*

Class: Existent Form

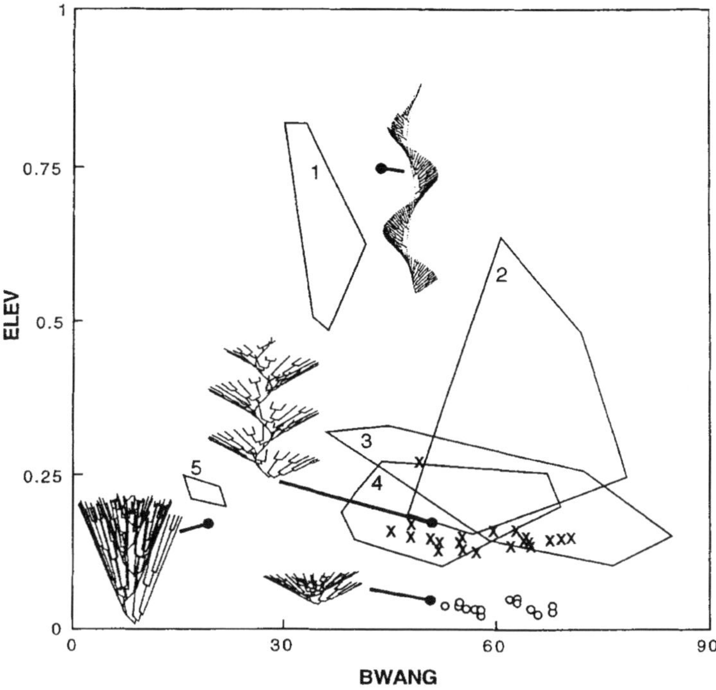

Figure 5.9
A two-dimensional slice through a three-dimensional theoretical morphospace of a potential helical colony form in bryozoans. Illustrated are four simulations of helical colony form that actually exist in fossil and living bryozoans. (Modified from McKinney and McGhee, 2003.)

and convergently evolved in the species *Crisidmonea archimediformis*; and the simulation in the top center region of the morphospace shows a form found in the aberrant species *Archimedes laxus*.

All of the simulations shown in the theoretical morphospace in figure 5.9 thus represent existent form. Note, however, that in figure 5.9 there are empty regions in which there are no data from real bryozoan colonies. What morphologies lie in these regions, morphologies that have never been evolved within the Bryozoa?

In figure 5.10 three simulations of nonexistent helical colony form are shown, illustrating the type of form that is geometrically possible (and can be computer-simulated) but has never been evolved by any bryozoan species. McGhee and McKinney (2000, 2002) and McKinney and McGhee (2003) have argued that the three nonexistent colony forms shown in figure 5.10 represent nonfunctional morphologies for filter-feeding organisms. For example, the simulated colony shown in

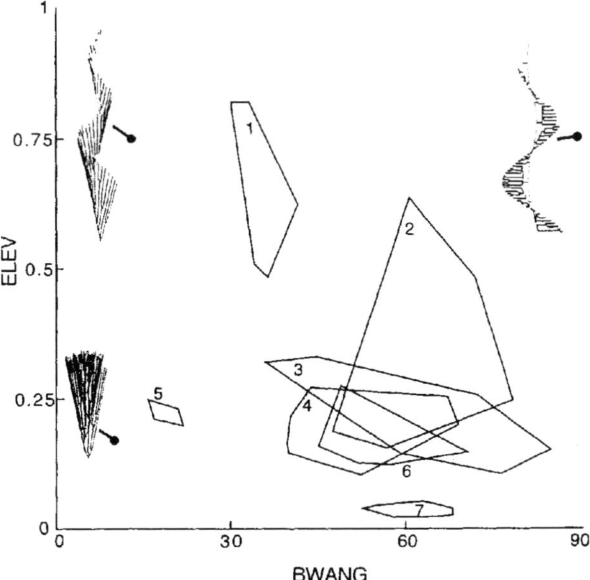

Figure 5.10
Three simulations of nonexistent helical colony form in fossil and living bryozoans, within the theoretical morphospace given in figure 5.9. Data polygons as in figure 5.9, polygons 6 and 7 represent *Crisidmonea archimediformis* and *Retiflustrea cornea*, respectively. (Modified from McKinney and McGhee, 2003.)

the upper right region of the morphospace has all of its branches oriented at a right angle to the central axial helix. Such a geometry would allow the majority of water passing through the colony to escape filtration by the colony zooids. In contrast, the simulated colony shown in the bottom left region of the morphospace has all of its branches folded close to the central axial helix. Such a geometry would produce a central dead-water zone in the center of the colony, a stagnant region of inert water lethal to the colony zooids. McKinney and McGhee (2003) argue that the boundaries of the data polygons shown in figure 5.10 represent the boundary between functional possible geometries and nonfunctional possible geometries for helical colony bryozoans. Thus the analytic techniques of theoretical morphology can actually map functional-constraint boundaries, a real demonstration of mapping the boundaries between the set of forms *NPF* and *FPF* in figure 5.2.

Analyzing Intrinsic Constraints

The analytic techniques of theoretical morphology are particularly powerful when applied to extrinsic constraints, those of geometry and function. They can also be used to help analyze the intrinsic constraints of phylogeny and development.

Phylogenetic constraints are best analyzed within the framework of phylogenetic hypothesis construction (McKitrick, 1993; Schwenk, 1995). In the example given in figure 5.10, the hypothesis could be put forth that the observed nonexistent forms are the result of phylogenetic constraint rather than functional constraint; that is, that these forms belong to the set $FPF - PPFx = \{f \mid f \in FPF, f \notin PPFx\}$, where x is the phylum Bryozoa. In this scenario, the Bryozoa have not yet evolved the genetic coding that would allow them to produce these nonexistent forms, but they might be able to evolve these forms at some time in the future, and these forms would be functional forms.

This argument can be refuted in two ways, one deterministic and the other probabilistic. First, functional analyses demonstrate that the nonexistent forms shown in figure 5.10 possess aspects of geometry that are unrealistic as fluid-filtering geometries for filter-feeding organisms (McGhee and McKinney, 2000; McKinney and McGhee, 2003).

Second, cladistic (phylogenetic) analyses reveal that helical colony geometries have convergently, independently, evolved in widely separate regions of the total cladogram for the phylum Bryozoa (figure 5.11). Thus the genetic coding that permits the development of helical colony form must be present near the very base of the cladogram of the Bryozoa, and had to be present some 495 million years ago. On probabilistic grounds one can argue that if the observed nonexistent colony forms were functional, the bryozoans would surely have discovered them in their 495 million years of evolution.

Last, the analytic techniques of theoretical morphology can be used to help the modeler to explore the phenomenon of developmental constraint (McGhee, 1999, 2001). Systematic theoretical morphological form simulations can reveal the constraints inherent in any given growth model and the limits to which those form simulations can be taken, and can contrast the constraints encountered in one growth model with those of another. Simulation of adaptive walks in theoretical morphospaces, where morphospace parameter combinations have been assigned relative fitness values, can explore the effect of developmental constraint on the evolution of a hypothetical group of organisms, versus the evolution of those same organisms in the absence of such constraint. For example, theoretical morphospace analyses of early land plant evolution show that developmental constraint does not limit the number of equally fit morphological variants that can be reached within the morphospace, but that the relative fitness of those variants is lower than those reached in unconstrained adaptive walks in the same morphospace (Niklas, chapter 14, this volume).

For the analyst, plotting growth-stage data for organisms within theoretical morphospaces can reveal the actual ontogenetic trajectories within that morphospace that nature has produced and, by comparison with the empty regions of the morphospace, reveal ontogenetic trajectories—developmental pathways—that are theoretically possible but have never evolved. Analysis of these nonexistent developmental

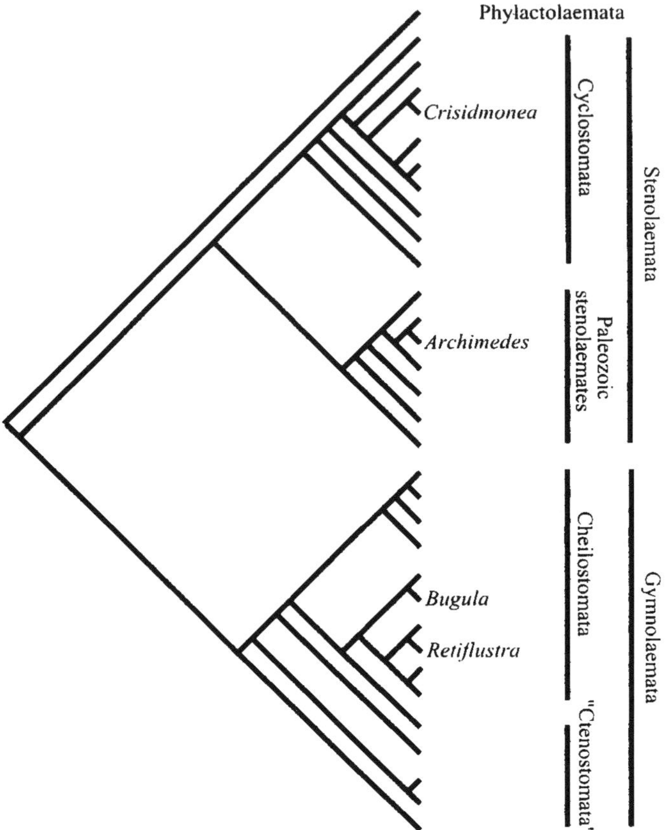

Figure 5.11
Cladogram of evolutionary relationships within the phylum Bryozoa. Helical colony forms have convergently and independently evolved within four separate genera of bryozoans widely scattered across the cladogram. (From McKinney and McGhee, 2003.)

pathways, similar to the analysis of nonexistent form, can be a powerful tool in the analysis of evolutionary constraint.

Modeling Biological Form in a Spatial Context

In summary, the analytic techniques of theoretical morphology go beyond the process of modeling any given particular biological form or growth process. In theoretical morphospace construction, the modeler seeks to simulate the entire spectrum of forms that biological entities could take and, in that process, not only to model those forms that evolution has produced but also to reveal nonexistent—but theoretically possible—forms that nature has never produced. Such techniques allow the analysis of evolutionary constraint to be taken into a spatial context, where actual geometric, functional, phylogenetic, and developmental constraint boundaries can be mapped within theoretical morphospaces.

References

Antonovics J, van Tienderen PH (1991) Ontoecogenophyloconstraints? The chaos of constraint terminology. Trends Ecol Evol 6: 166–168.

Blomberg SP, Garland T (2002) Tempo and mode in evolution: Phylogenetic inertia, adaptation and comparative methods. J Evol Biol 15: 899–910.

Eble GJ (2000) Theoretical morphology: State of the art. Paleobiology 26: 520–528.

Maclaurin J (2003) The good, the bad and the impossible. Biol Philos 18: 463–476.

Maynard Smith J, Burian R, Kauffman S, Alberch P, Campbell J, Goodwin B, Lande R, Raup D, Wolpert L (1985) Developmental constraints and evolution. Quart Rev Biol 60: 265–287.

McGhee GR (1999) Theoretical Morphology: The Concept and Its Applications. New York: Columbia University Press.

McGhee GR (2001) Exploring the spectrum of existent, nonexistent and impossible biological form. Trends Ecol Evol 16: 172–173.

McGhee GR, McKinney FK (2000) A theoretical morphologic analysis of convergently evolved erect helical colony form in the Bryozoa. Paleobiology 26: 556–577.

McGhee GR, McKinney FK (2002) A theoretical morphologic analysis of ecomorphologic variation in *Archimedes* helical colony form. Palaios 17: 556–570.

McKinney FK, McGhee GR (2003) Evolution of erect helical colony form in the Bryozoa: Phylogenetic, functional, and ecological factors. Biol J Linn Soc 80: 235–260.

McKinney FK, Raup DM (1982) A turn in the right direction: Simulation of erect spiral growth in the bryozoans *Archimedes* and *Bugula*. Paleobiology 8: 101–112.

McKitrick, MC (1993) Phylogenetic constraint in evolutionary theory: Has it any explanatory power? Ann Rev Ecol Sys 24: 307–330.

Raup DM, Michelson A (1965) Theoretical morphology of the coiled shell. Science 147: 1294–1295.

Schwenk K (1995) A utilitarian approach to constraint. Zoology 98: 251–262.

Stadler BMR, Stadler PF, Wagner GP, Fontana W (2001) The topology of the possible: Formal spaces underlying patterns of evolutionary change. J Theor Biol 213: 241–274.

6 Computational Models of Dendritic Morphology: From Parsimonious Description to Biological Insight

Alexei V. Samsonovich and Giorgio A. Ascoli

The study of neuronal morphology has a long history with three major stages: (1) early observation of neuronal shapes, their qualitative characterization, and comparative analysis (e.g. Ramón y Cajal, 1894–1904), (2) digital encoding of dendritic shapes and their quantitative statistical analysis, with gradual accumulation of digital cell databases (see Turner et al., 2002, for review), and (3) statistical reproduction of dendritic shapes with elegant computational algorithms (reviewed in Ascoli, 2002); these provide the capacity to generate an unlimited number of unique, realistic virtual shapes. Each of these stages is contingent on the development of a key technological base that becomes available only at a certain phase of scientific progress. For (1), the key was the technique of staining, beginning with the Golgi method. For (2), it was the technique of digital tracing based on computer-interfaced, high-resolution optical microscopy. For (3), the key is the combination of new computational powers with appropriate mathematical ideas and methods, such as L-systems, Markov models, and Bayesian analysis.

The goal of the current phase (3) is to describe very complex geometrical shapes with general, parsimonious rules and principles; in other words, to reduce the global complexity of individual structures to a concise, logical framework (Hillman, 1979; Ascoli et al., 2001a). This ambitious aim reflects the goal of science in general, and is also reminiscent of the principles of genetics. One can expect insights into developmental biology if the elements of models can be associated with their biophysical determinants or correlates. Although algorithms of virtual growth may not be considered as developmental or genetic models, they are likely to have an impact on developmental and/or genetic theories.

Recent efforts in computational neuroanatomy have aimed at accurately reproducing all relevant statistical details of dendritic morphology with stochastic models based on local rules (e.g., Markov models), using parameters measured from real neurons (e.g., Ascoli and Krichmar, 2000; Ascoli et al., 2001b). This chapter illustrates our implementation of hidden Markov algorithms (see "Classes of Algorithms and Their Limitations," below) to generate realistic dendritic trees, and discusses the biological implications of these results, using digitized rat CA3 and CA1 pyramidal and dentate granule cell dendrites (Samsonovich and Ascoli, 2003, 2005a, 2005b). In addition, we assess the basic limitations on classes of algorithms that can be used for statistical description of a given morphological class. In particular, we show that Markov algorithms may have significant limitations when they use only the geometric parameters locally available in the growing virtual structure.

Hidden Markov Modeling of Hippocampal Principal Cell Dendrites

Experimental Material

Digital files of 24 CA3 and 23 CA1 pyramidal cells (Ishizuka et al., 1995) were kindly provided by Dr. D. G. Amaral (University of California, Davis) for public distribution (http://krasnow.gmu.edu/L-Neuron). Reconstructed rat dentate gyrus granule cells (Carnevale et al., 1997; Rihn and Claiborne, 1990) were kindly made available by Dr. B. J. Claiborne (University of Texas, San Antonio) at www.utsa.edu/clairbornelab (36 of the 42 cells in this archive were used here). In both data sets, a dendritic tree consists of a system of segments (elementary pieces of dendritic structure being generated by the experimental reconstruction process) connected to each other at their extremities. Each segment is represented as an ideal cylinder characterized by two geometrical end points and the diameter. Diameter values are discrete (the minimum is 0.3 μm, with steps of 0.1 μm).

Algorithm Design

Our model of dendritic morphology consists of four steps. Each step produces a virtual structure with an increasing number of morphological features compatible with real neurons, starting from the structure produced in the previous step, and a set of measures taken from experimental data (table 6.1). We apply the resulting hidden Markov model description to the principal neurons of the rat hippocampus.

In this chapter we call an *embedding* the complete dendritic morphology of a neuron. The process of embedding implies allocating nodes and continuation points of a

Table 6.1
Steps of the dendritic morphology model

Step No.	Input	Output
1	Population statistics	Core data
2	Core data + the above	A tree skeleton
3	A tree skeleton + all the above	A dendrogram
4	A dendrogram + all the above	An embedding

given dendrogram in 3-D space. A *dendrogram* specifies all connections, lengths, and diameters of dendritic segments, but not the 3-D coordinates of each node. A *tree skeleton* includes the topology of branches (pieces of dendritic structure between consecutive branch points, or between branch point and consecutive terminal tips); that is, the point of origin, bifurcation, and termination nodes, and their connections, plus the lengths of internal branches, but no segment data. The lengths of terminal branches are also not specified at this level. The *core data* include all information about the soma and all dendritic trunks up to the second segment from the soma, plus the number of terminal tips for each truncated subtree, plus the density of bifurcations and the local termination probability, both taken as simple, few-parameter analytical functions of the path distance from the soma. Though all the above features can be taken from an individual cell, *population statistics* are derived from a given group of cells. Basal and apical dendrites are considered here as separate morphological classes.

Practically, the steps summarized in table 6.1 are implemented as four independent algorithmic modules that may work in sequence (1, 2, 3, 4) or in parallel. In contrast, our *search* for an algorithmic solution of this problem chained backward (4, 3, 2, 1). We therefore describe the four steps in reverse order.

Embedding a Dendrogram (Step 4)

Here the problem is to generate a realistic embedding from a given dendrogram, based on its core data and population statistics. The solution (Samsonovich and Ascoli, 2003) is a Markov process that starts from the core and proceeds toward the terminals. Specifically, given a growth node i (figure 6.1A), the preferred orientation $\mathbf{n}'_{i+1}$ of a new segment is determined by the sum of a vector parameter $\mathbf{a}$, the orientation $\mathbf{n}^r_i$ of the radius vector $\mathbf{r}_i$ multiplied by a scalar parameter b, and the orientation $\mathbf{n}_i$ of the parent segment ($\mathbf{n}'_i$, $\mathbf{n}_{i+1}$, and $\mathbf{n}^r_i$ are unit vectors). The sampled daughter orientation $\mathbf{n}_{i+1}$ is distributed around the preferred orientation $\mathbf{n}'_{i+1}$, exponentially over the angle α_{i+1} between $\mathbf{n}'_{i+1}$ and $\mathbf{n}_{i+1}$, and uniformly over the angle of rotation around $\mathbf{n}'_{i+1}$ (equations 6.1–6.3).

$$\mathbf{n}'_{i+1} \parallel \mathbf{a} + b\mathbf{n}^r_i + \mathbf{n}_i \tag{6.1}$$

$$\mathbf{n}_{i+1} = \mathbf{T}_{i+1}\mathbf{n}'_{i+1} \tag{6.2}$$

$$P(\mathbf{T}_{i+1}) \propto \exp -\frac{2\alpha_{i+1}}{\alpha_0}. \tag{6.3}$$

Here, $\mathbf{T}_{i+1}$ is an operator that deflects $\mathbf{n}'_{i+1}$ toward a random direction by an angle α_{i+1}, and $a_0 \ll 1$ is a parameter of the model, approximately equal to the mean $\langle a_i \rangle$.

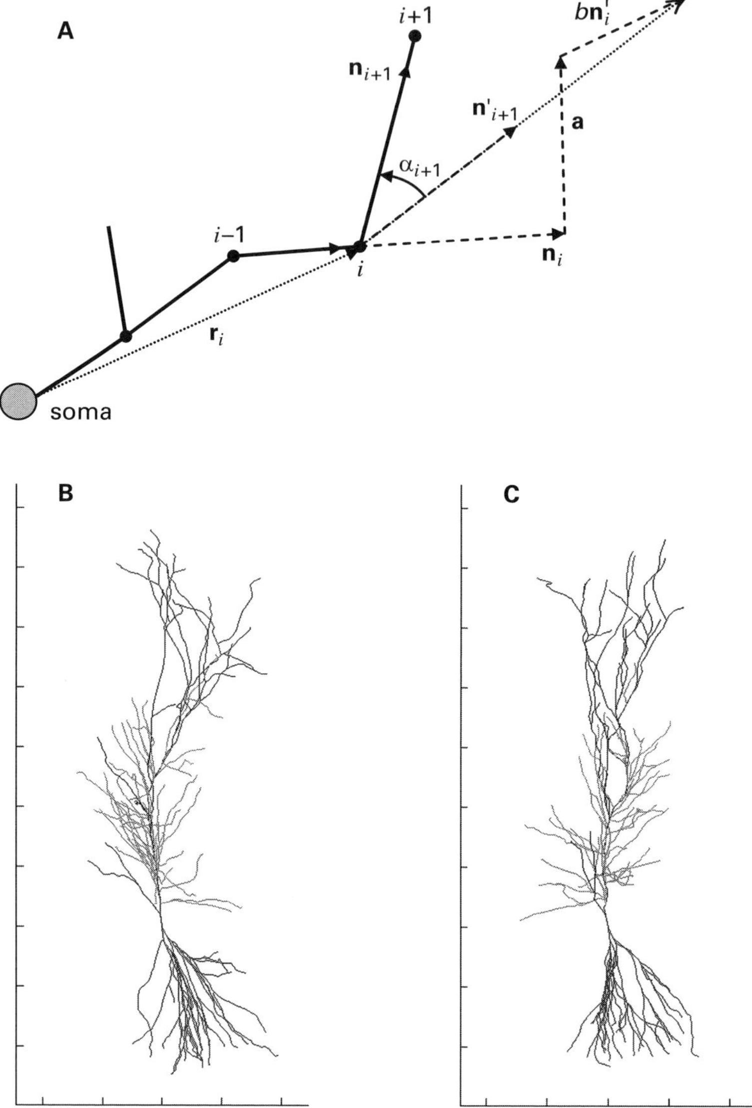

Figure 6.1
(A) The embedding algorithm (Samsonovich and Ascoli, 2003). Vectors labeled **n** have unit length; node i is the currently selected point of virtual growth; the angle a is measured in 3-D space. (B) Re-embedding of a CA1 neuron with parameter a given by eq. (6.4). Left: Amaral CA1 pyramidal cell c73166b. Right: the configuration of the same dendrogram re-embedded into 3-D space by the first run of the algorithm applied to the 23 CA1 cells. Twenty-two other results in this run were at the same qualitative level. Oblique apical branches[1] (gray), main apical branches, and basal branches were treated as separate classes. Each oblique subtree is related to its origin on the apical tree in the same way as each main or basal subtree is related to the soma. Tick spacing on all axes: 100 µm.

The embedding shown in figure 6.1B–C was carried out by orienting the vector **a** vertically (along the Y axis) and making its length dependent on the vertical distance y from the soma:

$$a_{\text{apical}}(y) = 0.7\frac{y_0^{\text{apical}} - |y|}{y_0^{\text{apical}}}, \quad y_0^{\text{apical}} = 400\ \mu\text{m}$$

$$a_{\text{basal}}(y) = 0.7y\frac{y_0^{\text{basal}} - |y|}{(y_0^{\text{basal}})^2}, \quad y_0^{\text{basal}} = 150\ \mu\text{m} \tag{6.4}$$

$$a_{\text{oblique}} = 0.$$

For all secondary oblique nodes,[1] $\mathbf{n}_i^r$ was drawn out of the corresponding primary oblique branching point (instead of from the soma). The constant parameter **b** was 0.5 for all basal and 0.35 for all apical and oblique dendrites (Samsonovich and Ascoli, 2003). The constant parameter α_0 was 0.15 (Samsonovich and Ascoli, 2003) except for primary oblique segments, terminal segments, and segments at a vertical distance from the soma > 400 μm. In these cases, α_0 corresponded to approximately 0.06.[2]

The resulting embedding is visually indistinguishable from the real cell (figure 6.1B–C). Similar results (also corroborated by extensive statistical morphometric comparison) can be obtained for CA3 pyramidal cells and dentate granule cells (Samsonovich and Ascoli, 2003, 2005b) by measuring all parameters $(\mathbf{a}, b, \alpha_0)$ from the experimental data. The results generally indicate that while **a** is not significantly different from zero in hippocampal principal cells, b always is. It is thus suggested that dendrites of hippocampal principal cells are repelled by their own cell bodies during growth (Samsonovich and Ascoli, 2003).

Transforming a Tree Skeleton into a Dendrogram (Step 3)

The task of generating a realistic dendrogram based on a given tree skeleton involves three distinct operations: dividing branches into individual segments, sampling the termination probability of segments in terminal branches, and assigning branch diameter (Samsonovich and Ascoli, 2005a).

The distribution of individual segment lengths in pyramidal cells is nearly independent of the cellular class (CA3 vs. CA1), type of branch (internal vs. terminal), or type of tree (basal vs. apical), and the overall experimental data are well fitted by gamma distributions (figure 6.2A–D). Dendritic branches can thus be algorithmically divided by sequentially sampling individual segment lengths. The local termination probability for each segment of terminal branches can be sampled from a fitted function proportional to the density of actual termination events in real neurons, normalized by the density of continuing segments (figure 6.2E–F).

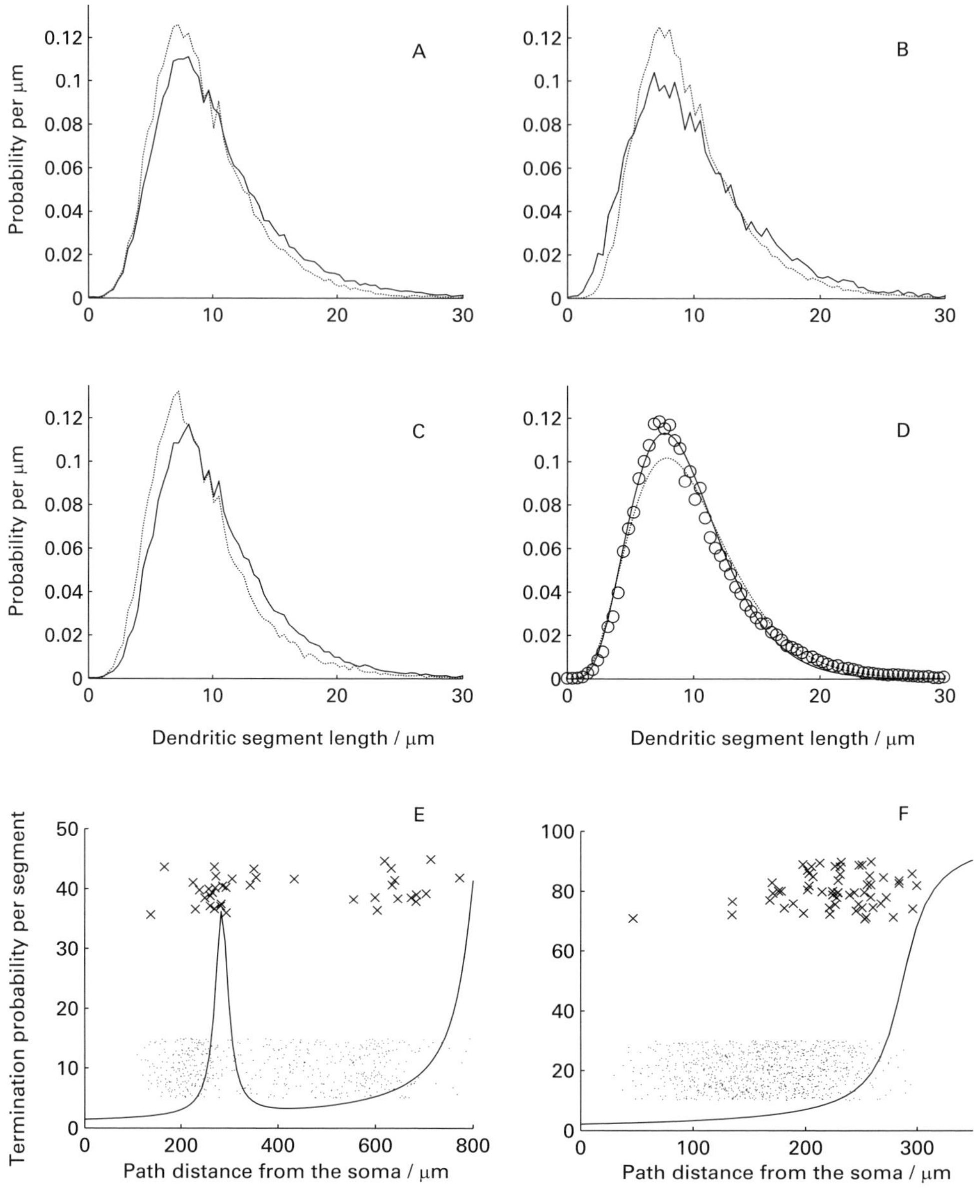

Figure 6.2
A–D: Histograms of dendritic segment lengths in the 23 CA1 and 24 CA3 pyramidal cells. (A) All 30,405 CA1 dendritic segments vs. all 33,692 CA3 dendritic segments (gray). (B) All 14,217 internal branch segments vs. all 49,880 terminal branch segments (gray). (C) All 36,971 apical dendritic segments vs. all 27,126 basal dendritic segments (gray). (D) All 64,097 dendritic segments (circles) vs. maximum likelihood gamma fit (gray: gamma parameters $a = 5.188$, $b = 1.877$) vs. least-square gamma fit (black: $a = 5.929$, $b = 1.561$). E–F: Examples of local termination probabilities per segment vs. path distance from the soma (CA3 Amaral cell C10861: E, apical; F, basal). The maximum-likelihood analytical fits (solid lines) are superimposed onto the vertically scattered actual events of segment continuation (dots) and termination (crosses). By construction, the fitted functions approach 100 percent at large distances from the soma (Samsonovich and Ascoli, 2005a.)

Previous studies on different morphological classes reported correlations of branch diameter between the bifurcation daughters (Burke et al., 1992) or between daughters and parent (Ascoli et al., 2001b). These trends, however, did not emerge from the experimental pyramidal cells considered here (figure 6.3A–B). In particular, except for the dense edges, the dots uniformly fill the plots, showing no significant correlation among the two daughter diameters or between daughter and parent diameters beyond a simple constraint, $0.3\ \mu m \le d_{daughter} \le d_{parent}$ (Pearson's coefficients: 0.073 for panel A and 0.187 for panel B). In addition, most points have zero taper (figure 6.3C–D).

We found a significant dependence of dendritic diameter on the degree (the number of terminals) of the subtree (figure 6.3E; correlation coefficient: 0.8). Therefore, in our model, branch diameter was assigned on the basis of degree (and limited by the parent diameter), and kept constant (zero taper) at all continuation points. In particular, the linear fits of mean and standard deviation of this dependence were used to calculate parameters of a gamma distribution, from which diameters were sampled.

Building a Tree Skeleton (Step 2)

A tree skeleton can be generated from the core and population-derived statistics by sampling bifurcation *partitions* (the distribution of the parent degree between the two daughters) and internal branch lengths. Consistent with the experimental data, our algorithm samples partition uniformly in all dendrites except CA1 apical trees (figure 6.4A). In this case, new partitions are randomly selected as oblique with a small probability, and then sampled from a Poisson distribution, to account for the dense edges (Samsonovich and Ascoli, 2005a).

After partitioning the parent degree, and selecting one of the daughters with a degree $m > 1$, a new branch length is generated as follows (figure 6.4B). The (a priori, fixed) distribution of bifurcation probability vs. path distance from the soma is sampled multiple times, until exactly m sampled events are accumulated to the right of the current branching event. The nearest sampled event is then selected as the new branch length, and the rest are discarded (for details and discussion of the algorithm, see Samsonovich and Ascoli, 2005a). Any resulting point that is too close to the current node is also rejected, consistent with the "gap" observed in the histogram of branch lengths in real neurons at very low values (Samsonovich and Ascoli, 2005a).

By putting together steps 2 and 3, dendrograms can be generated from a neuron core and population statistics. The results are qualitatively excellent for CA3 pyramidal cells and for CA1 basal trees (figure 6.5). Interestingly, when CA1 apical trees are generated with this algorithm, their dendrograms closely resemble those of CA3 cells, even if the bifurcation partitions are sampled directly from the CA1 experimental data (Samsonovich and Ascoli, 2005a). To obtain visually and statistically correct results based on this approach, CA1 apical dendrites need to be sampled on the basis

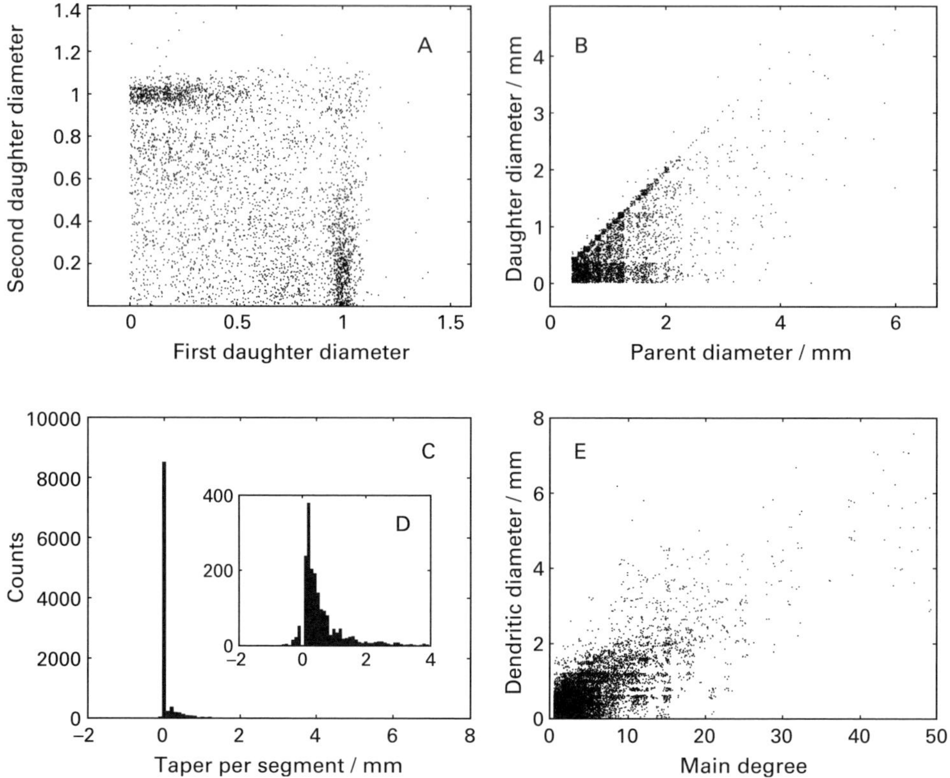

Figure 6.3
Dendritic diameters in all 47 pyramidal cells. (A) Scatter plot of daughter dendritic diameters at bifurcations, normalized by their parent diameter. Uniform artificial noise (± 0.05) was added to each daughter diameter before the normalization, for better visualization, and bifurcation points with the minimal parent diameter 0.3 μm were excluded. (B) Scatter plot of daughter diameter vs. parent diameter at bifurcations, with artificial noise (as in A) added to both parent and daughter values. (C) Histogram of taper at continuation points. The inset (D) shows the same distribution, without the peak at zero, scaled by 50. (E) Dependence of dendritic diameters on the main degree of the subtree (i.e., the degree calculated disregarding CA1 oblique branches). Artificial noise was added to discrete diameter values, as above. Solid and dashed lines are, respectively, maximum-likelihood linear fits to the mean and the standard deviation of diameter values as functions of the main degree.

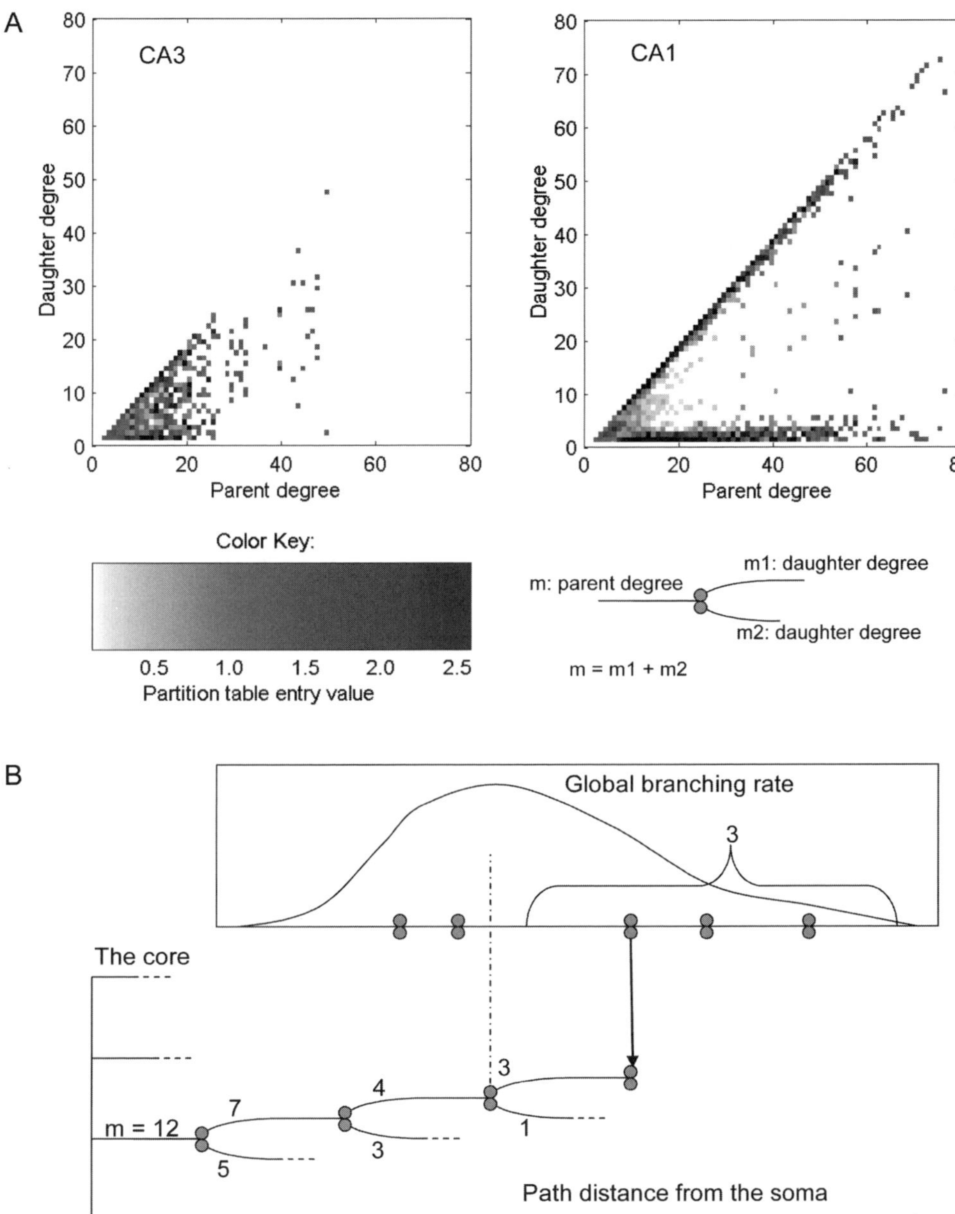

Figure 6.4
(A) Topological partitions of the number of terminals at branching points. The color key shows the gray shade for each value of the probability $P^{\text{part}}(m \mid m_{\text{parent}})$. Frequencies of partitions are normalized independently for each parent degree (as explained in Samsonovich and Ascoli, 2005a). (B) Sampling internal branches. Starting from the core, the length of each new branch with degree $m > 1$ is determined by repetitive sampling of the global branching rate, until there are m sampled bifurcations to the right of the current growth point.

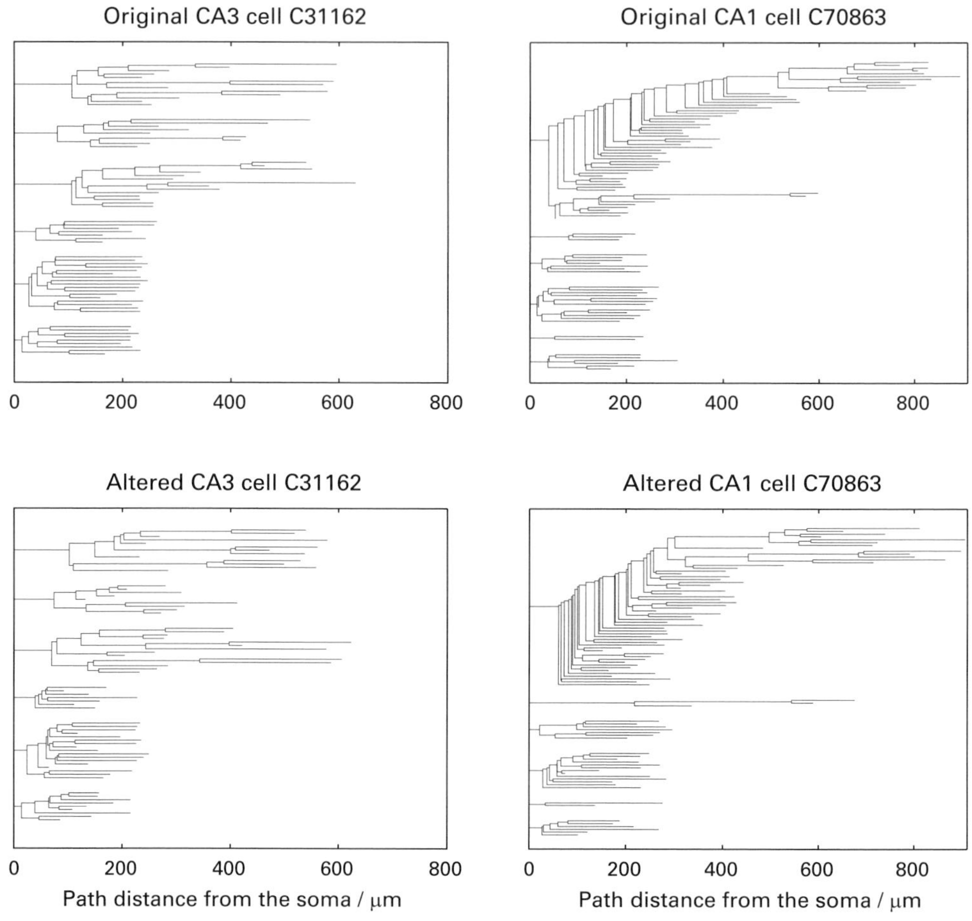

Figure 6.5
Examples of altering CA3 (left) and CA1 (right) pyramidal cell dendrograms (cells C31162 and C70863, respectively). The abscissa is the path distance from the soma in microns. Subtrees are vertically ordered, first by their kind (basal on top), then (given equal kind) by degree, and finally by total dendritic length.

of two kinds of branching—main and oblique—each separately described by its own branching and termination distributions.

We empirically define a branch of a real apical tree as *oblique* if its degree divided by the degree of the parent is <0.2, and if the maximal path distance within its subtree divided by the maximal path distance within the sibling subtree is <0.6 (Samsonovich and Ascoli, 2005a). Generation of virtual oblique branches is then performed with another instance of the same model, with the origin of the oblique subtree on the apical trunk playing the role of the soma. This "modified" algorithm produces higher quality CA1 dendrograms (figure 6.5, right). Quantitative morphometric comparison of real and virtual dendrograms (figure 6.6) confirms that the two populations are statistically consistent for both CA1 and CA3 cells (see Samsonovich and Ascoli, 2005a, for extensive analysis).

Adding the fourth step to the simulated dendrogram results in the generation of a complete virtual cell starting from the core data of a real neuron. Synthetic neurons of this type are qualitatively consistent with their real counterparts (figure 6.7).

Completing the Model: Sampling Core Data from Population Statistics (Step 1)

To generate completely virtual cells, the core data are sampled from population statistics (figure 6.8). Specifically, the number of trees and their degrees are sampled from Poisson distributions, and the parameters of basal and apical distributions of branching points and termination probabilities are sampled from normal and gamma distributions (Samsonovich and Ascoli, 2005b). The addition of the first step makes the algorithm complete and equally dependent on all individual real cells in the given population.

The resulting Markov algorithm to generate CA3 cells from population statistics can be summarized as follows. The number of dendritic trees, the lengths and spatial orientation of the first two segments and the degree of each tree, the basal and apical distributions of bifurcations and termination probabilities (both fitted by smooth curves)—all are sampled for each virtual cell. For each open node, the algorithm samples the branch length to the next bifurcation (if internal) or to termination (if terminal). At bifurcations, the parent degree is randomly and uniformly partitioned between the daughters, and each new branch is divided into segments based on the population distribution of segment lengths. The diameter of each daughter is sampled *independently* on the basis of degree, using linear regression from population data, and is inherited by the offsprings at continuation points. Finally, the spatial orientation of each segment is determined by adding a unit step along the direction of the parent, a step of constant length in the radial direction from the soma, and deflecting the resultant vector by an angle sampled from a fixed exponential distribution.

Modifications of the same algorithm are applied to CA1 pyramidal cells and dentate granule cells. In CA1, oblique dendrites need to be treated separately, possibly suggesting a different biophysical mechanism of branching during development (Samsonovich and Ascoli, 2005a). In granule cells, a maximum path distance from the soma is enforced on all dendrites (Samsonovich and Ascoli, 2005b). Results of the complete algorithm (steps 1–4) retain high quality levels (figure 6.9). In particular, we present a "Turing test" for specialists familiar with dendritic morphologies of the given neuronal classes. The task is to distinguish visually between real and virtual

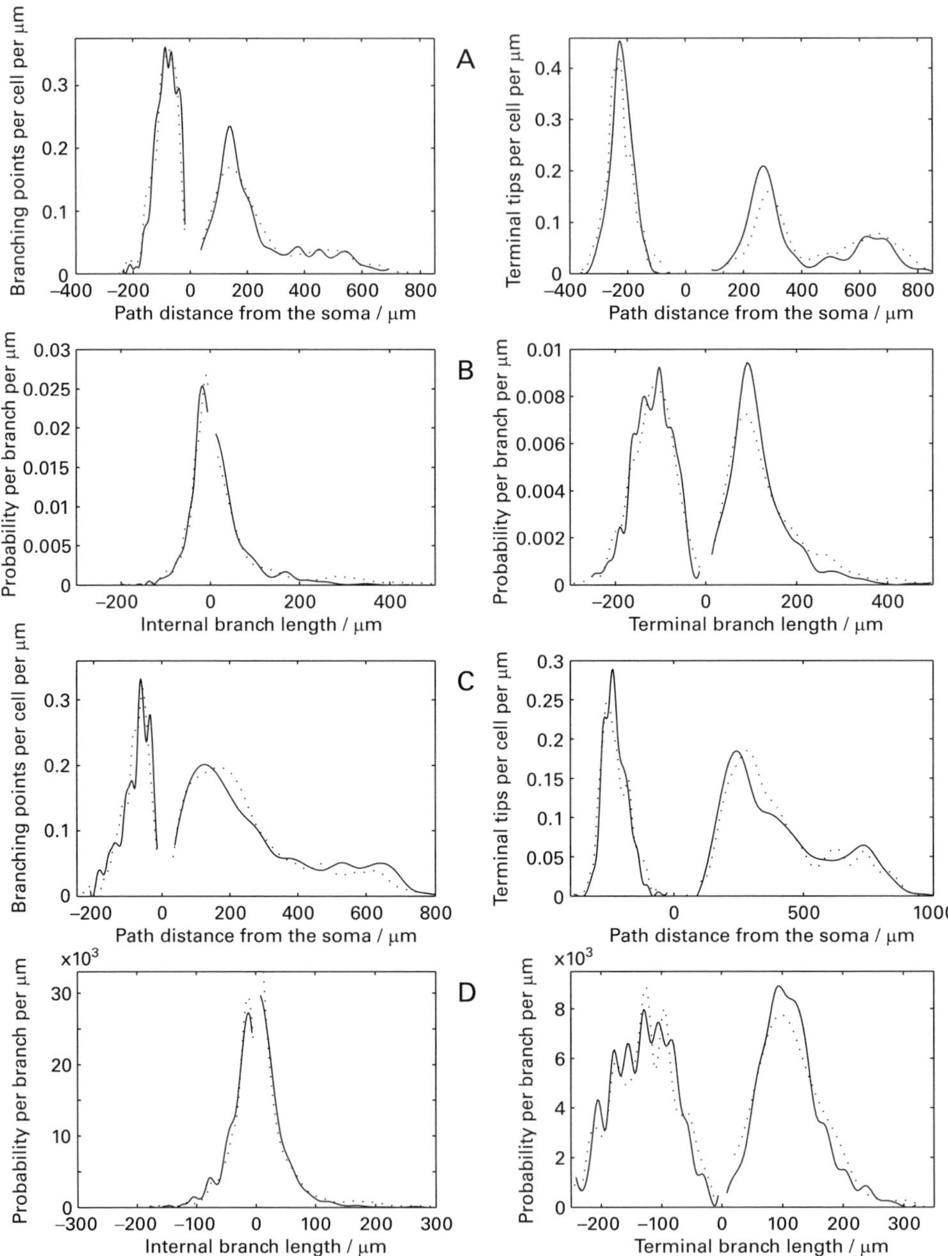

Figure 6.6
Node and branch length distributions in original (solid) vs. altered (dotted) CA3 (A, B) and CA1 (C, D) populations. (A, C) Distributions of bifurcations (left) and terminations (right) vs. path distance from the soma. (B, D) Histograms of branch lengths (left: internal; right: terminal) normalized as probability densities and interpolated by cubic splines. Negative abscissa corresponds to basal trees.

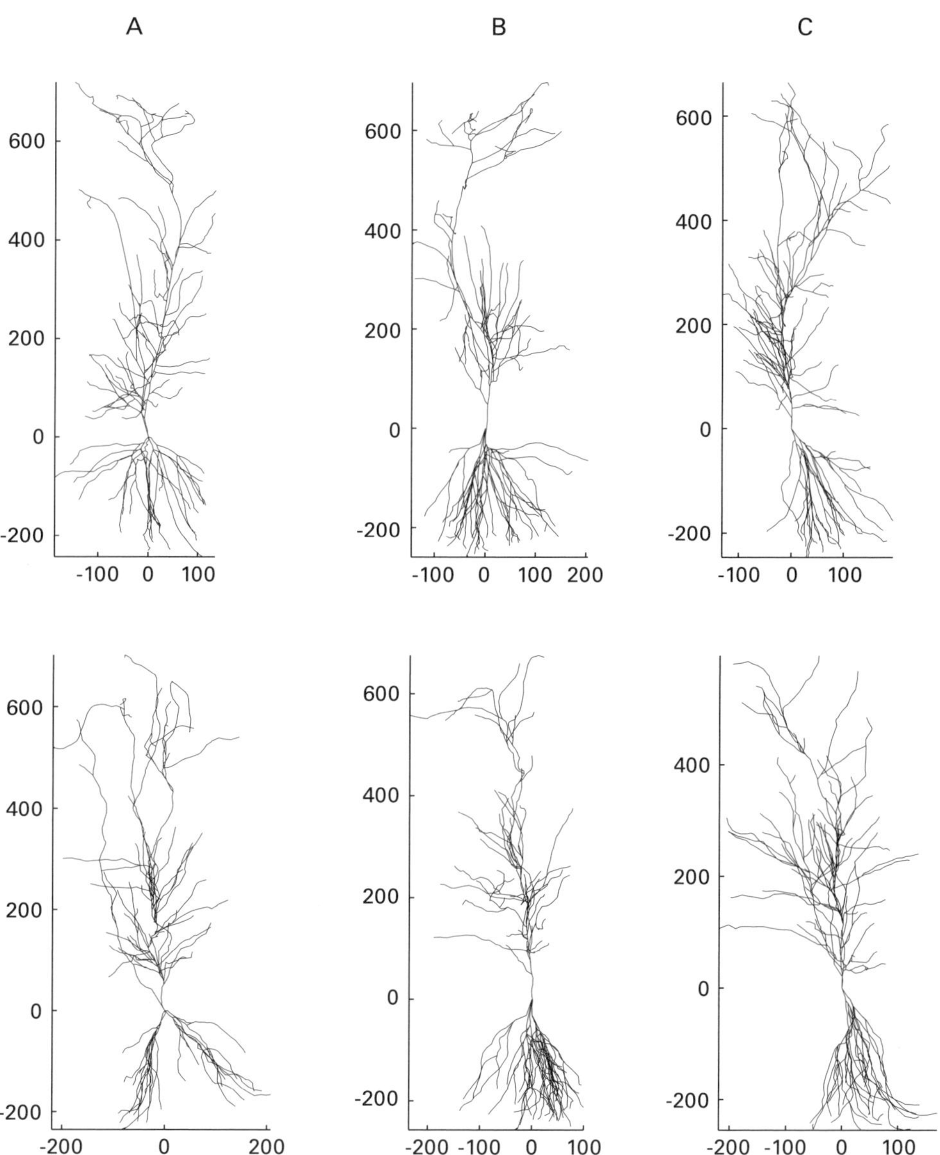

Figure 6.7
CA1 cell-by-cell remodeling, with only core data of a prototype (top) inherited by the virtual cell (bottom). (A–C) cells C70863, C8076e, and C73166b, respectively.

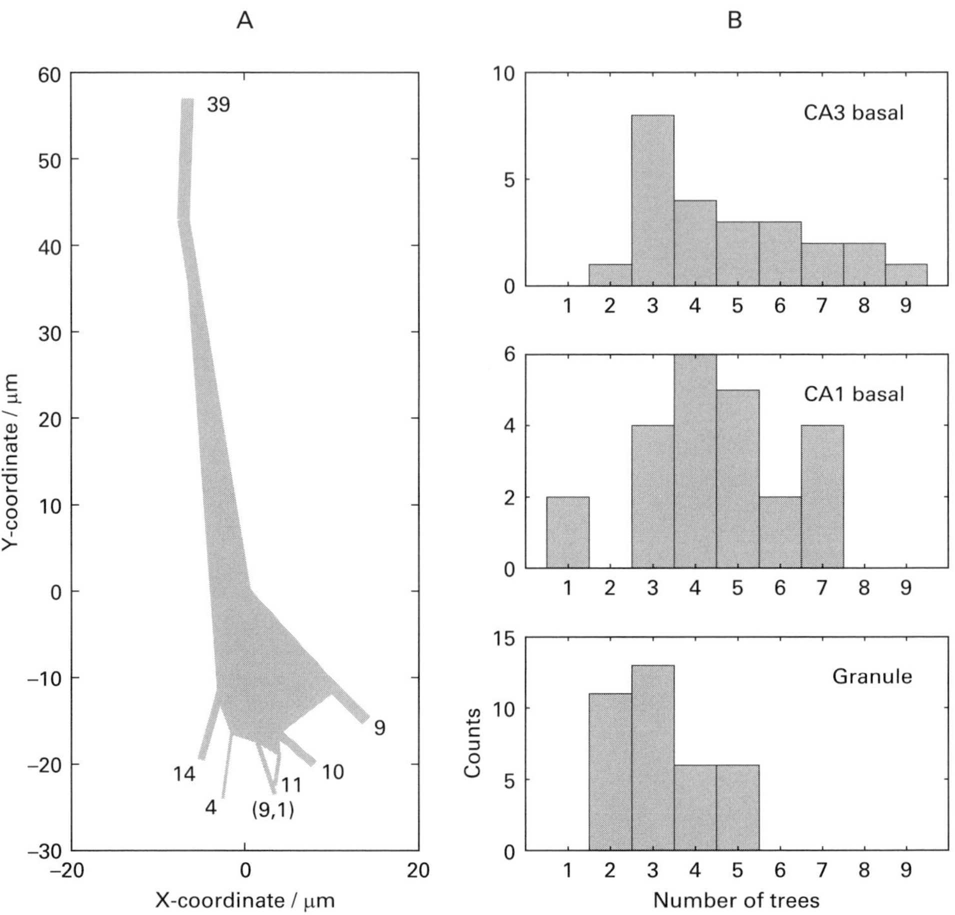

Figure 6.8
Sampling core data. (A) Core geometry of CA3 cell C10861. All stems are truncated at the second segment (apical pointing up). Line thickness measures diameter; numbers indicate the degrees of the truncated subtrees. A pair of numbers (9,1) indicates that the truncation node was a branching point. The choice of coordinates corresponds to a maximal-spread projection. (B) Histograms of the numbers of dendritic trees in real neurons.

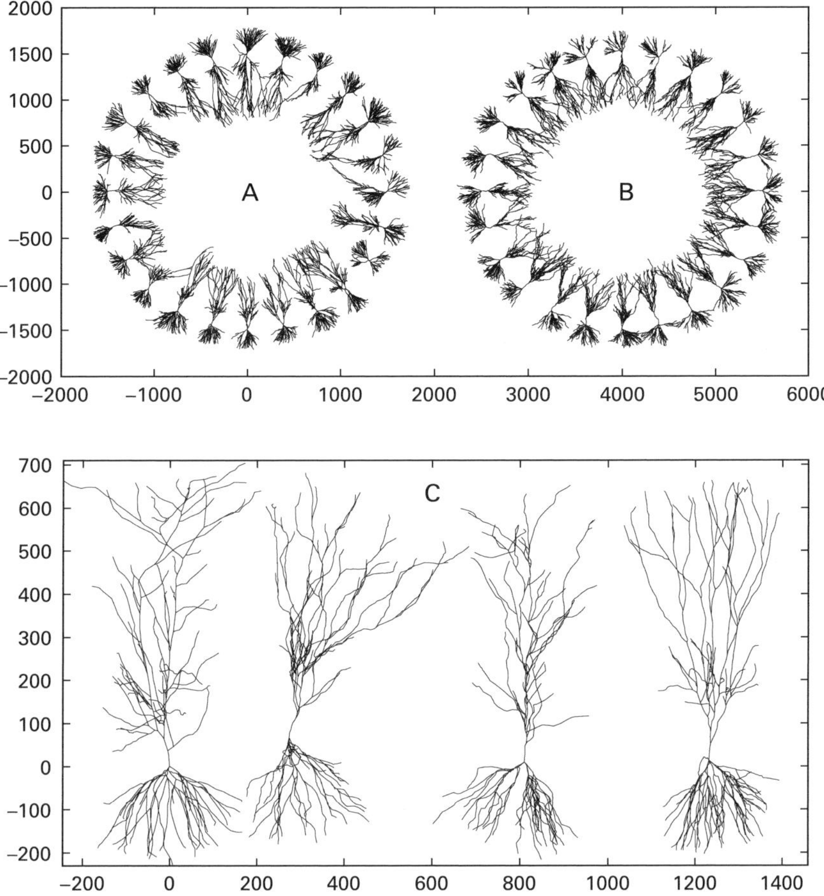

Figure 6.9
Populations of real (A) and entirely (steps 1–4) virtual (B) CA3 cells allocated on a circle. The units on the axes are microns. For related animations, see http://krasnow.gmu.edu/L-Neuron (case sensitive). (C) "Turing test": Two real and two virtual cells, one each from CA3 and one each from CA1. We ask the reader to guess which cells belong to which category.

cells. Preliminary results from this test offered to dozens qualified participants show the lack of confidence in discrimination of the two categories by domain experts (see also Samsonovich and Ascoli, 2005b).

These conclusions hold quantitatively upon statistical analysis of various morphometrics (figure 6.10), and similarly apply to granule cells (Samsonovich and Ascoli, 2005b). Comparison of figures 6.6A and 6.10A–D shows no substantial difference in the quality of results obtained with and without individual inheritance of the core data.

Classes of Algorithms and Their Limitations

General Considerations

It is almost always possible to find an "ugly" algorithmic solution of a given statistical descriptive task. Seldom does one find an elegant algorithmic solution in computational neuroscience at the level of complexity typical of dendritic arborizations. This challenge, however, has both practical and theoretical significance. If virtual neurons are to be assembled in anatomically realistic networks (Ascoli, 1999), their synthesis should be as computationally efficient as possible. Most important, in contrast to many other statistical studies, these numerical algorithms constitute constructive definitions of computational models. Stated differently, they are *theories rather than methods*. Therefore, they have to be parsimonious in order to allow for connections with the principles and mechanisms of real dendritic development.

We therefore restrict our consideration to *elegant algorithms*, understood as those that (a) operate in terms of the available digitized parameters of adult dendrites (such as segment lengths and diameters); (b) use a minimal number of constants, variables, and rules to achieve a given quality of results; and (c) are efficiently implemented for practical use. An example of an algorithm that, formally speaking, "solves" the task yet is not elegant could be a lookup table including the entire archive of available reconstructed cells.

Next, we consider two sorts of elegant algorithms: Markov and non-Markov. Consistent with the traditional notion of Markov processes (initially introduced by Markov, 1913), we adopt the working definition that an algorithm of dendritic tree generation is Markov if every step depends only on the immediately preceding step(s),

Figure 6.10
Population statistics resulting from the overall model (steps 1–4). Distributions of bifurcations (A, B) and terminations (C, D) vs. path distance from the soma in apical (A, C) and basal (B, D) CA3 dendrites (solid: real; dashed: virtual). (E) Topological asymmetry (1–4: CA3; 5–8: CA1; 1,2,5,6: apical; 3,4,7,8: basal; odd: real, even: virtual). Each box has lines at the lower quartile, median, and upper quartile values. Numbers show Wilcoxon rank-sum test *P*-values of virtual vs. (corresponding) real data.

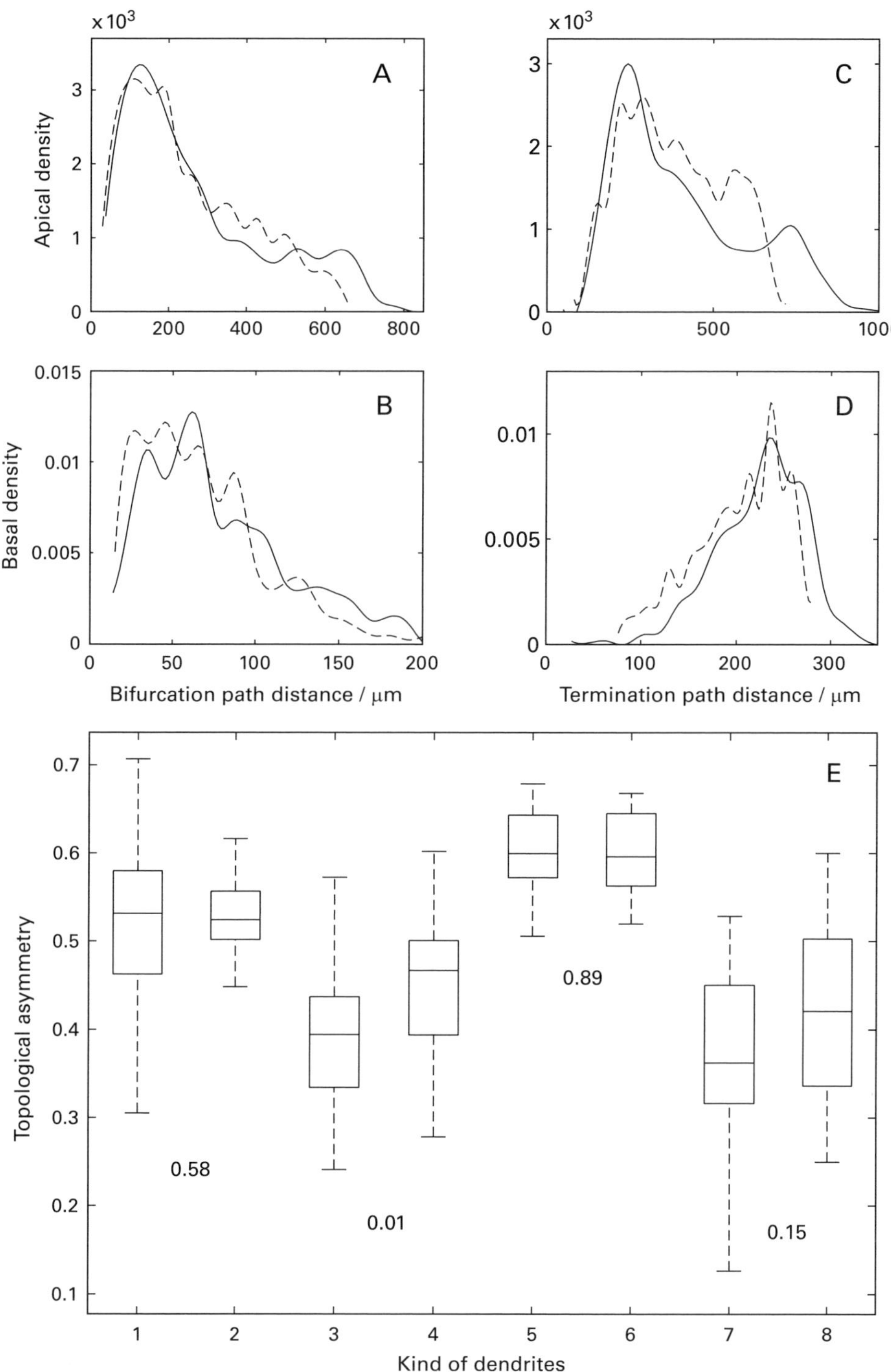
x 10^3
A
C
B
D
E
Apical density
Basal density
Bifurcation path distance / μm
Termination path distance / μm
Topological asymmetry
Kind of dendrites
0.58
0.01
0.89
0.15

with the order of precedence determined by locations on the tree. Here we count an act of attachment of a new dendritic branch B to the growing tree T as one step, meaning that B is regarded as an elementary unit.

More formally, a Markov algorithm must be consistent with the joined probability $P_0(B_1, B_2 \mid B_P) \equiv P_0(B_2, B_1 \mid B_P)$ in order to observe the configurations of two sibling branches B_1 and B_2, conditioned by their parent branch configuration B_P, such that P_0 is the same function for all bifurcations, regardless of their position in the tree. Therefore, specifying the parent and the sibling configurations makes a given branch statistically independent of the rest of the tree. Practically, this definition implies the following constraint (equations 6.5 and 6.6) on the transition probabilities (i.e., sampling rules):

$$P(T + B \mid T, i) = \begin{cases} P_1(B \mid B_P), & \text{if the sibling is missing,} \\ P_2(B \mid B_P, B_S), & \text{if the sibling is present.} \end{cases} \tag{6.5}$$

Here $P(T + B \mid T, i)$ is the probability of attachment of a branch B to an incomplete tree structure T at a selected growth node i (B and T label the sets of parameters associated with the corresponding objects rather than the objects themselves), B_P is the parent branch, and B_S is the sibling branch at node i. P_1 and P_2 are the conditional probabilities that satisfy the relations

$$P_1(B \mid B_P) = \sum_{B_S} P_0(B, B_S \mid B_P), \quad P_2(B \mid B_P, B_S) = \frac{P_0(B, B_S \mid B_p)}{P_1(B_S \mid B_P)}. \tag{6.6}$$

The sum in the first relation is taken over all possible configurations of the sibling branch B_S. The second relation in (6.6) is the Bayes formula.

The constraint (equations 6.5, 6.6) is very coarse and holds for a variety of dendritic growth models that are based on local rules. One may refine the constraint (equations 6.5 and 6.6) by changing the choice of elementary unit and, accordingly, of elementary step. For instance, a unit could be an individual dendritic segment, or a bit of segment, instead of a branch. These fine-grained Markov models (with bits of segments regarded as elementary units) would still satisfy equations (6.5) and (6.6) understood in the original sense (i.e., with branches taken as elementary units).

The model and the related algorithm presented in the preceding section belong to the class of *hidden Markov* models (Baum and Petrie, 1970; Rabiner, 1989) because they operate on "hidden" variables (e.g., m) that are locally associated with the growing virtual object without being determined by the local geometry. Therefore, we distinguish this approach from a case of Markov models that operate in terms of locally measurable geometrical parameters only, which we call *explicit Markov* models.

In looking for an elegant and accurate algorithmic description of the morphometric statistics of a given dendritic class, one can therefore choose among three model categories: non-Markov, explicit Markov, and hidden Markov. Above, we summarized one hidden Markov solution that proves to be efficient and accurate. Now we hypothesize that a solution of this problem cannot be found among explicit Markov algorithms. Below we prove this hypothesis for fine-gained algorithms, in which elementary units are bits of segments rather than whole branches. The proof involves constructing a best-possible, fine-grained explicit Markov algorithm and demonstrating its inability to generate realistic neurons. The constructed algorithm is best-possible in the sense that it exactly reproduces the local probabilities P_0 defined (in terms of bits of segments) by the given population of real cells. A formal possibility of a suitable coarse explicit Markov algorithm (with branches regarded as elementary units) still remains.

Best-Possible Explicit Markov Algorithm

A Markov algorithm describing digitized dendritic morphology can be generally visualized as a process of assembling a tree from a set of available building blocks. At each step, a new block is attached to an available site in the growing structure. According to equations (6.5) and (6.6), the sequential selection of each block can depend only on properties of the immediately preceding block(s), (i.e., previously attached block[s] that can be "touched" at the given site). Limiting the set of inherent block properties to local geometrical measures yields the explicit Markov paradigm defined above. The dendritic model originally described by Hillman (1979) belongs to this category.

Each possible definition of a set of building blocks specifies a subclass of explicit Markov algorithms. The task of finding best representatives within a given subclass can be precisely formulated as follows: Given a set of blocks, find a Markov algorithm that provides maximal statistical consistency of all local geometrical properties (comprised by the probability P_0 above) across the populations of virtual and digitized experimental cells. Below we specify a particular set of building blocks and describe a best-possible explicit Markov algorithm for this set.

We first select a finite scale ε that limits the minimal size of building blocks. Next, we generate a set of blocks by decomposing every digitized cell into fragments. This is done by cutting all dendritic segments that are not immediately adjacent to the soma, and have diameter $d \leq \varepsilon$ and length $l > 2\varepsilon$. Each of these segments is cut twice, at distances ε and 2ε from the origin of the segment (figure 6.11A). All fragments resulting from this procedure form the set of building blocks. In particular, there are five categories of blocks: a core (the soma together with all attached trunks), basal and apical node blocks, and basal and apical links. A link is the middle fragment of each cut segment, a cylinder of length ε. Node blocks include bifurcations, terminations, continuation points, and combinations of the above.

Finally, we arrange all building blocks into rows, sorted by their parent diameters and by their categories (figure 6.11B). The first row contains all cores; the second row

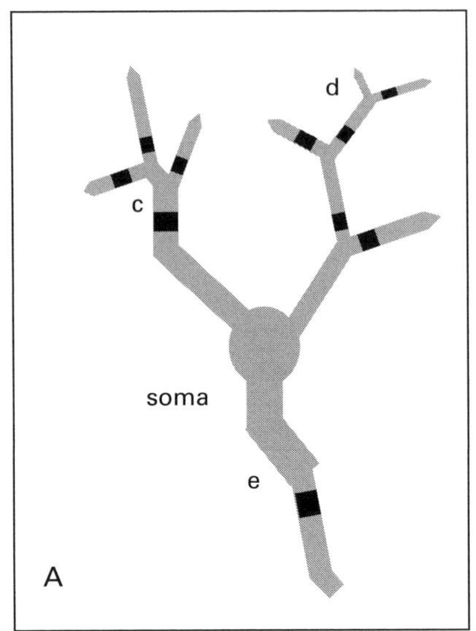
d
c
soma
e
A

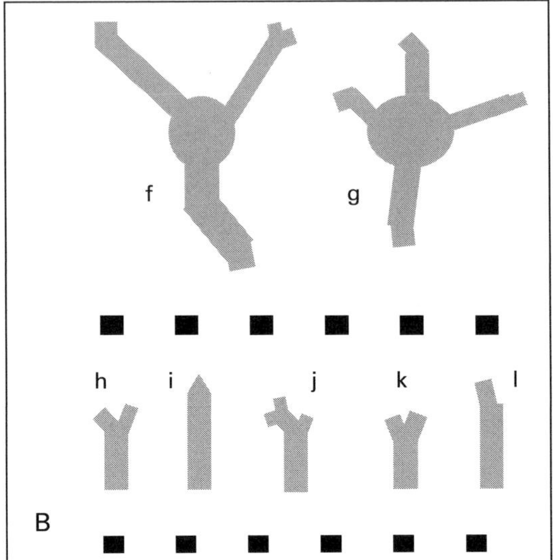
f
g
h
i
j
k
l
B

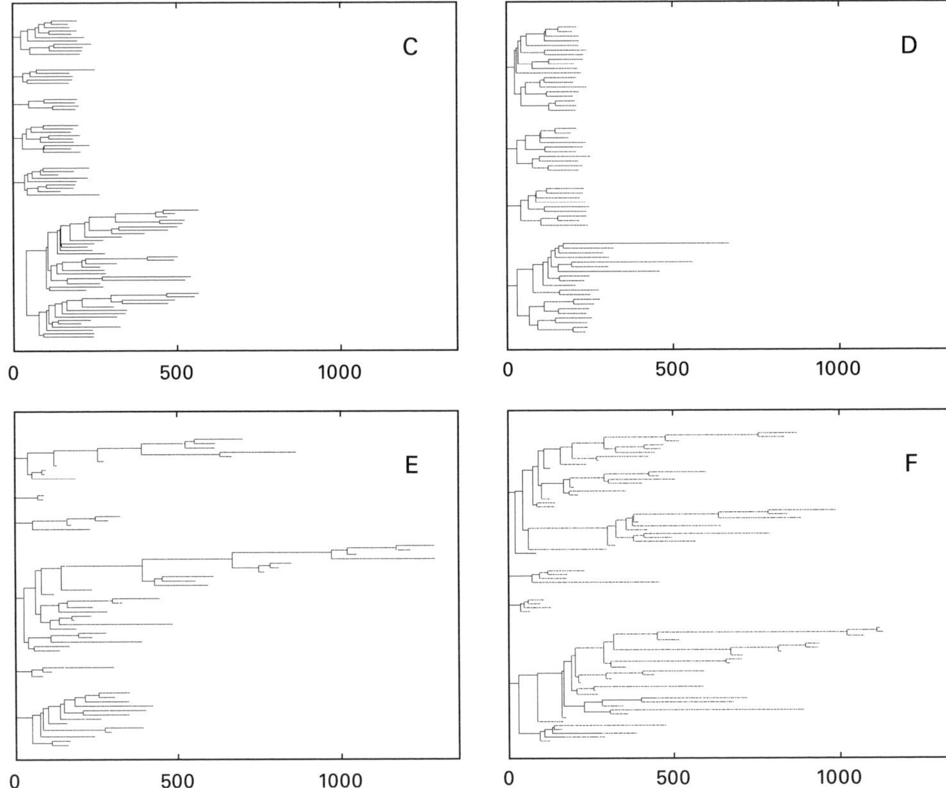
C
D
E
F
0
500
1000

includes all links with the maximal diameter; and so on, up to the last row, containing all node blocks of the minimal parent diameter value (0.3 μm in the data set considered here). Given this set of building blocks, the following constitutes a best-possible explicit Markov algorithm.

Virtual trees grow step by step, starting from a randomly selected core and moving toward the terminals. Each step consists of attaching a new fragment to a randomly selected free site. The new fragment is randomly selected from a row of the table with matching category and parent diameter. The process ends when there are no remaining free sites in the tree.

This is an explicit Markov algorithm, because each step uses only information available from the preceding step within the same branch (diameter matching). The algorithm is best-possible because it preserves P_0 (defined by the set of original trees), and therefore all possible statistical measures defined for a pair of adjacent blocks. This is because one of any two adjacent blocks is always a link of length ε, its only variable parameter is the diameter, and matching the diameters is a rule.[3] At the same time, the algorithm randomly shuffles the order of building blocks in the original morphologies as much as the local constraints allow, thereby erasing all nonlocal correlations that are not entailed by P_0.

Figure 6.11C–F illustrates the failure of this algorithm applied to the population of CA3 cells in the limit of zero ε (similar results are expected for a small finite ε; e.g., ~1 μm). Despite the two cells in each vertical pair sharing the same core, all their morphological characteristics are substantially different, with the virtual cells (6.11E–F) far out of normal range. Therefore, typical virtual cells constructed according to this best-possible explicit Markov algorithm are not qualitatively compatible with real cells.

Conclusions

The structural complexity and natural variability of individual neurons make it exceedingly difficult to achieve a complete characterization of dendritic morphology.

Figure 6.11
Best possible explicit Markov algorithm. (A) The decomposition of dendritic trees into building blocks is illustrated with an oversimplified structure made of a soma and a set of cylindrical segments. Sharpened ends indicate termination points. At the first step, a link fragment (black) of fixed length ε is cut out of every dendritic segment that satisfies certain conditions (see text). (B) Building blocks are sorted into rows by their parent diameter and by their category. Core fragments (f, g) are in the top row, followed by the row of (all identical) links of diameter ε. These are followed by a row of node blocks with parent diameter ε (h, k: bifurcations; i: termination; l: continuation; j: a two-node fragment). Links of the next diameter value follow below, and so on, down to the smallest parent diameter value (0.3 μm). The process of virtual tree generation consists of random attachments of diameter matching building blocks of alternating color, starting from a core, until all branches terminate. (C–F) Typical examples of real (C–D) and virtual (E–F) CA3 dendrograms created with the best possible Markov algorithm. Each vertical pair (C/E, D/F) shares the same core. The scale (in microns) is the same in all panels; apical dendrites are at the bottom.

In recent years, we proposed a novel solution to this open problem. We have implemented computational algorithms to simulate dendritic morphology based on distributions of parameters measured from the experimental data. Because the quantitative characterization of dendritic morphology of a neuronal class is necessarily statistical, its algorithmic description ought to be intrinsically stochastic. Therefore, resulting virtual (computer-generated) neurons are each different and unique, yet statistically and visually compatible with real cells of the corresponding morphological class. The algorithm that generates these synthetic neurons sets an upper limit on the amount of morphological information present in real cells that can be effectively captured by a given set of measurements.

Here we have described a series of algorithms to model both dendrogram properties and three-dimensional orientation of neuronal trees. In particular, this chapter has illustrated how the dendritic morphologies of rat CA3 and CA1 pyramidal cells and dentate granule cells can be accurately described in all functionally relevant statistical details by an elegant hidden Markov algorithm. The total amount of information used by the algorithm is substantially smaller than that representing a single digitized neuron. The hidden variables correspond to the expected number of terminal tips of a subtree and the path distance from the soma. An equivalent result cannot be obtained with a best explicit Markov algorithm (one using all local information available in the digitized experimental data, but not relying on hidden variables). The model successfully captures all major morphometrics of basal and apical dendrites of rat CA3 pyramidal cells, basal CA1 dendrites, and dentate granule cells, including distributions of bifurcations and terminations, and topological asymmetry. Interestingly, in order to obtain similar results with CA1 apical dendrites, it is necessary to distinguish between main and oblique apical branches, which suggest possibly different underlying developmental mechanisms. In addition, the model suggests that dendrites of all classes in the principal cells of the hippocampus may selectively repel their own somata.

Acknowledgments

This work was supported by Human Brain Project Grant R01-NS39600, funded jointly by NIMH, NINDS, and NSF.

Notes

1. Oblique branches are not labeled as such in the original data. Rather, they are defined algorithmically, as explained in "Building a Tree Skeleton (Step 2) below." (See also Samsonovich and Ascoli, 2005a.)

2. Technically, *alpha* in this case was sampled from the gamma probability distribution function with parameters derived from the experimental primary oblique segments by maximization of the likelihood ($a = 1.41$, $b = 23.6$).

3. Thus, in an infinite virtual population, the statistical averages of all local geometrical properties (within the range ε) are necessarily preserved from the original data.

References

Ascoli GA (1999) Progress and perspectives in computational neuroanatomy. Anat Rec 257(6): 195–207.

Ascoli GA (2002) Neuroanatomical algorithms for dendritic modelling. Network 13(3): 247–260.

Ascoli GA, Krichmar JL (2000) L-Neuron: A modeling tool for the efficient generation and parsimonious description of dendritic morphology. Neurocomputing 32–33: 1003–1011.

Ascoli GA, Krichmar JL, Nasuto SJ, Senft SL (2001a) Generation, description and storage of dendritic morphology data. Phil Trans R Soc Lond B356(1412): 1131–1145.

Ascoli GA, Krichmar JL, Scorcioni R, Nasuto SJ, Senft SL (2001b) Computer generation and quantitative morphometric analysis of virtual neurons. Anat Embryol 204(4): 283–301.

Baum LE, Petrie T (1970) Statistical inference for probabilistic functions of finite state Markov chains. Ann Math Stat 41: 164.

Burke RE, Marks WB, Ulfhake B (1992) A parsimonious description of motor neurons' dendritic morphology using computer simulation. J Neurosci 12: 2403–2416.

Carnevale NT, Tsai KY, Claiborne BJ, Brown TH (1997) Comparative electrotonic analysis of three classes of rat hippocampal neurons. J Neurophys 78: 703–720.

Hillman DE (1979) Neuronal shape parameters and substructures as a basis of neuronal form. In: The Neurosciences, Fourth Study Program (Schmitt F, ed), 477–498. Cambridge, Mass.: MIT Press.

Ishizuka N, Cowan WM, Amaral DG (1995) A quantitative analysis of the dendritic organization of pyramidal cells in the rat hippocampus. J Comp Neurol 362: 17–45.

Markov AA (1913) An example of statistical investigation in the text of "Eugene Onegin" illustrating coupling of "tests" in chains. Proc Acad Sci St Petersburg 7: 153–162.

Rabiner LR (1989) A tutorial on hidden Markov models and selected applications in speech recognition. Proc IEEE 77(2): 257–286.

Ramón y Cajal S (1894–1904) *Textura del sistema nervioso del hombre y los vertebrados.* 1994 English trans. *Histology of the Nervous System of Man and Vertebrates* by N. Swanson and L. W. Swanson. New York: Oxford University Press.

Rihn LL, Claiborne BJ (1990) Dendritic growth and regression in rat dentate granule cells during late postnatal development. Brain Res Dev 54: 115–124.

Samsonovich AV, Ascoli GA (2003) Statistical morphological analysis of hippocampal principal neurons indicates cell-specific repulsion of dendrites from their own cell. J Neurosci Res 71(2): 173–187.

Samsonovich AV, Ascoli GA (2005a) Statistical determinants of dendritic morphology in hippocampal pyramidal neurons: A hidden Markov model. Hippocampus 15: 166–183.

Samsonovich AV, Ascoli GA (2005b) Algorithmic description of hippocampal granule cell dendritic morphology. Neurocomputing 65–66: 253–260.

Turner DA, Cannon RC, Ascoli GA (2002) Web-based neuronal archives: Neuronal morphometric and electrotonic analysis. In: Neuroscience Databases—A Practical Guide (Kotter R, ed), 81–98. Amsterdam: Elsevier.

7 Geometric Modeling and Quantitative Visualization of Virus Ultrastructure

Chandrajit Bajaj

Viruses are among the smallest parasitic nano-objects that are agents of human disease (White and Fenner, 1994). They have no systems for translating RNA, generating ATP, or synthesizing proteins or nucleic acids, and therefore need the subsystems of a host cell to sustain and replicate (White and Fenner, 1994). It would be natural to classify these parasites according to their eukaryotic or prokaryotic cellular hosts (e.g., plants, animals, bacteria, fungi, etc.), but viruses exist which have more than one sustaining host species (White and Fenner, 1994). Currently, viruses are classified simultaneously via the host species (algae, archaea, bacteria, fungi, invertebrates, mycoplasma, plants, protozoa, spiroplasma, vetebrates), the host tissues that are infected, the method of virial transmission, the genetic organization of the virus (single- or double-stranded, linear or circular, RNA or DNA), the protein arrangement of the protective closed coats housing the genome (helical, icosahedral symmetric nucleocapsids), and whether the virus capsids have a further outer envelope covering (the complete virion) (White and Fenner, 1994). Table 7.1 summarizes a small yet diverse collection of viruses and virions (ICTV Database).

The focus of this chapter is on the computational geometric modeling and visualization of the nucleocapsid ultrastructure of plant and animal viruses exhibiting the diversity and geometric elegance of the multiple protein arrangements. Additionally, a regression relationship between surface area and enclosed volume is computed for spherical viruses with icosahedral symmetric protein arrangements. The computer modeling and quantitative techniques for virus capsid shells' ultrastructure that we review here are applicable to atomistic, high-resolution (less than 4Å) model data, as well as medium (5Å–15Å) resolution map data reconstructed from cryoelectron microscopy.

The Morphology of Virus Structures

Minimally, viruses consist of a single nucleocapsid made of proteins for protecting their genome, as well as for facilitating cell attachment and entry. The capsid

Table 7.1
Helical and icosahedral viruses and viral subunit structures

Name	Family	Host	NA	Capisd Sym. (# T)	#Shell (E?)	Modality (res in A) (pdbid)
Tobacco mosaic [15, 65]	Tobamoviridae	P	sR (L)	He	1(n)	X(2.45) (1ei7)
Ebola [93]	Filoviridae	V	sR (L)	He	1(E)	X(3) (1ebo)
Vaccinia [27]	Poxviridae	V	dD (L)	He	1(E)	X(1.8) (1luz)
Rabies [60]	Rhabdoviridae	V	sR (L)	He	1(E)	X(1.5) (1vyi)
Satellite tobacco necrosis [56]	Tombusviridae	P	sR (L)	Ic (1)	1(n)	X(2.5) (2stv)
L-A (Saccharomyces cerevisiae) [63]	Totiviridae	F	dR (L)	Ic (1)	1(n)	X(3.6) (1m1c)
Canine parvovirus-Fab complex [97]	Parvoviridae	V	sD (L)	Ic (1)	1(n)	X(3.3) (2cas)
T1L reovirus core [76]	Reoviridae	V	dR (L)	Ic (1,1)	2(n)	X(3.6) (1ej6)
T3D reovirus core [87]	Reoviridae	V	dR (L)	Ic (1,1)	2(n)	X(2.5) (1muk)
P4 (Ustilago maydis) [53]	Totiviridae	P	dR (L)	Ic (1)	1(n)	X(1.8) (1kp6)
Tomato bushy stunt [43]	Tombusviridae	P	sR (L)	Ic (3)	1(n)	X(2.9) (2tbv)
Cowpea chlorotic mosaic [83]	Bromoviridae	P	sR (L)	Ic (3)	1(n)	X(3.2) (1cwp)
Cucumber mosaic [83]	Bromoviridae	P	sR (L)	Ic (3)	1(n)	X(3.2) (1f15)
Norwalk [71]	Caliciviridae	V	sR (L)	Ic (3)	1(n)	X(3.4) (1ihm)
Rabbit hemorrhagic disease VLP-MAb-E3 complex [68]	Caliciviridae	V	sR (L)	Ic (3)	1(n)	X(2.5) (1khv)
Galleria mellonella denso [82]	Parvoviridae	I	sD (L)	Ic (1)	1(n)	X(3.7) (1dnv)
Semiliki Forest [59]	Togaviridae	I,V	sR	Ic (4,1)	2(E)	C(9) (1dyl)
Polyoma [23]	Papovaviridae	V	dD (C)	Ic (7D)	1(n)	X(2.2) (1cn3)
Simian [85]	Papovaviridae	V	dD (C)	Ic (7D)	1(n)	X(3.1) (1sva)
Papillomavirus initiation complex [33]	Papovaviridae	V	dD (C)	Ic (7D)	1(n)	X(3.2) (1ksx)
Blue tongue [37]	Reoviridae	V	dR (L)	Ic (1,13L)	2(n)	X(3.5) (2btv)
Rice dwarf [64]	Reoviridae	P	dR (L)	Ic (1,13L)	2(n)	X(3.5) (1uf2)
T1L reovirus virion [55]	Reoviridae	V	dR (L)	Ic (1,13L)	2(n)	X(2.8) (1jmu)
Simian rotavirus (SA11-4F) TLP [39]	Reoviridae	V	dR (L)	Ic (1,13L)	2(n)	X(2.38) (1lj2)
Rhesus rotavirus [31]	Reoviridae	V	dR (L)	Ic (1,13L)	2(n)	X(1.4) (1kqr)
Reovirus [101]	Reoviridae	V	dR (L)	Ic (1,13L)	2(n)	C(7.6)
Nudaurelia capensis w [42]	Tetraviridae	I	sR (L)	Ic (4)	1(n)	X(2.8) (1ohf)
Herpes simplex [18]	Herpesviridae	V	dD (L)	Ic (7L)	1(E)	X(2.65) (1jma)
Chilo iridescent [98]	Iridoviridae	I	dD (C)	Ic (147)	1(E)	C(13)
Paramecium bursaria chlorella [98]	Phycodnaviridae	P	dD (L)	Ic (169D)	1(E)	C(8)
HepBc (human liver) (nHBc) [92]	Hepadnaviridae	V	dD (C)	Ic (4)	1(E)	X(3.3) (1qgt)

Table 7.1
(continued)

Notes: Structure reference is in square brackets, following name.
Family nomencleature from the ICTV database.
Host types are P (plant), V (vertebrate), I (invertebrate), and F (fungi).
Virus nucleic acid (NA) type is single-stranded RNA (sR) or DNA (sD), double-stranded RNA (dR) or DNA (dD), and linear (L) or circular (C).
Capsid symmetry is helix (He) or icosahedral (Ic), with the triangulation number of each capsid shell in parentheses.
The number of capsid shells is followed by whether enveloped (E) or not (n).
The acquisition modality X-ray is followed by feature resolution and PDBid, each in parentheses.

proteins magically self-assemble, often into a helical or icosahedral symmetric shell (henceforth referred to as capsid shells). There are several examples of capsid shells which do not exhibit any global symmetry (ICTV Database), but we focus on the symmetric capsid shells in the remainder of this chapter.

Different virus morphologies that are known (a small sampling is included in table 7.1) are distinguished by optional additional outer capsid shells, the presence or lack of a surrounding envelope for these capsid shells (often derived from the host cell's organelle membranes), and additional proteins within these capsids and envelopes, that are necessary for the virus life cycle (figure 7.1). The complete package of proteins, nucleic acids, and envelopes is often termed a virion.

The asymmetric structural subunit of a symmetric capsid shell may be further decomposable into simpler and smaller protein structure units termed protomers. Protomers can be a single protein in monomeric form (for example, TMV) or form homogeneous dimeric or trimeric structure units (for example, RDV). These structure units also often combine to form symmetric clusters, called capsomers, and are predominantly distinguishable in visualizations at even medium and low resolution virus structures. The capsomers and/or protomeric structure units pack to create the capsid shell in the form of either a helical or an icosahedral symmetric arrangement (figure 7.2).

The subsequent subsections of the chapter dwell on the geometry of the individual protomers and capsomers as part of a hierarchical arrangement of symmetric capsid shells.

The Geometry of Helical Capsid Shells

Helical symmetry can be captured by a 4×4 matrix transformation $H_{(\vec{a},\phi,L)}$ parameterized by $\vec{a} = (a_x, a_y, a_z)$, a unit vector along the helical axis; by θ, an angle in the plane of rotation; and by the pitch L, the axial rise for a complete circular turn.

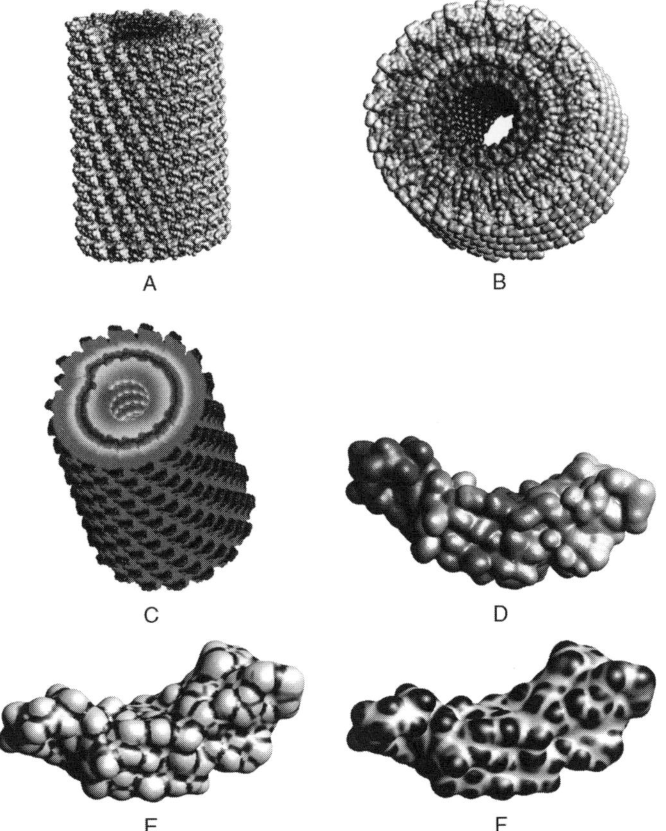

Figure 7.1
Organization of the tobacco mosaic virus (1EI7) with its helical nucleocapsid shown in (A), (B), and (C). (A) and (B) are surface-rendered; (C) is volume-rendered. The asymmetric protomeric structure unit is visualized in (C) as an implicit solvation molecular surface colored by distance from the helix symmetry axis (D), with a transparent molecular surface and the protein backbone showing helix secondary structures. (E) Molecular surface of protomer with the mean curvature function; red shows positive mean curvature and green shows negative mean curvature. (F) Gaussian curvature function on the protomer molecular surface, with green showing positive Gaussian curvature and red signifying negative curvature, displayed on the molecular surface.

Figure 7.2
Organization of rice dwarf virus (1UF2) with icosahedral capsid shells. (A) 2-D, texture-based visualization of the outer capsid shell showing a single sphere per nonhydrogen atom, and colored to distinguish individual protein subunits. (B) The outer capsid shell is shown as a smooth, analytic molecular surface, and the inner capsid surface is displayed using 2-D texture maps of a union of spheres and colors. (C) shows the outer capsid. (D) displays the inner capsid. (E) shows the icosahedral asymmetric structure unit of the outer unit. (F) displays the icosahedral asymmetric structure unit of the inner unit. (G) shows the protein backbone of the structure unit in (E). (H) shows the protein backbone of the structure unit in (F).

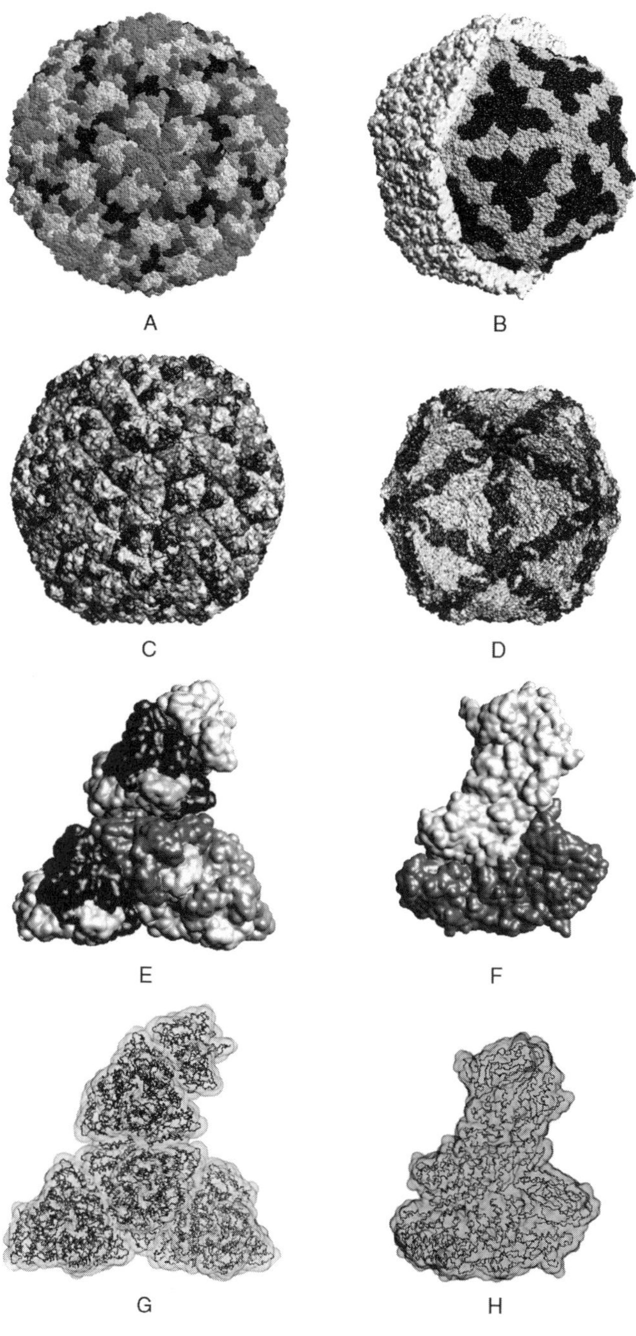
A
B
C
D
E
F
G
H

$$H_{(\vec{a},\phi,L)} = \begin{bmatrix} a_x^2(1-\cos\theta)+\cos\theta & a_xa_y(1-\cos\theta)-a_z\sin\theta & a_xa_z(1-\cos\theta)+a_y\sin\theta & \frac{a_xL\theta}{2\pi} \\ a_xa_y(1-\cos\theta)+a_z\sin\theta & a_y^2(1-\cos\theta)+\cos\theta & a_ya_z(1-\cos\theta)-a_x\sin\theta & \frac{a_yL\theta}{2\pi} \\ a_xa_z(1-\cos\theta)-a_y\sin\theta & a_ya_z(1-\cos\theta)+a_x\sin\theta & a_z^2(1-\cos\theta)+\cos\theta & \frac{a_zL\theta}{2\pi} \\ 0 & 0 & 0 & 1 \end{bmatrix}$$

If P is the center of any atom of the protomer, then P' is the transformed center, and $P' = H^*P$. Repeatedly applying this transformation to all atoms in a protomer yields a helical stack of protomeric units. The desired length of the helical nucleocapsid shell is typically determined by the length of the enclosed nucleic acids. The capsid shell of the tobacco mosaic virus (TMV) exhibits helical symmetry (figures 7.1 and 7.3), with the asymmetric protein structure unit or the protomer consisting of a single protein (pdb id 1EI7).

The Geometry of Icosahedral Capsid Shells

In numerous cases the virus structure, a regular icosahedron with edge length l, will have a radius, measured from center to a vertex, $r = l/2 \cdot V(\phi Vs)$. The advantage over helical symmetry is the efficient construction of a capsid of a given size using the smallest protein subunits. An icosahedron has twelve vertices, twenty equilateral triangular faces, and thirty edges, and exhibits 5:3:2 symmetry. A fivefold symmetry axis passes through each vertex, a threefold symmetry axis passes through the center of each face, and a twofold axis passes through the midpoint of each edge (see figure 7.4).

A rotation transformation around an axis $\vec{a} = (a_x, a_y, a_z)$ by an angle θ is described by the 4×4 matrix

$$R_{(\vec{a},\phi)} = \begin{bmatrix} a_x^2(1-\cos\theta)+\cos\theta & a_xa_y(1-\cos\theta)-a_z\sin\theta & a_xa_z(1-\cos\theta)+a_y\sin\theta & 0 \\ a_xa_y(1-\cos\theta)+a_z\sin\theta & a_y^2(1-\cos\theta)+\cos\theta & a_ya_z(1-\cos\theta)-a_x\sin\theta & 0 \\ a_xa_z(1-\cos\theta)-a_y\sin\theta & a_ya_z(1-\cos\theta)+a_x\sin\theta & a_z^2(1-\cos\theta)+\cos\theta & 0 \\ 0 & 0 & 0 & 1 \end{bmatrix}$$

The vertices of a canonical icosahedron are given by $\{(0,\pm1,\pm\Phi), (\pm1,\pm\Phi,0), (\pm\Phi,0,\pm1)\}$, where $\Phi = (1+\sqrt{5})/2$ is the golden ratio. For a fivefold symmetry transformation around the vertex $(0,\pm1,\pm\Phi)$, the normalized axis of rotation is $\vec{a} = (0, 0.52573, 0.85064)$ and the angle of rotation is $\theta = 2\pi/5$, yielding a fivefold symmetry transformation matrix:

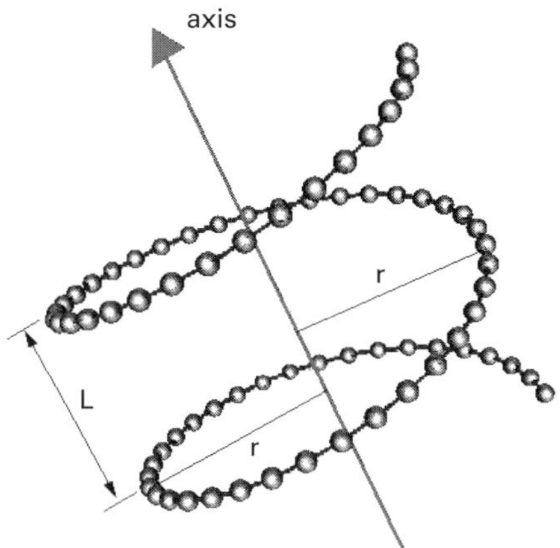

Figure 7.3
Helical symmetry axis.

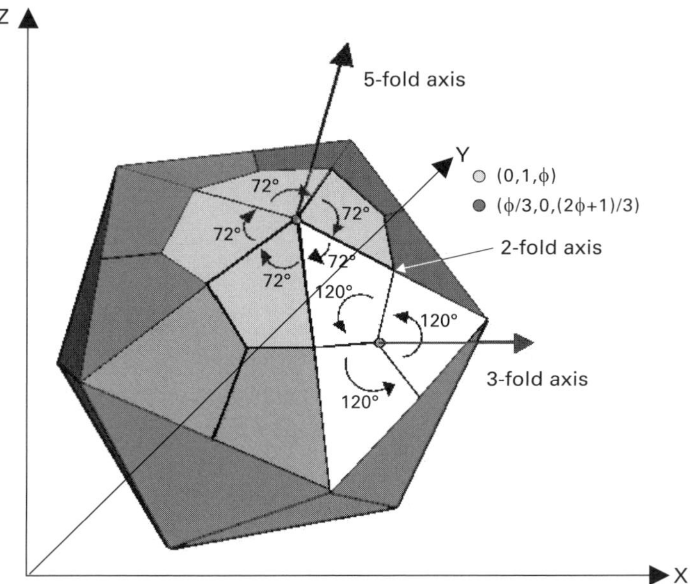

Figure 7.4
Icosahedral transformations showing fivefold and threefold symmetry axes.

$$R_{(5-fold)} = \begin{bmatrix} 0.30902 & -0.80902 & 0.5000 & 0 \\ 0.80902 & 0.5000 & 0.30902 & 0 \\ -0.5000 & 0.30902 & 0.80902 & 0 \\ 0 & 0 & 0 & 1 \end{bmatrix}$$

Similarly, one is able to construct fivefold symmetry transformation matrices for the other icosahedron vertices. Using the generic rotational transformation matrix $R_{(\vec{a},\phi)}$, one can construct the threefold transformation matrices via the rotation axis that passes through the centroid of the triangular faces of the icosahedron and an angle of rotation of $\theta = 2\pi/3$. Consider the triangular face with corners at $(0, 1, \Phi)$, $(0, -1, \Phi)$ and $(\Phi, 0, 1)$. The centroid is at $(\Phi/3, 0, (2\Phi + 1)/3)$; the normalized axis of rotation is $\vec{a} = (0.356822, 0, 0.934172)$; and the transformation matrix is

$$R_{(3-fold)} = \begin{bmatrix} 0.30902 & -0.80902 & 0.5000 & 0 \\ 0.80902 & -0.5000 & -0.30902 & 0 \\ 0.5000 & 0.30902 & 0.80902 & 0 \\ 0 & 0 & 0 & 1 \end{bmatrix}$$

A polyhedron with faces that are all equilateral triangles is called a deltahedron. Deltahedra with icosahedral symmetry are classified as icosadeltahedra. Any icosadeltahedron has 20 T facets, where T is the *triangulation number* given by $T = Pf^2$, where $P = h^2 + hk + k^2$, for all pairs of integers h and k which do not have a common factor, and f is any integer (Caspar and Klug, 1962). The possible values of P are $1, 3, 7, 13, 19, 21, 31, 37, \ldots$. Figure 7.5(A) shows triangles with different triangulation numbers for icosahedral virus structures.

With a fixed-size asymmetric unit, the greater the T number, the larger the size of the virus capsid. Each triangular portion of the icosahedral virus capsid is easily subdivided into its three asymmetric units, with each unit containing some combination of protein structure units (protomers). In total, an icosahedral virus capsid has 60 T asymmetric units with numerous protein structures intertwined to form a spherical mosaic. In figure 7.5 we see that when $T = 1$, each vertex is at the center of a pentagon, and the capsid proteins are in equivalent environments (i.e., five neighbors cluster at a common vertex). However, for icosadeltahedra with larger triangulation numbers (e.g., $T = 13$), there are pentagons and hexagons in the capsid mosaic (figure 7.5). Therefore, even though the capsid proteins (protomers) may be chemically identical, some cluster into a fivefold neighborhood and the others into a sixfold neighborhood. Such locally symmetric clusterings of protomers are alternatively termed capsomers. In these situations, the proteins are no longer globally symmetrically equivalent, but only *quasi-equivalent* (Caspar and Klug, 1962).

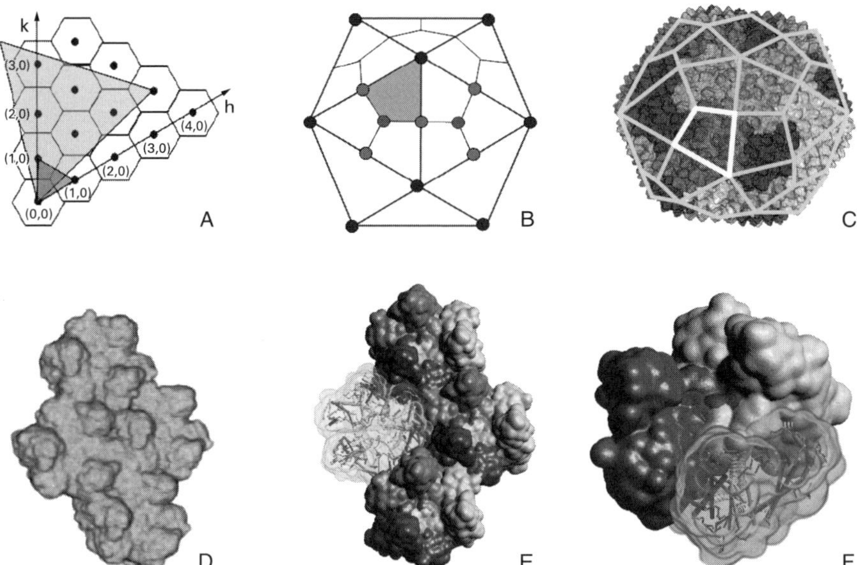

Figure 7.5
Architecture of icosahedral viruses. (A) Caspar-Klug triangulation number (T) via a hexagonal lattice. Dark gray triangle is T = 1, and light gray triangle represents T = 13. (B) shows the asymmetric unit of an icosahedron. (C) Asymmetric structure units of the capsid shell. (D) A single asymmetric structure unit. (E) Asymmetric unit colored by protein and also showing protein backbone. (F) A capsomere consisting of three proteins. (VIRUS PDB: 1GW8.)

Surface and Volumetric Modeling and Visualization

Atomistic Resolution Model Structures

Numerous schemes have been used to model and visualize biomolecules and their properties (Zhang et al., 2006; Bajaj et al., 2004; Bajaj, Pascucci, et al., 2003:47). These different visual representations are often derived from an underlying geometric model constructed from the positions of atoms, bonds, chains, and residues—information deposited as part of an atomic resolution structure of the protein or nucleic acid in the Protein Data Bank (PDB). Hence, structural models are designed to represent the primary (sequence), secondary (e.g., α-helices, β-sheets), tertiary (e.g., $\alpha - \beta$ barrels) (subparts), and quaternary structures of the entire protein or nucleic acid.

An early approach to molecular modeling is to consider atoms as hard spheres, and their union as an attempt to capture shape properties as well as spatial occupancy of the molecule. This is similar to our perception of surfaces and volume occupancy of macroscopic objects. Figure 7.2A and 7.2B show hard-sphere model visualizations of the twin rice dwarf capsid shells, with individual proteins colored differently. Solvated versions of these molecular surfaces have been proposed by Lee, Richards (Lee and Richards, 1971) and Connolly (1983) for use in computational biochemistry and biophysics. Much of the preliminary work, along with later extensions, focused on finding fast methods of triangulating this molecular surface, is sometimes referred to as the solvent contact surface. Additionally, see figure 7.6 for several examples of molecular surface approximations of protein capsid shells of virus structures. Two prominent obstacles in modeling are the correct handling of surface self-intersections (singularities) and the high communication bandwidth needed when sending tessellated surfaces to the graphics hardware.

A more analytic and smooth description of molecular surfaces (without singularities) is provided by a suitable level set of the electron density representation of the molecule. Isotropic Gaussian kernels have traditionally been used to describe atomic electron density due to their ability to approximate electron orbitals. The electron density of a molecule with M atoms, centered at $\vec{x_j}$, $j \in \{1 \dots M\}$ can thus be written as

$$F_{elec\_dens}(\vec{x}) = \sum_{j=1}^{M} \gamma_j K(\vec{x} - \vec{x_j}),$$

where γ_j and K are typically chosen from a quadratic exponential description of atomic electron density:

$$\begin{aligned} Atom(\vec{x}) &= e^{-(d/r^2)((\vec{x}-\vec{y})^2-r^2)} = e^d e^{-(d/r^2)(\vec{x}-\vec{y})^2} \\ &= AK^q(\vec{x}-\vec{y})\gamma_{elec\_dens}(\vec{x}) \\ &= e^{-(d/r^2)((\vec{x}-\vec{x_j})^2-r^2)} = \gamma_j K(\vec{x}-\vec{x_j}). \end{aligned}$$

The atomic electron density kernels are affected by the radius r of individual atoms and the decay parameter d. Smooth and molecular surface models for individual proteins, structure units, and entire capsid shells can easily be constructed as a fixed level set of

$$F_{elec\_dens}(\vec{x}) = \sum_{j=1}^{M} \gamma_j K(\vec{x} - \vec{x_j}).$$

An array of such structural molecular model visualizations is shown as figures 7.1–7.5 as well as figure 7.6. Some of them use transparency on the solvated molecular surface and show the protein backbone structure (folded chains of α-helices and β-sheets).

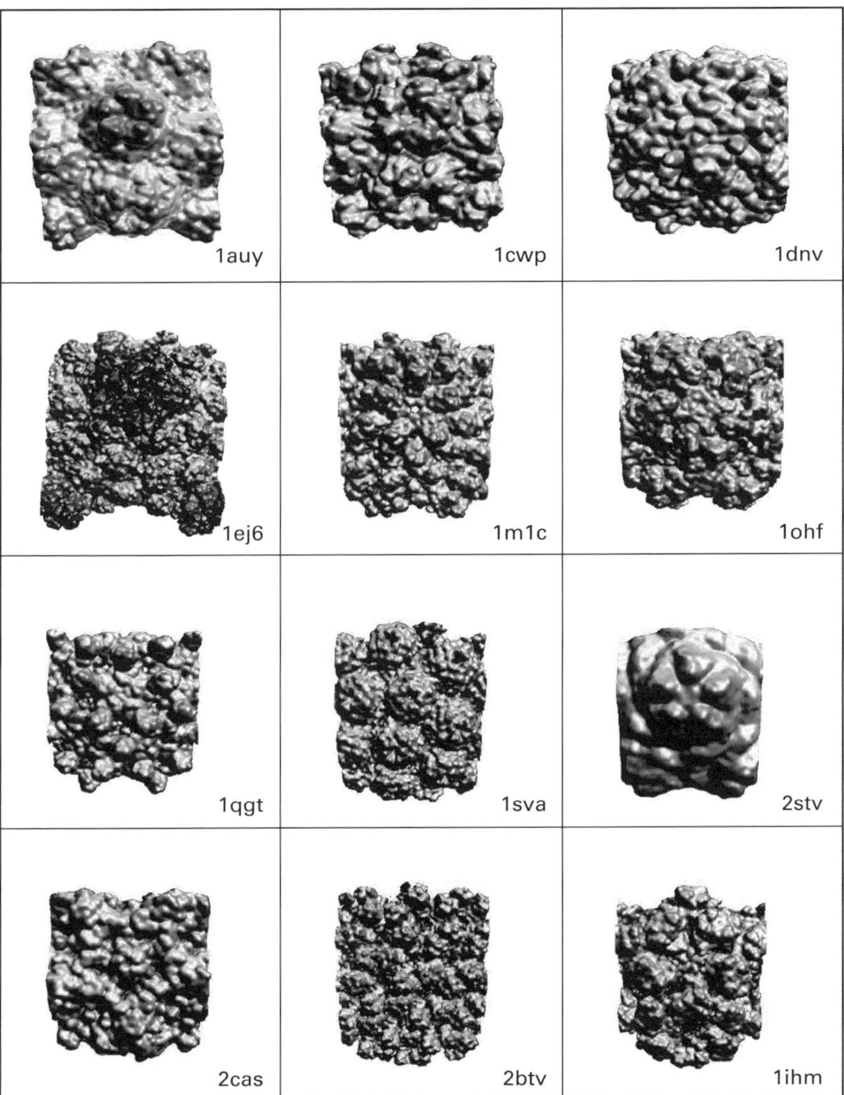

Figure 7.6
Portions of capsid shells of icosahedral viruses showing a significant portion of the protein capsid which properly includes the asymmetric subnit. Note that the isosurface is selected to provide a good molecular surface approximation while maintaining topological equivalence to a sphere. This makes the surface area and enclosed volume computation directly amenable to the calculations reported in the contour spectrum paper.

Structure Elucidation from 3-D Maps

Electron microscopy (EM), in particular single-particle reconstruction using cryo-EM, has rapidly advanced over recent years, such that many virus structures can be resolved routinely at low resolution (10–20Å), and in some cases at subnanometer (intermediate) resolution (7–10Å) (Baker et al., 1999; Belnap et al., 1999).

Symmetries within the virus capsid shells are exploited both in the 3-D map reconstructions from raw 2-D EM images and in structure elucidation in the 3-D map. In many cases, the 3-D maps are of spherical viruses, with protein capsid shells exhibiting icosahedral symmetry. In these cases, the global symmetry detection can be simplified to computing the location of the fivefold rotational symmetry axes passing through the twelve vertices of the icosahedron, from which the threefold symmetry axes for the twenty icosahedron faces and the twofold symmetry axes for the thirty icosahedron edges can be easily derived.

However, determining the local symmetries of the capsomers (structure units) is more complicated, because they exhibit varied k-fold symmetry, and their detection requires a modified correlation-based search algorithm (Yu and Bajaj, 2005). Volumetric segmentation methods are additionally utilized to partition, color, and thereby obtain a clearer view into the macromolecules' architectural organization. Furthermore, electronically dissecting the local structure units from a 3-D map allows for further structural interpretation (tertiary and secondary folds). Visualizations from the aforementioned local symmetry detection and automatic segmentation, applied to a 3-D volumetric map of the turnip yellow mosaic virus (PDB id 1AUY), are shown in figure 7.7.

Quantitative Visualization

The geometric modeling of virus capsids and the individual virus structure units can be further augmented by the computation of several global and local shape metrics (Bajaj and Yu, 2006). Although integral, topological, and combinatorial metrics capture global shape properties, differential measures such as mean and Gaussian curvatures have also proved useful to an enhanced understanding and quantitative visualization of macromolecular structures.

Integral Properties

Integral shape metrics include the area of the molecular capsid surface defining the capsid, the volume enclosed by closed capsid shells, and the gradient integral on the molecular capsid surface. Given our smooth analytic level set definition of the molecular surface from the aforementioned section on "Atomistic Resolution Model Structures" above,

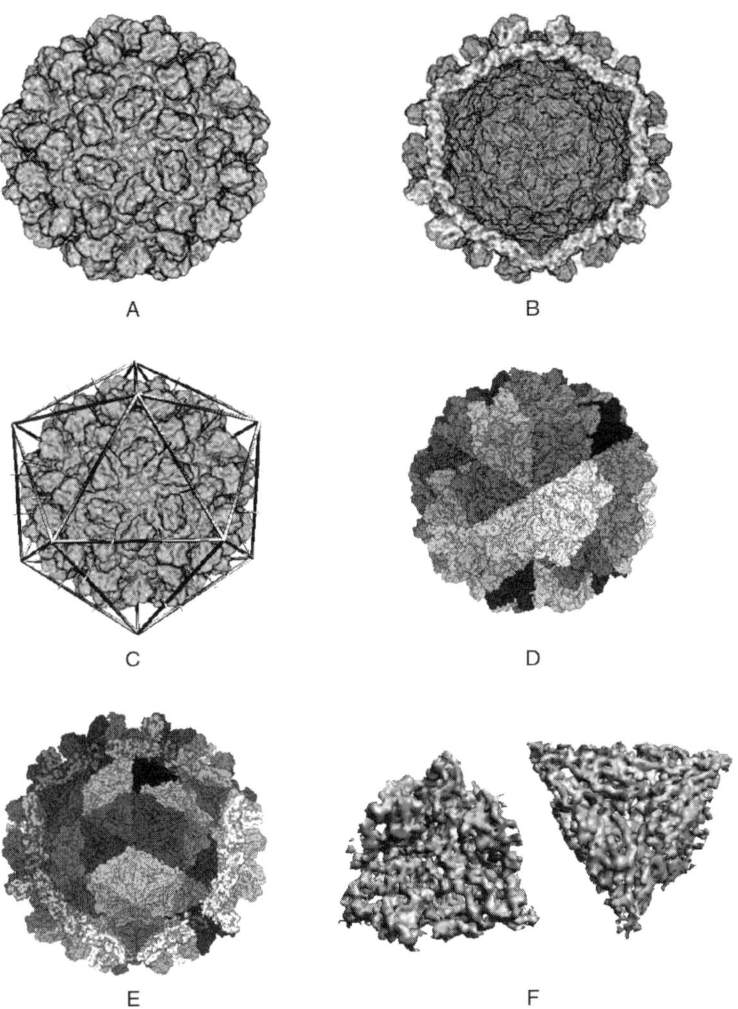

Figure 7.7
(PDB id = 1AUY. Size: 256^3. Resolution: ~4Å). (A) Gaussian blurred map (outside view) from the nonhydrogen atom locations given in the PDB. (B) Gaussian blurred map (inside view). (C) Symmetry detected, including global and local threefold symmetry axes. (D) Segmented trimers (outside view), with randomly assigned colors. (E) Segmented trimers (inside view). (F) One of the segmented trimers (left, outside view; right, inside view).

$$F_{elec\_dens}(\vec{x}) = \sum_{j=1}^{M} \gamma_j K(\vec{x} - \vec{x_j}) = const,$$

for all the atoms that make up either an individual structure unit or the entire virus capsid, an efficient and accurate integration computation for these metrics is given by the contour spectrum (Bajaj et al., 1997). The surface integrations can be performed by adaptively sampling the capsid surface using a technique known as contouring (Bajaj et al., 1997). Contouring is often performed by first decomposing (meshing) the space surrounding the capsid surface into either a rectilinear Cartesian grid mesh or a tetrahedral or a hexahedral mesh. For a tetrahedral mesh, the surface area for the portion of the level set inside a tetrahedron can be represented by a quadratic polynomial B-spline (Bajaj et al., 1997). Summing these B-splines over all of the tetrahedra containing the capsid surface yields the capsid surface area. The volume enclosed by a closed capsid surface is determined by the definite integration of the surface area polynomial B-splines.

Figure 7.8 displays the results of surface area and volume calculations, and a regression relationship between the two, for a selection of spherical icosahedral capsids for virus structures summarized in table 7.2. The analytic molecular surfaces were first computed, and then surface area and enclosed volume were estimated through B-spline evaluation.

Differential Properties

The gradient function of our smooth analytic capsid surface is simply

$$\nabla F_{elec\_dens}(\vec{x}) = \sum_{j=1}^{M} \gamma_j \nabla K(\vec{x} - \vec{x_j}),$$

the summation of the vector of first derivatives of the atomic electron density function. This gradient function is non-zero everywhere on the virus capsid surface (i.e., no singularity). The second derivatives of the molecular surface capture additional differential shape properties and provide suitable metrics. Popular metrics are the magnitudes of *mean curvature* **H** and the *Gaussian curvature* **G**. These are given directly as $\mathbf{H} = \frac{1}{2}(k_{min} + k_{max})$ and $\mathbf{G} = k_{min}k_{max}$ and are, respectively, the average and the product of the twin principal curvatures—namely, k_{min} and k_{max}, also sometimes known as the minimum and maximum curvatures at a point on the surface. Again, for our level set-based analytic molecular surface $F_{elec\_dens}(\vec{x}) = cons = f$, the twin curvatures **H** and **G** can be evaluated as $\mathbf{H} = (\sum(f_x^2(f_{yy} + f_{zz})) - 2 * \sum(f_x f_y f_{xy}))/(2 * (\sum(f_x^2))^{1.5})$ and $\mathbf{G} = (2 * \sum(f_x f_y(f_{xz} f_{yz} - f_{xy} f_{zz})))/((\sum(f_x^2))^2)$, where $\sum$ represents a cyclic summation over x, y, and z, and where, additionally, $f_{x,etc.}$ denotes partial differentiation with respect to those variables.

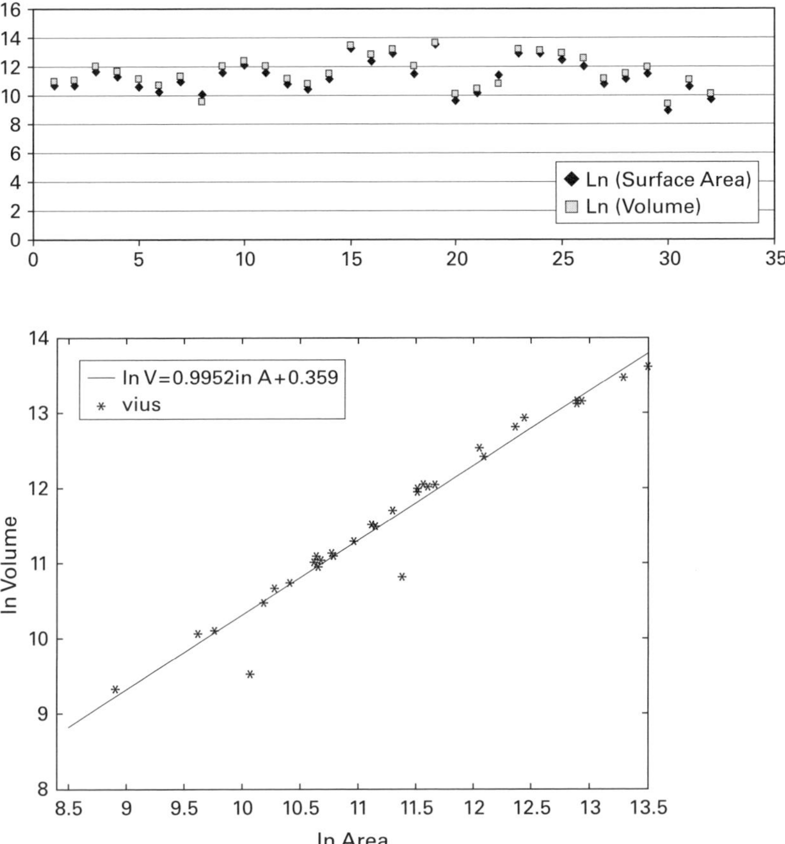

Figure 7.8
Surface area and volume relationship for icosahedral viruses in table 7.1. The area and volume units are square Ångstrom and cube Ångstrom, respectively.

Displaying the magnitude of the gradient function and its variation, as expressed by the mean and Gaussian curvature functions over a molecular surface, helps in quantitative visualization of the bumpiness or lack thereof of an individual protomer, a structure unit, or the entire viral capsid. In figure 7.1 the bottom two pictures display the mean and Gaussian curvature functions of the tobacco mosaic virus's asymmetric protomer surface, exhibiting and enhancing the bumpiness of the surface.

Topological and Combinatorial Properties

Affine invariant topological structures of volumetric functions f, such as our smooth analytic electron density function as mentioned before, include the Morse complex

Table 7.2
Icosahedral viruses and viral components

Virus	Surf Area	Vol.	Ln (Surf Area)	Ln (Vol.)
Satellite tobacco necrosis	17401.22	24419.51	9.7643	10.1031
L-A (Saccharomyces cerevisiae)	99643.60	155223.15	11.5094	11.9526
Canine parvovirus-Fab complex	48482.09	66028.46	10.7889	11.0978
T1L reovirus core	412654.67	517093.80	12.9304	13.1560
T3D reovirus core	99627.14	161424.33	11.5092	11.9918
P4 (Ustilago maydis)	7362.92	11269.59	8.9042	9.3299
Tomato bushy stunt	69600.33	98169.33	11.1505	11.4944
Cowpea chlorotic mosaic	42523.74	56607.48	10.6578	10.9439
Cucumber mosaic	43317.17	61885.43	10.6763	11.0330
Norwalk	116674.31	170940.17	11.6671	12.0491
Rabbit hemorrhagic disease VLP-MAb-E3 complex	80585.54	121611.94	11.2971	11.7086
Galleria mellonella densovirus	33251.61	46216.49	10.4119	10.7411
Human rhino	67337.70	99964.30	11.1175	11.5126
HepBc (human liver) (nHBc)	41669.23	65963.21	10.6375	11.0969
Nudaurelia capensis w	170957.88	278225.27	12.0492	12.5362
Semiliki Forest	47586.60	68392.18	10.7703	11.1330
Polyoma	104897.53	171532.31	11.5607	12.0525
Simian	177557.44	246603.02	12.0870	12.4155
Herpes simplex virus glyco-protein	29035.25	43098.51	10.2763	10.6712
Blue tongue	590265.25	711692.59	13.2883	13.4754
Rice dwarf	727228.58	820906.09	13.4970	13.6182
T1L reovirus virion	412654.67	517093.80	12.3640	12.8133
Simian rotavirus (SA11-4F) TLP	26451.69	35311.17	10.1831	10.4720
Rhesus rotavirus	15093.03	23469.18	9.6220	10.0634

Area units are square Angstrom; volume units are cube Angstrom; logarithm entries are illustrated in figure 7.8.

(Edelsbrunner et al., 2001; Milnor, 1963) and the contour tree (CT) (van Kreveld et al., 1997). Both the Morse complex and the contour tree are related to the critical points of the volumetric function f (i.e., those points in the domain M where the function gradient vanishes ($\nabla f = 0$). The *functional range* of f is the interval between the minimum and maximum values of the function f: $[f_{min}, f_{max}]$. For a scalar value $w \in [f_{min}, f_{max}]$, the level set of the field f at the value w is the subset of points $L(w) \subset M$ such that $f(x) = w \forall x \in L(w)$.

A level set may have several connected components, called contours. The topology of the level set $L(w)$ changes only at the critical points in M, whose corresponding functional values are called critical values. A *contour class* is a maximal set of continuous contours which have the same topology and do not contain critical points. Without loss of generality, the critical points are assumed to be nondegenerate (i.e., only isolated critical points). This assumption can be enforced by small perturbations of the function values. If the critical points are nondegenerate, then the *Hessian* $H(a)$ at a critical point a has non-zero real eigenvalues. The *index* of the critical point a is the number of negative eigenvalues of $H(a)$. For a 3-D volumetric function, there are four types of critical points: index 0 (minima), indices 1 and 2 (saddle points), and index 3 (maxima).

The contour tree (CT) was introduced by van Kreveld et al. (1997) to find the connected components of level sets for contour generation. The CT captures the topological changes of the level sets for the entire functional range $[f_{min}, f_{max}]$ of f; each node of the tree corresponds to a critical point and each arc corresponds to a contour class connecting two critical points.

As an example, the contour tree for a virus capsid is shown in figure 7.9. Each leaf node of the CT represents the creation or deletion of a component at a local minimum or maximum, and each interior node represents the joining and/or splitting of two or more components or topology changes at the saddle points. A cut on an arc of the tree $(v_1, v_2) \in T$ by an isovalue $v_1 \leq w \leq v_2$ represents a contour of the level set $L(w)$. Therefore, the number of connected components for the level set $L(w)$ is equal to the number of cuts to the CT at the value w.

The CT can be enhanced by tagging arcs with topological information such as the Betti numbers of the corresponding contour classes (van Kreveld et al., 1997). Betti numbers β_k $(k = 0, 1, \ldots)$ intuitively measure the number of k-dimensional holes of a virus capsid surface or of any individual structure unit. Only the first three Betti numbers $(\beta_0, \beta_1, \beta_2)$ of a smooth surface are non-zero: β_0 corresponds to the number of connected components; β_1 corresponds to the number of independent tunnels; β_2 represents the number of voids enclosed by the surface. For example, a sphere has the Betti numbers $(\beta_0, \beta_1, \beta_2) = (1, 0, 1)$, whereas a torus has $(\beta_0, \beta_1, \beta_2) = (1, 2, 1)$. Betti number computations for virus capsid surfaces provide useful topological and combinatorial structural information.

Conclusion

Ultrastructure modeling and visualization of virus capsids are clearly just two of the steps in a computational modeling pipeline for determining structure-to-function relationships for such nano-sized objects (Bajaj and Yu, 2006). Efforts are under way by several research groups for virus structure modeling and visualization (Shepherd et al., 2006), computation of virus energetics in solvated environments, atomistic and

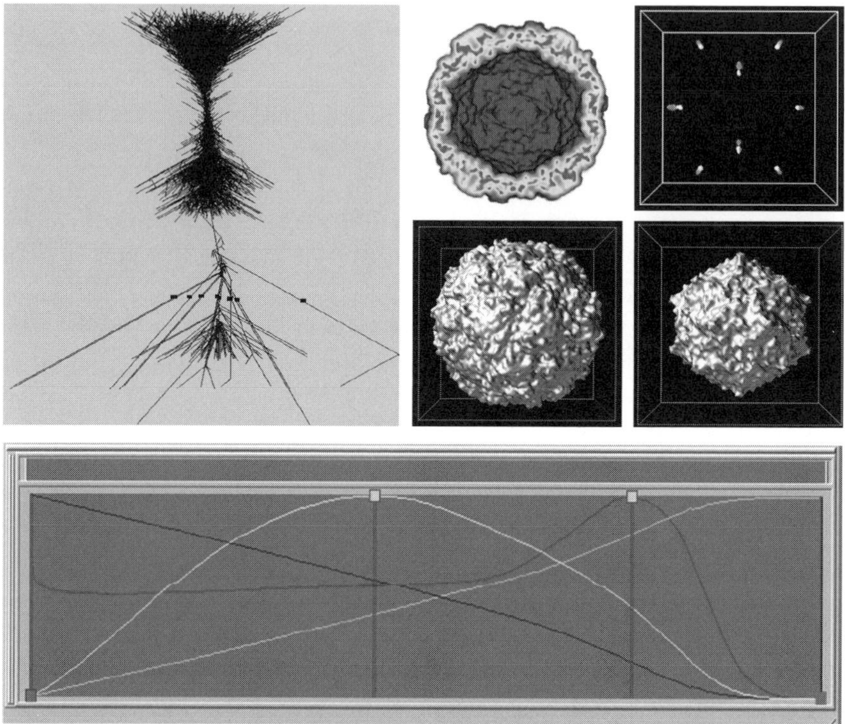

Figure 7.9
The contour tree (upper left) and the contour spectrum (bottom) for the human rhinovirus serotype 2 (PDB id: 1FPN). The red (dark gray) spectrum curve is the graph of molecular surface area; the blue and green (black, light gray) curves are the volume excluded and enclosed by the various level surfaces of the volumetric density map. The horizontal axis of the plot above is the density map, and the vertical axis is spectrum function value.

coarse-grained virus dynamics, and interactions and binding of various ligands and proteins to the nucleocapsids.

Acknowledgments

Sincere thanks to my students S. Goswami, S. Siddahanvalli, J. Wiggins, and W. Zhao for their help with this manuscript. Thanks also to invaluable discussions with and many helpful suggestions from Dr. Tim Baker at the University of California, San Diego. All pictures in this manuscript were generated by our in-house TexMol (Bajaj et al., 2004), VolRover (Bajaj, Yu, et al., 2003), and LBIE-mesher (Zhang et al., 2006). The software is freely downloadable from http://ccvweb.csres.utexas

.edu/software. This work was supported in part by NSF-ITR grant EIA-0325550, and grants from the NIH: 0P20 RR020647, R01 GM074258, and R01 GM073087.

References

Aloy P, Russell RB (2002) The third dimension for protein interactions and complexes. Trends Biochem Sci 27(12): 633–638.

Anderer FA, Schlumberger HD, Koch MA, Frank H, Eggers HJ (1967) Structure of simian virus 40. II. Symmetry and components of the virus particle. Virology 32(3): 511–523.

Arnold E, Rossmann MG (1988) The use of molecular-replacement phases for the refinement of the human rhinovirus 14 structure. Acta Crystallogr A44: 270–282.

Aroul-Selvam R, Hubbard T, Sasidharan R (2004) Domain insertions in protein structures. J Mol Biol 338(4): 633–641.

Bajaj C, Djeu P, Siddavanahalli V, Thane A (2004) Interactive visual exploration of large flexible multicomponent molecular complexes. Proc IEEE: Visualization, 243–250.

Bajaj C, Pasucci V, Schikore D (1997) The contour spectrum. Proc IEEE: Visualization, 167–173.

Bajaj C, Pasucci V, Schikore D (1998) Visualization of scalar topology for structural enhancement. Proc IEEE: Visualization, 51–58.

Bajaj C, Pascucci V, Shamir A, Holt R, Netravali A (2003) Dynamic maintenance and visualization of molecular surfaces. Discr Appl Math 127: 23–51.

Bajaj C, Yu Z (2006) Geometric processing of reconstructed 3D maps of molecular complexes. In: Handbook of Computational Molecular Biology (Aluru S, ed). Boca Raton, FL: Chapman & Hall/CRC Press.

Bajaj C, Yu Z, Auer M (2003) Volumetric feature extraction and visualization of tomographic molecular imaging. J Struct Biol 144(1–2): 132–143.

Baker TS, Olson NH, Fuller SD (1999) Adding the third dimension to virus life cycles: Three-dimensional reconstruction of icosahedral viruses from cryo-electron micrographs. Microbiol Mol Biol Rev 63(4): 862–922.

Belnap DM, Kumar A, Folk JT, Smith TJ, Baker TS (1999) Low-resolution density maps from atomic models: How stepping "back" can be a step "forward." J Struct Biol 125(2–3): 166–175.

Belnap DM, Olson NH, Cladel NM, Newcomb WW, Brown JC, Kreider JW, Christensen ND, Baker TS (1996) Conserved features in papillomavirus and polyomavirus capsids. J Mol Biol 259(2): 249–263.

Berger B, Shor PW, Tucker-Kellogg L, King J (1994) Local rule-based theory of virus shell assembly. Proc Natl Acad Sci USA 91(16): 7732–7736.

Bhyravbhatla B, Watowich SJ, Caspar DL (1998) Refined atomic model of the four-layer aggregate of the tobacco mosaic virus coat protein at 2.4-Å resolution. Biophys J 74: 604–615.

Burley SK, Bonanno JB (2002) Structuring the universe of proteins. Ann Rev Genom Hum Genet 3: 243–262.

Canady MA, Larson SB, Day J, McPherson A (1996) Crystal structure of turnip yellow mosaic virus. Nat Struct Biol 3: 771–781.

Carfi A, Willis SH, Whitbeck JC, Krummenacher C, Cohen GH, Eisenberg RJ, Wiley DC (2001) *Herpes simplex* virus glycoprotein D bound to the human receptor HveA. Mol Cell 8: 169–179.

Caspar DL, Klug A (1962) Physical principles in the construction of regular viruses. Cold Spring Harb Symp Quant Biol 27: 1–24.

Caston JR, Trus BL, Booy FP, Wickner RB, Wall JS, Steven AC (1997) Structure of L-A virus: A specialized compartment for the transcription and replication of double-stranded RNA. J Cell Biol 138(5): 975–985.

Ceulemans H, Russell RB (2004) Fast fitting of atomic structures to low-resolution electron density maps by surface overlap maximization. J Mol Biol 338(4): 783–793.

Chen WN, Oon CJ (1999) Human hepatitis B virus mutants: Significance of molecular changes. FEBS Lett 453(3): 237–242.

Chen XS, Stehle T, Harrison SC (1998) Interaction of polyomavirus internal protein VP2 with the major capsid protein VP1 and implications for participation of VP2 in viral entry. EMBO J 17: 3233–3240.

Cheng RH, Caston JR, Wang GJ, Gu F, Smith TJ, Baker TS, Bozarth RF, Trus BL, Cheng N, Wickner RB, Steven AC (1994) Fungal virus capsids, cytoplasmic compartments for the replication of double-stranded RNA, formed as icosahedral shells of asymmetric Gag dimers. J Mol Biol 244(3): 255–258.

Connolly M (1983) Analytical molecular surface calculation. J Appl Crystallog 16: 548–558.

Crowther RA (1971) Procedures for three-dimensional reconstruction of spherical viruses by Fourier synthesis from electron micrographs. Philos Trans Roy Soc Lond B261(837): 221–230.

Crowther RA, DeRosier DJ, Klug A (1970) The reconstruction of a three-dimensional structure from projections and its application to electron microscopy. Proc Roy Soc Lond A317: 319–340.

Dar AC, Sicheri F (2002) X-ray crystal structure and functional analysis of *Vaccinia* virus K3L reveals molecular determinants for pkr subversion and substrate recognition. Mole Cell 10: 295–305.

de Haas F, Paatero AO, Mindich L, Bamford DH, Fuller SD (1999) A symmetry mismatch at the site of RNA packaging in the polymerase complex of dsRNA bacteriophage phi6. J Mol Biol 294(2): 357–372.

DeRosier DJ, Klug A (1968) Reconstruction of three-dimensional structures from electron micrographs. Nature 217: 130–134.

Dong G, Lu G, Chiu W (2001) Electron cryomicroscopy and bioinformatics suggest protein fold models for rice dwarf virus. Nat Struct Biol 8(10): 868–873.

Dormitzer PR, Sun Z-YJ, Wagner G, Harrison SC (2002) The rhesus rotavirus VP4 sialic acid binding domain has a galectin fold with a novel carbohydrate binding site. EMBO J 21: 885–897.

Edelsbrunner H, Harer J, Zomorodian A (2001) Hierarchical Morse complexes for piecewise linear 2-manifolds. Symp Comput Geom 70–79.

Enemark EJ, Stenlund A, Joshua-Tor L (2002) Crystal structures of two intermediates in the assembly of the papillomavirus replication initiation complex. EMBO J 21: 1487–1496.

Fuller SD, Butcher SJ, Cheng RH, Baker TS (1996) Three-dimensional reconstruction of icosahedral particles—the uncommon line. J Struct Biol 116(1): 48–55.

Goldsmith-Fischman S, Honig B (2003) Structural genomics: Computational methods for structure analysis. Prot Sci 12(9): 1813–1821.

Grimes J, Basak AK, Roy P, Stuart D (1995) The crystal structure of bluetongue virus VP7. Nature 373(6510): 167–170.

Grimes JM, Burroughs JN, Gouet P, Diprose JM, Malby R, Zientara S, Mertens PP, Stuart DI (1998) The atomic structure of the bluetongue virus core. Nature 395(6701): 470–478.

Grimes JM, Jakana J, Ghosh M, Basak AK, Roy P, Chiu W, Stuart DI, Prasad BV (1997) An atomic model of the outer layer of the bluetongue virus core derived from X-ray crystallography and electron cryomicroscopy. Structure 5(7): 885–893.

Groft CM, Burley SK (2002) Recognition of eIF4G by rotavirus NSP3 reveals a basis for mRNA circularization. Mol Cell 9: 1273–1283.

Grunewald K, Desai P, Winkler DC, Heymann JB, Belnap DM, Baumeister W, Steven AC (2003) Three-dimensional structure of herpes simplex virus from cryo-electron tomography. Science 302(5649): 1396–1398.

He J, Schmid MF, Zhou ZH, Rixon F, Chiu W (2001) Finding and using local symmetry in identifying lower domain movements in hexon subunits of the herpes simplex virus type 1B capsid. J Mol Biol 309(4): 903–914.

Helgstrand C, Munshi S, Johnson JE, Liljas L (2004) The refined structure of *Nudaurelia capensis* omega virus reveals control elements for a T = 4 capsid maturation. Virology 318: 192–203.

Hopper P, Harrison SC, Sauer RT (1984) Structure of tomato bushy stunt virus. V. Coat protein sequence determination and its structural implications. J Mol Biol 177: 701–713.

Jiang W, Li Z, Zhang Z, Baker ML, Prevelige PE Jr, Chiu W (2003) Coat protein fold and maturation transition of bacteriophage P22 seen at subnanometer resolutions. Nat Struct Biol 10(2): 131–135.

Johnson JE (1996) Functional implications of protein-protein interactions in icosahedral viruses. Proc Natl Acad Sci USA 93(1): 27–33.

Johnson JE, Speir JA (1997) Quasi-equivalent viruses: A paradigm for protein assemblies. J Mol Biol 269(5): 665–675.

Kang BS, Devedjiev Y, Derewenda U, Derewenda ZS (2004) The PDZ2 domain of syntenin at ultra-high resolution: Bridging the gap between macromolecular and small molecule crystallography. J Mol Biol 338(3): 483–493.

Kidd-Ljunggren K, Zuker M, Hofacker IL, Kidd AH (2000) The hepatitis B virus pregenome: Prediction of RNA structure and implications for the emergence of deletions. Intervirology 43(3): 154–164.

Klug A, Finch JT (1965) Structure of viruses of the papilloma-polyoma type. I. Human wart virus. J Mol Biol 11: 403–423.

Kreveld, M van, van Oostrum R, Bajaj C, Pascucci V, Schikore D (1997) Contour trees and small seed sets for isosurface traversal. Proc 13th Ann Sympos Comput Geom 212–220.

Leach AR (1996) Molecular Modeling: Principles and Applications. London: Longman.

Lee MS, Feig M, Salsbury FR Jr, Brooks CL III (2003) New analytic approximation to the standard molecular volume definition and its application to generalized Born calculations. J Comput Chem 24(11): 1348–1356.

Lee B, Richards F (1971) The interpretation of protein structures: Estimation of static necessibility. J Mol Biol 55(3): 379–400.

Li N, Erman M, Pangborn W, Duax WL, Park CM, Bruenn J, Ghosh D (1999) Structure of *Ustilago maydis* killer toxin KP6 alpha-subunit. A multimeric assembly with a central pore. J Biol Chem 274: 20425–20431.

Liddington RC, Yan Y, Moulai J, Sahli R, Benjamin TL, Harrison SC (1991) Structure of simian virus 40 at 3.8-Å resolution. Nature 354(6351): 278–284.

Liemann S, Chandran K, Baker TS, Nibert ML, Harrison SC (2002) Structure of the reovirus membrane-penetration protein, Mu1, in a complex with its protector protein, Sigma3. Cell 108: 283–295.

Liljas L, Strandberg B (1984) The structure of satellite tobacco necrosis virus. Bio Macromol Assem 1: 97–104.

Lu G, Zhou ZH, Baker ML, Jakana J, Cai D, Wei X, Chen S, Gu X, Chiu W (1998) Structure of double-shelled rice dwarf virus. J Virol 72(11): 8541–8549.

Lwoff A, Horne R, Tournier P (1962) A system of viruses. Cold Spring Harb Symp Quant Biol 27: 51–55.

Mancini EJ, Clarke M, Gowen BE, Rutten T, Fuller SD (2000) Cryo-electron microscopy reveals the functional organization of an enveloped virus, Semiliki Forest virus. Mol Cell 5(2): 255–266.

Mavrakis M, McCarthy AA, Roche S, Blondel D, Ruigrok RWH (2004) Structure and function of the C-terminal domain of the polymerase cofactor of rabies virus. J Mol Biol 343: 819.

McKenna R, Xia D, Willingmann P, Ilag LL, Krishnaswamy S, Rossmann MG, Olson NH, Baker TS, Incardona NL (1992) Atomic structure of single-stranded DNA bacteriophage phi X174 and its functional implications. Nature 355: 137–143.

Milnor J (1963) Morse Theory. Princeton, N.J.: Princeton University Press.

Naitow H, Tang J, Canady M, Wickner RB, Johnson JE (2002) L-A virus at 3.4Å resolution reveals particle architecture and mRNA decapping mechanism. Nat Struct Biol 9: 725–728.

Nakagawa A, Miyazaki N, Taka J, Naitow H, Ogawa A, Fujimoto Z, Mizuno H, Higashi T, Watanabe Y, Omura T, Cheng RH, Tsukihara T (2003) The atomic structure of rice dwarf virus reveals the self-assembly mechanism of component proteins. Structure 11: 1227–1238.

Namba K, Pattanayek R, Stubbs G (1989) Visualization of protein-nucleic acid interactions in a virus: Refined structure of intact tobacco mosaic virus at 2.9Å resolution by X-ray fiber diffraction. Mol Biol 208: 307–325.

Natarajan P, Johnson JE (1998) Molecular packing in virus crystals: Geometry, chemistry, and biology. J Struct Biol 121(3): 295–305.

Navaza J, Lepault J, Rey FA, Alvarez-Rua C, Borge J (2002) On the fitting of model electron densities into EM reconstructions: A reciprocal-space formulation. Acta Crystallogr D58: 1820–1825.

Ng KK, Cherney MM, Vazquez AL, Machin A, Alonso JM, Parra F, James MN (2002) Crystal structures of active and inactive conformations of a caliciviral RNA-dependent RNA polymerase. J Biol Chem 277: 1381–1387.

Padilla JE, Colovos C, Yeates TO (2001) Nanohedra: Using symmetry to design self-assembling protein cages, layers, crystals, and filaments. Proc Natl Acad Sci USA 98(5): 2217–2221.

Poranen MM, Tuma R (2004) Self-assembly of double-stranded RNA bacteriophages. Virus Res 101(1): 93–100.

Prasad BV, Hardy ME, Dokland T, Bella J, Rossmann MG, Estes MK (1999) X-ray crystallographic structure of the Norwalk virus capsid. Science 286: 287–290.

Qu F, Morris TJ (1997) Encapsidation of turnip crinkle virus is defined by a specific packaging signal and RNA size. J Virol 71(2): 1428–1435.

Qu F, Ren T, Morris TJ (2003) The coat protein of turnip crinkle virus suppresses posttranscriptional gene silencing at an early initiation step. J Virol 77(1): 511–522.

Rapaport DC, Johnson JE, Skolnick J (1999) Supramolecular self-assembly: Molecular dynamics modeling of polyhedral shell formation. Comput Phys Communic 121: 231–235.

Reddy VS, Natarajan P, Okerberg B, Li K, Damodaran KV, Morton RT, Brooks CL III, Johnson JE (2001) Virus Particle Explorer (VIPER), a website for virus capsid structures and their computational analyses. J Virol 75(24): 11943–11947.

Reinisch KM, Nibert ML, Harrison SC (2000) Structure of the reovirus core at 3.6Å resolution. Nature 404: 960–967.

Roberts MM, White JL, Grutter MG, Burnett RM (1986) Three-dimensional structure of the adenovirus major coat protein hexon. Science 232(4754): 1148–1151.

Rossmann MG (2000) Fitting atomic models into electron-microscopy maps. Acta Crystallogr D56(pt 10): 1341–1349.

Rossmann MG, Blow DM (1962) The detection of sub-units within the crystallographic asymmetric unit. Acta Crystallogr 15: 24–31.

Schwartz R, Prevelige PE Jr, Berger B (1998) Local rules modeling of nucleation-limited virus capsid assembly. Cambridge, Mass.: MIT Technical Report. MIT-LCS-TM 584.

Shepherd CM, Borelli IA, Lander G, Natarajan P, Siddavanahalli V, Bajaj C, Johnson JE, Brooks CL, Reddy VS (2006) VIPERdb: A relational database for structural virology. Nucl Acids Res 34 (database issue): D386–D389.

Simpson AA, Chipman PR, Baker TS, Tijssen P, Rossmann MG (1998) The structure of an insect parvovirus (*Galleria mellonella* densovirus) at 3.7Å resolution. Structure 6: 1355–1367.

Smith TJ, Chase E, Schmidt T, Perry KL (2000) The structure of cucumber mosaic virus and comparison to cowpea chlorotic mottle virus. J Virol 74: 7578–7586.

Speir JA, Munshi S, Wang G, Baker TS, Johnson JE (1995) Structures of the native and swollen forms of cowpea chlorotic mottle virus determined by X-ray crystallography and cryo-electron microscopy. Structure 3: 63–77.

Stehle T, Gamblin SJ, Yan Y, Harrison SC (1996) The structure of simian virus 40 refined at 3.1Å resolution. Structure 4(2): 165–182.

Stehle T, Harrison SC (1997) High-resolution structure of a polyomavirus VP1-oligosaccharide complex: Implications for assembly and receptor binding. EMBO J 16(16): 5139–5148.

Tao Y, Farsetta DL, Nibert ML, Harrison SC (2002) RNA synthesis in a cage—structural studies of reovirus polymerase lambda3. Cell 111: 733–745.

Tendulkar AV, Joshi AA, Sohoni MA, Wangikar PP (2004) Clustering of protein structural fragments reveals modular building block approach of nature. J Mol Biol 338(3): 611–629.

Trus BL, Heymann JB, Nealon K, Cheng N, Newcomb WW, Brown JC, Kedes DH, Steven AC (2001) Capsid structure of Kaposi's sarcoma-associated herpesvirus, a gammaherpesvirus, compared to those of an alphaherpesvirus, *Herpes simplex* virus type 1, and a betaherpesvirus, cytomegalovirus. J Virol 75(6): 2879–2890.

Van Heel M, Gowen B, Matadeen R, Orlova EV, Finn R, Pape T, Cohen D, Stark H, Schmidt R, Schatz M, Patwardhan A (2000) Single-particle electron cryo-microscopy: Towards atomic resolution. Quart Rev Biophys 33(4): 307–369.

Wang J, Simon AE (2000) 3′-end stem-loops of the subviral RNAs associated with turnip crinkle virus are involved in symptom modulation and coat protein binding. J Virol 74(14): 6528–6537.

Weissenhorn W, Carfi A, Lee KH, Skehel JJ, Wiley DC (1998) Crystal structure of the ebola membrane-fusion subunit, GP2, from the envelope glycoprotein ectodomain. Mol Cell 2: 605–616.

White DO, Fenner FJ (1994) *Medical Virology (4th ed)*. San Diego: Academic Press.

Wikoff WR, Johnson JE (1999) Virus assembly: Imaging a molecular machine. Curr Biol 9(8): 296–300.

Wodak SJ, Mendez R (2004) Prediction of protein-protein interactions: The CAPRI experiment, its evaluation and implications. Curr Opin Struct Biol 14(2): 242–249.

Wu H, Rossmann MG (1993) The canine parvovirus empty capsid structure. J Mol Biol 233: 231–244.

Wynne SA, Crowther RA, Leslie AG (1999) The crystal structure of the human hepatitis B virus capsid. Mol Cell 3: 771–780.

Yan X, Olson NH, van Etten JL, Begoin M, Rossman MG, Baker TS (2000) Structure and assembly of large lipid-containing dsDNA viruses. Nat Struct Biol 7: 101–103.

Yu Z, Bajaj C (2005) Automatic ultra-structure segmentation of reconstructed cryo-EM maps of icosahedral viruses. IEEE Transactions on Image Proc 14(9): 1324–1337. Special issue on molecular and cellular bioimaging.

Zhang X, Walker SB, Chipman PR, Nibert ML, Baker TS (2003) Reovirus polymerase $\lambda 3$ localized by cryo-electron microscopy of virions at a resolution of 7.6Å. Nat Struct Biol 10: 1011–1018.

Zhang Y, Xu G, Bajaj C (2006) Quality Meshing of Implicit Solvation Models of Biomolecular Structures. Special issue of Comput Aided Geom Design on geom modeling in the life sci.

Zhou ZH, Baker ML, Jiang W, Dougherty M, Jakana J, Spencer JV, Trus BL, Booy FP, Steven AC, Newcomb WW, Brown JC (1997) Structure of the *Herpes simplex* virus capsid: Peptide A862-H880 of the major capsid protein is displayed on the rim of the capsomer protrusions. Virology 228(2): 229–235.

Zhou ZH, Chen DH, Jakana J, Rixon FJ, Chiu W (1999) Visualization of tegument-capsid interactions and DNA in intact *Herpes simplex* virus type 1 virions. J Virol 73(4): 3210–3218.

ICTV Database http://www.ncbi.nlm.nih.gov/ICTVdb/Ictv/index.htm.

IV DEVELOPMENT

Developmental biology is experiencing a major trend toward computation and modeling. This is prompted not only by the increasing availability of bioinformatic tools but primarily by the progression from the descriptive and experimental study of embryological forms to the causal and mechanistic analysis of developmental processes. Since the dynamical component inherent in these processes cannot be directly measured and analyzed in live specimens, models fulfill essential representational, analytical, and heuristic roles in developmental biology. This requirement was perceived long before computation was around, with physical models providing visualizations of consecutive stages of organ formation, and also with the mathematical capture of processes such as growth and patterning, as pioneered by Hans Przibram, Ludwig von Bertalanffy, D'Arcy Thompson, and Alan Turing. The modern computational techniques lend unprecedented power to these methodologies and, more important, afford completely new kinds of quantitative data about the dynamics of development, such as those derived from three- and four-dimensional representations of gene activity, molecule concentrations, diffusion rates, and cell and tissue parameters.

The crucial problem of modern-day developmental biology is the genotype–phenotype relation. Large quantities of data emerge from genomics and other high-throughput molecular techniques, but a mere statistical association of gene data with developmental events does not solve the topological and organizational questions related to cells building bodies. Even though gene activity and regulation are among the best-understood aspects of development, they do not suffice to present a full account of the genotype-phenotype relation. As more data of molecular epigenetic gene regulation, cell-cell, and cell-environment interaction become available, it is necessary to include aspects of the dynamic interactions of cells with other cells and cell products, and with the environment that can account for the physical aspects of collective cell behaviors and developing cell masses. This cannot be achieved without measurement, abstraction, and computational modeling, in particular when the evolution of these processes should also be considered, as in EvoDevo (see part VI). The modeling approaches discussed here in part IV reflect this requirement for the understanding of cell behavioral processes and epigenesis by uniting processes from different levels of complexity and organization.

Diego Rasskin-Gutman et al. (chapter 8) explore three modeling strategies regarding the early specification of left-right asymmetry in vertebrates. Asymmetry is a fundamental problem of development because the bilaterian body plan is basically symmetric, and deviations from symmetry must involve early differential behavior of cells that share the same genetic background. Some organ-level asymmetries can be traced to very early left-right differences in gene activation. The first model concerns the expression of the gene Nodal in the perinodal cells exclusively on the left side of the anterior embryonic pole of gastrulating chick embryos, a conserved feature of left-right asymmetry formation in vertebrates. Modeling six elements of a gene-protein network associated with the expression of Nodal characterizes and predicts the behavior of perinodal cells under different initial conditions.

Rasskin-Gutman's second model looks at the dynamics of ciliary movement and extraembryonic fluid flow in the node area of the early mouse embryo. Leftward-directed fluid flow is a necessary condition for the specification of the left-right axis. Data from high-resolution videomicroscopy shows that left-right asymmetry in the cila-beating of each individual nodal cell collectively generates the unidirectional fluid flow that represents the epigenetic trigger for left-right axial asymmetry formation of the embryo.

The third model concerns the looping of the heart tube, one of the first manifestations of organ asymmetry in vertebrate embryos. Here the dynamics of differential adhesiveness of heart cells are modeled, providing a quantitative basis for the looping phenomenon of the heart tube. The key parameter is the density of the cell-to-cell junctions on the left and right sides of the tube. A number of interesting predictions concerning the molecular regulation of differential cell behavior are derived from the model.

Eirikur Palsson (chapter 9) models "cell" behavior in the cellular slime mold *Dictyostelium discoideum*. Slime molds are superorganisms composed of individual amoebae that exhibit various behaviors during their life cycles that are akin to the developmental processes of metazoans (i.e., the amoebae coalesce, adhere, differentiate, sort out, and respond to signaling cues). Because of the relatively large size and small numbers of the amoebae that compose one organism, the simple lab conditions for keeping the colonies, and the easy access for experiment, the slime molds have become one of the exemplary "model organisms" of developmental biology. Palsson uses a three-dimensional cell model to explore *in silico* a number of features of the *Dictyostelium* life cycle, such as wave patterns and spiral formation, in order to understand how these processes, in combination with chemotaxis, cell adhesion, and stiffness, orchestrate the movement of cells in different phases of *Dictyostelium* formation. This 3-D modeling of physical cell behavior permits, apart from various kinds of mathematical analyses of the behavior of the model itself, conjectures about the types of chemical or physical signals that could underlie cell movement in specific phases of structural differentiation of the biological organisms.

Jaap van Pelt and Harry Uylings (chapter 10) study mammalian nerve cell formation, using models of neuronal morphogenesis. Nerve cells show enormous diversity and variation in shape originating from complex and dynamic developmental processes. Since a quantitative relationship must exist between the shape of the neurons and the developmental processes generating these shapes, theoretical and computational modeling approaches can be applied to elucidate this relationship. The authors employ dendritic growth models, using stochastic rules of dendrite formation for studying the shapes of dendrites. Elongation and branching are found to represent the cruical parameters of neuronal shape variation. Quantitative descriptions and several predictions regarding the biological nature of intrinsic and extrinsic factors influencing neuronal growth parameters can be derived from the model.

8 Modeling Developmental Asymmetries

Diego Rasskin-Gutman, Marta Ibañes Miguez, and Juan Carlos Izpisúa-Belmonte

A certain lack of confidence in the ability of theoreticians to contribute to biology is summed up in the tale of a team of mathematicians who were given ten years by a group of biologists to contribute something to their subject. Summing up a decade's work by his group, the chief mathematician opened his lecture with the phrase, "Consider a spherical elephant . . .". Biologists will not benefit much from theoretical work until they understand that this story is not really funny. Sometimes it is perfectly reasonable to regard an elephant as spherical, for the error is immaterial and a simple and insightful view of the situation is thereby obtained.
—Lee Segel (1980)

Models as a Way of Knowing in Developmental Biology

Models bring together three elements that are very much needed in order to deal with the overwhelming complexity of biology: abstraction, simplicity, and heuristic power. When building a model of a biological system, the most relevant aspects of its numerous components and their complex interactions are set apart, leaving behind unimportant details that often obscure a proper understanding of the system: the spherical elephant alluded to by Lee Segel (figure 8.1). Moreover, the same abstraction and simplicity often call for the existence of "hidden variables" or components that were not part of the model but that are needed in order to explain what the model itself cannot simulate (Rasskin-Gutman and Buscalioni, 2001). In this line of thought, William Wimsatt favors the use of false, "neutral" models as a way to arrive at a better knowledge of reality in biology, building a convincing case to suggest that proposing a false model can "lead to the detection and estimation of other relevant variables" (Wimsatt, 1987).

Adding and fine-tuning parameters and variables until resemblance to reality is best approximated makes modeling a continuous work in progress. In addition, they are heuristic, that is, once the system has been abstracted and the parameters and variables carefully chosen and delimited, they can be used to explore the possible outcomes of specific conditions, an extreme that can be experimentally tested. Even when the model has enough complexity so that it is not possible to reach analytical

Figure 8.1
The spherical elephant. Modeling requires a thorough understanding of the problem at hand, something not always achieved in published papers. A spherical elephant provides a good approximation to the mass, volume, and surface relationship of the animal, all important physiological indicators that can be used to infer the metabolism of the animal and even its longevity prospects. However, it says nothing of interest about many other biological aspects, such as gene expression patterns or cognitive abilities.

solutions, there are numerical methods that, along with the computational and display power of computers, can offer approximate solutions to the long-term behavior of the system.

Modeling in developmental biology has a long tradition, spanning about a century of ideas, concepts, and, of course, models. It is fair to say that D'Arcy W. Thompson was the father of this broad discipline with his famous book *On Growth and Form*, first published in 1917, in which he systematized the use of mathematical and physical principles as seats for causal and mechanistic explanation of natural phenomena (Thompson, 1942). However, his efforts in showing this relationship between an abstract science and a profound materialistic biology, the path biology followed during the twentieth century, was better welcomed by physicists and mathematicians than by biologists themselves.

It was an embryologist who—even before Thompson—first employed a mathematical description as a biological abstraction. Hans Driesch, in the nineteenth century (see Oppenheimer, 1967:74), worked on the idea of the existence of a "coordinate system" that cells within the developing embryo are able to read. Much later, in the 1960s, this theoretical insight was given operational status by Lewis Wolpert within the concept of "positional information," incorporating the whole idea into an elegant model dubbed "the French flag model" (Wolpert, 1969; but see also Weiss, 1939; Crick, 1970).

Mathematicians have also contributed greatly to formulating concepts that have sparked interest in the biological community. Alan Turing, also known as one of the fathers of digital computers, in 1952 published a seminal paper that has given rise to a theoretical approach to understanding the possible physicochemical basis of cell differentiation. More or less independently, Gierer and Meinhardt (1972) proposed similar mathematical approaches to account for what is now called reaction-diffusion mechanisms (Nicholas Rashevsky seems to have discovered the same sort of mathematical system even before Turing; see Fox-Keller, 2002 and Graham et al., 1993). Rashevsky, founder of the journal *Bulletin of Mathematical Biophysics*, touched upon many biological themes mathematically, but his efforts were a bit too advanced and abstract for his time. Nevertheless, the spirit of the Biomathematics Committee he founded at the University of Chicago is still thriving.

As pointed out by Gerald Edelman in a very important, albeit underrated, book *Topobiology* (1988), any model in developmental biology needs to address at least two kinds of questions. First, it needs to account for the topobiological problem, that is, to relate multiscale events—molecules, cells, tissues, organs—to the formation of tissues. Second, it needs to account for the epigenetic problem, that is, the nonlinear mapping between gene expression and phenotype. In most instances it is virtually impossible to answer both questions completely. A decomposition of the problem at hand is then warranted, and each level of this multiscale process is treated separately. We shall proceed in the present article accordingly.

Modeling Developmental Asymmetries

An impressive amount of knowledge has been generated since the 1990s regarding the early establishment of left-right asymmetries in the body plan of metazoans. There are three kinds of insights that have appeared in the literature in this relatively young field: (1) theoretical insights concerning possible theories about the initial mechanisms responsible for left-right asymmetry, such as chirality of a theoretically postulated protein (Brown and Wolpert, 1990); (2) epigenetic mechanisms such as cilia movement in the nodal cells of mice embryos (Nonaka et al., 1998) and ion flux through cellular gap junctions in early *Xenopus* gastrulation (Levin and Mercola, 1998); and (3) gene expression data and some correlations about the spatiotemporal actions of these genes (Capdevila et al., 2000). Despite this, we still know very little about how side-specific signaling pathways are regulated, and even less about the mechanisms that these pathways use to coordinate the generation of the normal body plan.

We present here three modeling strategies to tackle early specification of left-right asymmetries. First, at the molecular level, the early expression of Notch is necessary in order to differentially elicit the expression of Nodal in the left side of the perinodal region in the chick. Modeling critical steps of the gene network associated with the Notch pathway characterizes and predicts the behavior of perinodal cells under different conditions (Raya et al., 2004). Second, at the cellular organelle level, we will look at the dynamics of ciliary movement in the node of the early mouse embryo, which is paramount for the early specification of the left-right axis (Buceta et al., 2005). Finally, the looping of the heart tube is a cellular event that marks one of the first manifestations of organ asymmetry in vertebrates. A model of cooperation of adhesion membrane molecules provides an entry point to understand this phenomenon (Rasskin-Gutman and Izpisúa-Belmonte, 2004).

Modeling Notch in the Chick Perinodal Region

Henson's node is a transient embryonic structure that regulates proper formation of organ primordia during gastrulation at many levels. In chick development, the node appears initially in the rostral end of the embryo at stage HH4 (18–19 hours), progressing caudally until it reaches the tailbud zone. As the node shifts posteriorly, the notochord is formed behind it. At the same time, cells from the paraxial mesoderm enter the node and migrate, arriving at locations of the embryo where organs such as the heart will form. It has been observed repeatedly that disturbing the normal development of this area changes the normal asymmetric layout of the gene expression pattern. At the molecular level, the left-sided expression of Nodal in the lateral plate mesoderm is a conserved feature necessary for the establishment of normal left–right asymmetry during vertebrate embryogenesis. Experimental evidence has demonstrated that the activity of the Notch pathway is necessary to elicit Nodal expression around the node, as well as to specify the establishment of the left–right axis (Raya et al., 2004, 2003; see also Krebs et al., 2003). This model simulates the dynamical expression pattern of relevant genes in chick embryos during stages HH4–HH6, highlighting the importance of key elements that play a role in the expression of Nodal in the left perinodal region at stage HH6 (Raya et al., 2004).

Modeling always involves simplification. The first step in simplifying our present molecular problem has been to reduce the number of elements participating in the process. According to the dynamic expression patterns, the well-known interactions among them, and their effects on left-right asymmetry (Raya et al., 2003, 2004), the model has focused on six different molecular elements: Notch receptor (N); two ligands, Delta 1 (Dll1) and Serrate1 (S); a glycoprotein that modifies Notch, Lunatic Fringe (Lfng); the activation pathway of Notch (NA); and Nodal (Nd) as a target of the Notch pathway. We have considered both the mRNAs and proteins of these elements, as well as the receptor-ligand complexes (ND and NS).

Therefore, in our model every single cell has been characterized by the amount of the mRNA, protein, and protein complexes of the above-mentioned elements (figure 8.2). The dynamics of each molecular species is a balance of its production (mRNA transcription and protein translation), its decay rate, and the binding rate of complex formation, and the interaction among macromolecular species is described according to mass-action kinetics. Cell-to-cell interactions result from receptor-ligand binding

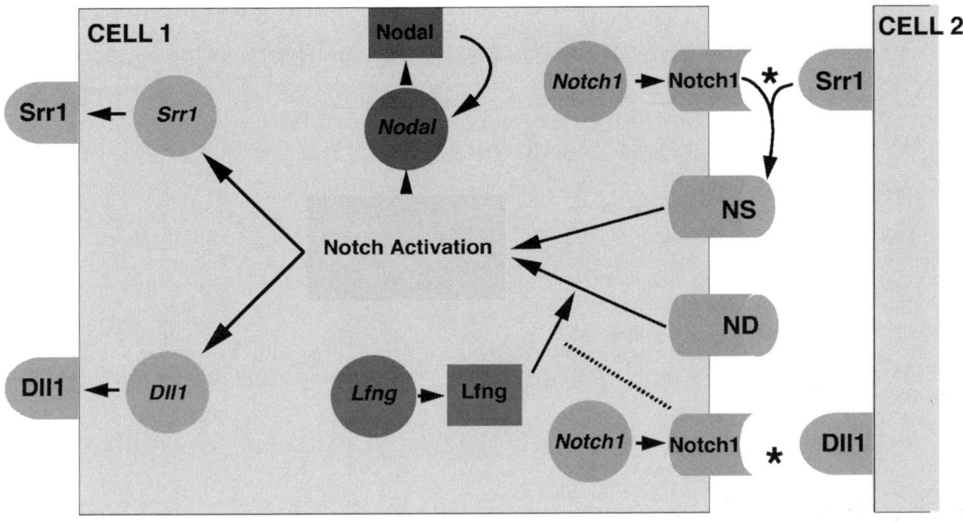

Figure 8.2
Perinodal regulatory network. The mathematical model characterizes the activation dynamics of the Notch regulatory pathway in the chick node. Circles symbolize mRNA and boxes represent proteins or protein complexes. Only two of the total 32×16 cells that form the lattice for each half of the embryo in the simulation are shown. In cell 1, the full set of interactions can be seen. Asterisks denote the action of calcium as an external LR modulator in the formation of Notch complexes with Delta1 and Serrate. The broken line indicates the regulation of Lfng action by Notch. (After Raya et al., 2004.)

(Bosenberg and Massague, 1993; Collier et al., 1996; Owen et al., 2000). Thus, the model is a set of thirteen coupled nonlinear delay and ordinary differential equations for each cell (Raya et al., 2004).

The model also assumes that, at the level of the above-mentioned molecular elements, the left and right sides of the embryo proceed dynamically in an independent way. Indeed, in the case of Lfng, it has been shown that it sweeps the embryo bilaterally and independently throughout the left and right halves in a caudorostral direction, producing a dynamic pattern of waves of expression (Panin et al., 1997; Jouve et al., 2002).

The gene-protein network that constitutes the model is as follows. Notch receptor binds to its ligands, Dll1 and Serrate1, in neighboring cells, eliciting the activation of the Notch pathway in the receiving cell. This cascade activates the mRNAs of Dll1, Serrate1, and Nodal (Krebs et al., 2003; Raya et al., 2003; Shi and Stanley, 2003), which in turn elicit the production of their specific proteins. The four waves of Lfng that sweep the embryo caudorostrally up to the most anterior part of the node, influence the activation of the Notch pathway cell autonomously (Panin et al., 1997; Jouve et al., 2002). The model does not underscore which molecular mechanisms are the basis of these waves of expression, and focuses only on the modification of the Notch pathway as a result of Lfng. Specifically, the model proposes that the activity of Notch resulting from the binding to its ligand Delta increases in the presence of Lfng. As the node regresses posteriorly, at stage HH4 there is a *Dll1-Serrate1* interface posterior to the node, whereas at stage HH6 it is anterior.

By way of example, the equations for *Dll1(Dll)* (the mRNA), Dll1 (Dll) (the protein), and NA are

$$\frac{dDll_i(t)}{dt} = a_d\psi_3(\mathrm{NA}_i(t), h_d) - \mu_d Dll_i(t) \tag{8.1}$$

$$\frac{d\mathrm{Dll}_i(t)}{dt} = t_D Dll_i(t - T_D) - b_{DN} \sum_{j \in nn(i)} \mathrm{Dll}_i(t)\mathrm{Notch}_j(t) - \mu_D \mathrm{Dll}_i(t) \tag{8.2}$$

$$\frac{d\mathrm{NA}_i(t)}{dt} = a_{NA}\psi_2(A(\mathrm{Lfng}_i(t)/\mathrm{Notch}_i(t)) \cdot \mathrm{ND}_i(t) + B \cdot \mathrm{NS}_i(t), h_{NA}) - \mu_{NA}\mathrm{NA}_i(t), \tag{8.3}$$

where ψ_κ is a saturating function with stiffness controlled by the Hill or cooperative coefficient κ, and parameter h is a measure of the threshold value above which activation is greater than 50 percent. We will see the Hill function again when modeling heart tube formation. According to equation (8.1), *Dll* expression is enhanced above a threshold of NA and it saturates at a maximum stationary value of $Dll_{\max} = a_d/\mu_d$, where a_d and μ_d are the transcription and decay rates, respectively. *Dll* expression elicits the translation of its protein, which requires a period of time to reach the cell membrane and be able to interact with other transmembrane molecules such as Notch. The delay term in equation (8.2) accounts for this period of time (T_D).

The above-described interactions, in the presence of domains of expression of Notch, Delta, and Serrate, are able to elicit the expression of *Nodal* through a high activation of Notch achieved only when several waves of Lfng have swept the embryo (HH6). In addition, this strong activation elicits the expansion of the domain of Delta expression at slightly later stages (HH5). Therefore, the model reproduces the dynamic gene expression patterns (Raya et al., 2004).

Left-right asymmetries in the chick embryo have been observed at very early stages, such as HH4 (Raya and Izpisúa-Belmonte, 2004). However, none of these previously known asymmetries has yet been related to the pathway of Notch. We have hypothesized that a left-right asymmetry might be present in our gene-protein network such that it enables the expression of Nodal only on the left side of the embryo, through an enhancement of the activity of Notch. To identify potential modulators of Notch activity on the left or right side of Hensen's node, we mathematically explored the response generated by our model under different parameter values for specific initial conditions. Notably, our simulations show that Dll1 and Nodal can be expressed asymmetrically by varying the binding rates of the complexes Notch1–Dll1 and

Notch1–Serrate1. If the binding rates are higher, the activation of the Notch pathway increases and elicits Nodal expression.

Thus, we find that a left-right asymmetry in the binding rates is the most parsimonious way to reproduce the left and right expression patterns during stages HH4–HH6. This narrows the search space in which molecular elements might be eliciting this left-right asymmetry to those involved in modifying the binding rates of Notch to its ligands. We have experimentally shown that a left-right asymmetric concentration of extracellular calcium, which can elicit a differential binding of Notch receptor with its ligands, is responsible for the correct situs (organ positioning) in chick embryos (Raya et al., 2004).

Cilia Dynamics and Fluid Flow in the Mouse Node

In recent years a number of reports have suggested that a leftward extraembryonic fluid flow on the surface of the node of mouse embryos is involved in the consistent specification of the left-right axis (Nonaka et al., 1998, 2002; Okada et al., 1999, 2005). The induction of such a leftward flow has been related to the clockwise (ventral view) rotation-like motion of the 9+0 cilia present in the monociliated cells that form the node. Specifically, mutant mouse embryos either lacking cilia or with immotile cilia in the node show left-right randomization as well as absence of directional fluid flow (Nonaka et al., 1998; Okada et al., 1999; Takeda et al., 1999; Marszalek et al., 1999).

Importantly, recent experiments have shown that such leftward fluid flow dynamics is conserved among different species and occurs in different ciliated organs: the node in mouse embryos, the notochord in rabbit embryos, and the Kupffer's vesicle in medaka and zebra fish embryos (Okada et al., 2005; Kawakami et al., 2005). Zebra fish embryos with immotile cilia or disrupted cilia structure show absence of directional fluid flow and alterations of left-right axis specification (Essner et al., 2005; Kramer-Zucker et al., 2005; Kawakami et al., 2005). In addition, elegant experiments have shown that the phenotype in mutant mouse embryos with immotile cilia can be rescued if an external artificial fast leftward flow is imposed (Nonaka et al., 2002).

These experimental results highlight the importance of cilia dynamics to induce an asymmetric flow that subsequently establishes the proper situs in vertebrate embryos. But how can the rotation-like dynamics of cilia induce such a leftward flow? This challenging question involving both hydrodynamics and several aspects of cilia biology needs to be addressed from a multidisciplinary point of view. We describe the theoretical framework we have proposed to address this question, based on and supported by data taken from high-resolution videomicroscopy imaging (Buceta et al., 2005).

Cartwright and colleagues were the first to address from a theoretical point of view how the leftward flow is induced from the clockwise rotational motion of cilia. Their simple yet appealing approach hinted at the right direction: the motion of cilia occurs around a posteriorly tilted axis (Cartwright et al., 2004). Recent experimental data show that their prediction is correct and that this feature, together with the clockwise direction of motion, is a conserved element of cilia dynamics (Okada et al., 2005; Kramer-Zucker et al., 2005). The theoretical approach they used was focused on the steady fluid flow induced by rotational dynamics, and it involved neither the modeling of nodal cilia nor the time-dependent motion of latex beads used in experimental setups to visualize the fluid flow (Cartwright et al., 2004). Therefore, although their pioneering approach provided an insightful prediction, a more detailed modeling strategy able to translate different features of cilia dynamics was required to gain a full characterization of the experimental observations.

The nodal fluid flow is in the low Reynolds number ($LRN \approx 10^{-3}, 10^{-4}$) regime where motion follows Aristotelian dynamics rather than Newtonian dynamics, and viscosity dominates over inertia in the node (Cartwright et al., 2004; Buceta et al., 2005). It has been established that due to the lack of inertia in this regime, and disregarding possible boundary effects, at least two degrees of freedom are needed to induce a directional motion, which is commonly known as the *scallop theorem* (Purcell, 1977). Thus, a rotating stiff cilium can hardly generate an *asymmetric* flow in the LRN regime. Indeed, directional fluid motion at low Reynolds numbers has been thoroughly studied in the context of 9+2 cilia and flagella, which show a characteristic two-phase beating dynamics.

This dynamics involves a power or effective stroke during which the cilium is extended and moves fast, and a recovery stroke that corresponds to the slower return motion of the bent cilium. In many cases this motion is predominantly planar (i.e., the phases of motion occur in the same plane). This two-phase dynamics, even if planar, is described by more than one degree of freedom (the flagellum or cilium alters its shape during the motion), and thus it is able to induce a unidirectional fluid flow in the direction of the power stroke (Blake, 1974; Sleigh, 1974).

However, this two-phase beating dynamics has not been described for 9+0 cilia. These primary cilia with a 9+0 architecture are mostly immotile mechanosensors widely found in the surface of many cells in both embryos and adults of many species (Wheatley, 2004). The primary cilia in the mouse embryo node, in the rabbit embryo notochord, and in the medaka and zebra fish embryos' Kupffer's vesicle are a notable exception, showing an apparent rotational motion (Okada et al., 2005). However, experiments with mutant *Chlamydomonas* with 9+0 phenotype in their flagella instead of their normal 9+2 structure (Wakabayashi et al., 1997) confirm that while the 9+2 flagella have a planar beating motion, mutant flagella lacking the central pair of microtubules perform a more irregular asymmetric beating motion when demembranated and in the presence of a low concentration of calcium ($<10^{-6}$ M). Thus, planar and three-dimensional beating patterns seem to be interchangeable, given the right external conditions, such as concentration of calcium and viscosity (Woolley and Vernon, 2001).

Given this evidence, we have hypothesized that the periodic motion of nodal cilia is not just a rotation but a two-phase beating dynamics involving a power stroke and a recovery stroke (Buceta et al., 2005). Indeed, high-resolution in vivo videomicroscopy data depict a two-phase motion with slow and fast strokes and axonemal bending when the cilium is moving more slowly close to the cell surface (Okada et al., 2005). In order to address the question of whether such beating dynamics is able to induce a leftward fluid flow, we have used the formalism of polymer hydrodynamics to construct a model of the nodal cilia dynamics.

Specifically, we have modeled each cilium as a string of n moving spheres of radius a connected by massless rods plus one static sphere at the base (figure 8.3). Thus, our approach allows setting the specific length (L) and width (a) of cilia. We have also defined the two-phase dynamics with a novel formalism, in which the stiffness of cilia, as well as the frequency (f), apex angle (Ψ), and axis of motion (Θ), are free parameters (Buceta et al., 2005). The spatial configuration and dynamics of a cilium are described in terms of the vector $\vec{R}_{i,k}(t) = (x_{i,k}(t), y_{i,k}(t), z_{i,k}(t))$, which indicates the spatial position of sphere i of cilium k. The periodic two-phase beating dynamics are characterized by the function $\beta_i(t)$ such that the bending of the cilium is included (Buceta et al., 2005). Therefore, our model allows a detailed characterization of the dynamics of cilia.

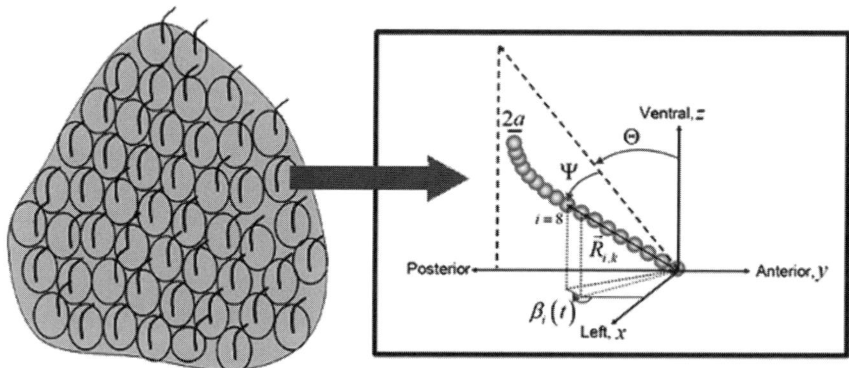

Figure 8.3
Model for nodal cilia dynamics. Left: The node is a pear-shaped structure composed of monociliated cells that harbor a 9 + 0 cilium about 5 microns in length. Right: Each cilium is modeled as a string of n moving spheres. For mouse embryos, the radius and the length of cilia are $a \approx 150$ nm and $L \approx 5$ μm, respectively, which leads to $n = 16$ moving spheres. The function $\beta_i(t)$ characterizes the nonplanar beating dynamics with power and recovery strokes, as well as the bending of the cilium along its length. (After Buceta et al., 2005.)

We have computed the trajectory of the tip of a cilium performing nonplanar beating around an axis tilted to the posterior ($\Theta \approx 40^\circ$) with frequency $f \approx 10$ Hz and apex angle $\Psi \approx 40^\circ$, which are the parameter values characterizing the dynamics of cilia in the mouse node (Okada et al., 2005). We have included the time-dependent bending of cilia by choosing the parameter value that leads to the experimentally measured ratio between angular speeds in the recovery and power strokes ($\dot{\xi}_{rec}/\dot{\xi}_{pow} \approx 0.6$–$0.7$). Importantly, we have set that the recovery stroke occurs while the cilium is moving close to the cell surface, which corresponds to a rightward motion for cilia moving clockwise around an axis tilted to the posterior, and which is supported by in vivo microscopy data (Okada et al., 2005). The ventral view of the numerically computed trajectory of the tip of the cilium for these parameter values fits perfectly the experimental data concerning the tip trajectory of mouse nodal cilia (Buceta et al., 2005; Okada et al., 2005).

The nodal fluid flow has been experimentally visualized by the motion of small, fluorescent latex beads added to the embryo culture medium (Nonaka et al., 1998; Okada et al., 2005). Thus, in order to have a realistic comparison of our theoretical results with the experimental data, we have formulated the equations describing the dynamics of such beads (modeled as spherical particles) embedded in the nodal fluid. The equation of motion for a small, spherical particle in a nonuniform flow can be derived from first principles (Maxey and Riley, 1983). An analysis of such an equation for low Reynolds number regimes reveals that the beads used in the nodal experiments are truly passive tracers: their velocities in the bulk of the node correspond to the fluid velocity. Thus, the equation of motion of a bead in the node simply reads (Buceta et al., 2005)

$$\frac{d\vec{r}}{dt} = \vec{u}(\vec{r}, t), \tag{8.4}$$

where $\vec{r}$ is the position of the bead, $\vec{u}(\vec{r}, t)$ is the velocity field that at position $\vec{r}$, and time t is produced by the dynamics of the cilia.

Moreover, since in the LRN regime the Navier-Stokes equations describing the dynamics of the extraembryonic fluid become linear, the flow generated by an ensemble of cilia can be approximated as the sum of the flow induced by each single cilium. Thus, the velocity field $\vec{u}(\vec{r}, t)$ appears as a result of the action of the N cilia present within the node (Buceta et al., 2005):

$$\vec{u}(\vec{r}, t) = \sum_{k=1}^{N} \vec{u}_k(\vec{r}, \{\vec{R}_{i,k}\}, t), \tag{8.5}$$

where $\vec{u}_k(\vec{r}, \{\vec{R}_{i,k}\}, t)$ represents the velocity fields that at point $\vec{r}$ and time t are induced by the dynamics of cilium k that has its basis at $\vec{R}_{0,k}$ and is formed by a set of n moving spheres with centers located at spatial positions $\{\vec{R}_{i,k}\}_{i=1,\dots,n}$. The velocity field $\vec{u}_k(\vec{r}, \{\vec{R}_{i,k}\}, t)$ induced by cilium k is computed in terms of the coordinates and velocities of the spheres that define the cilium (equation 8.4) by means of the well-known hydrodynamic expression that accounts for the velocity field induced by a moving sphere in a quiescent fluid (Landau and Lifshitz, 1959).

Specifically, $\vec{u}_k(\vec{r}, \{\vec{R}_{i,k}\}, t)$ is the sum of the velocity fields generated by the spheres of the cilium (Buceta et al., 2005),

$$\vec{u}_k(\vec{r}, \{\vec{R}_{i,k}\}, t) = \frac{3}{4}\sum_{i=0}^{n} \frac{a_{i,k}(t)}{|\vec{r} - \vec{R}_{i,k}|^3} \left[\dot{\vec{R}}_{i,k} \left(|\vec{r} - \vec{R}_{i,k}|^2 + \frac{a_{i,k}^2(t)}{3} \right) + (\vec{r} - \vec{R}_{i,k}) \left((\dot{\vec{R}}_{i,k} \cdot (\vec{r} - \vec{R}_{i,k})) \left(1 - \frac{a_{i,k}^2(t)}{|\vec{r} - \vec{R}_{i,k}|^2} \right) \right) \right], \tag{8.6}$$

where the effect of conformational changes has been included by means of a new time-dependent radii $a_{i,k}(t)$ approach (Buceta et al., 2005).

The *in silico* trajectories described by beads embedded in a node with many two-phase beating cilia resembles the motion of latex beads in the murine node: beads move leftward, performing some swirls and with an average speed $\langle u_x \rangle = 14 \pm 6$ μm/s, which is in agreement with experimentally available data (figure 8.4; Buceta et al., 2005). Therefore, a two-phase beating cilia dynamics is able to explain the nodal leftward fluid flow.

Our theoretical framework allows underscoring the dynamics of cilia and latex beads in the fluid flow of homozygous *inv* mouse mutants. *Inv*/*inv* mutants show a leftward flow, although it is slower and less directional, as evidenced by the more swirling, slower motion of latex beads (Okada et al., 1999). However, *inv*/*inv* mutants always exhibit situs reversal (*situs inversus*). Thus, *inv*/*inv* mutants exemplify that the establishment of the left-right axis requires not just an asymmetric signal which is directly translated to define the situs, but a host of different signals (direction and speed of the flow, for instance), not all of which need to be left-right asymmetric.

In order to understand the fluid dynamics in *inv*/*inv* mutants, we have hypothesized that if cilia always perform the recovery stroke when moving close to the cell surface, then a more swirly trajectory of beads, as observed in experiments, will emerge if few of the cilia in the *inv*/*inv* mutant have a reversed beating with a power stroke toward the right. This reversed beating might arise either from a clockwise motion around an anteriorly tilted axis or a counterclockwise motion around a posteriorly tilted axis. Beads move locally leftward or rightward, depending on the beating motion of the cilia that are closer. Thus, if there are few cilia with a reversed beating, the overall motion of beads will still be leftward, although more swirly. Indeed, *inv*/*inv* mutants have 20 percent of nodal cilia anteriorly tilted with clockwise

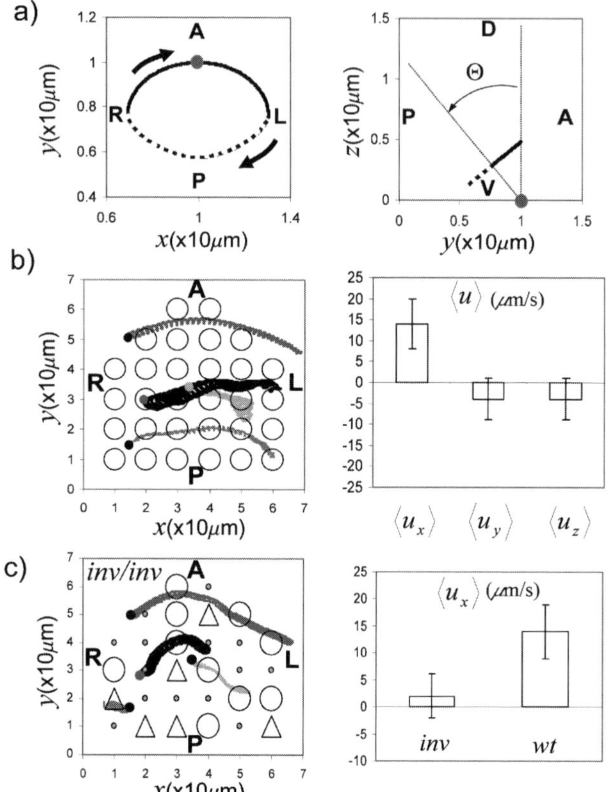

Figure 8.4
In silico results for the dynamics of cilia and of beads embedded in the *wt* and *inv* mutant embryonic fluid flows. (a) Trajectory (ventral view and right view) of the tip of a cilium tilted to the posterior with a non-planar beating motion having power stroke toward the left (orange) and recovery stroke (black) toward the right ($n = 16$, $\theta = 3\pi/2$, $\tau_b = 0.5\pi/\omega$, $\Psi = \Theta = 0.22\pi \approx 40°$). (b, right) Motion of beads embedded in *wt* embryos with beating cilia having the dynamics shown in (a). The beads are tracked during sixty periods of beating. Dots indicate the initial positions of the beads ($z = 5$ μm). All cilia are posteriorly tilted with their bases located at the center of the circles. (b, right) Average fluid flow velocities for *wt* embryos along the three anatomical axes. The bars stand for the statistical deviation. (c, left) Motion of beads embedded in the *inv* mutant node. 20 percent of cilia are anteriorly tilted (triangles) and thus perform a rightward power stroke, and 50 percent of the cilia are immobile (small circles). (c, right) Average leftward fluid flow velocities for *wt* and *inv* mutant embryos. Anterior (A), posterior (P), left (L), and right (R) sides are indicated. (After Buceta et al., 2005.)

motion, and 50 percent immotile (Okada et al., 2005). A slight increase in the number of counterclockwise-moving cilia has also been observed (Okada et al., 2005).

Importantly, an analysis of the speed of motion of anteriorly tilted cilia, compared with posteriorly tilted cilia, reveals that our hypothesis of a reversed beating is correct and that cilia always slow their motion when moving closer to the cell surface (when moving leftward or rightward for anteriorly or posteriorly tilted cilia, respectively (Buceta et al., 2005; Okada et al., 2005). Our *in silico* motion of beads in such a mutant node depicts a slower ($\langle u_x \rangle = 2 \pm 4$ μm/s) and more swirly motion, as expected (figure 8.4).

Therefore, our work proposes that the leftward fluid flow arises from the left-right asymmetric two-phase beating dynamics of cilia. This left-right asymmetric cilia dynamics arises from previously broken symmetries along the anteroposterior (A-P) and dorsoventral (D-V) axes: the tilting of the axis (A-P asymmetry) coupled to the ventral protrusion and clockwise motion of cilia (D-V asymmetries) reveals that the power stroke occurs when the tip of the cilium is moving leftward. Hence, in these ciliated organs (murine node, notochord, and Kupffer's vesicle), the left-right symmetry is broken at the level of each single cilium and cell. Since these ciliated structures are located along the midline, the left-right asymmetry at the level of each single cell is translated to an axial left-right asymmetry in the embryo through the fluid flow.

Cellular Level: Differential Adhesiveness in the Heart Tube Formation

The heart is the first functional organ in the anatomy of the vertebrates, and it is also the first anatomical phenomenon that breaks the initial symmetry of the embryo. We can distinguish the following stages in the formation of the heart in chick embryos: (1) bilateral heart field formation in the anterior lateral plate mesoderm; (2) migration of both heart epithelia toward the midline of the embryo; (3) epithelial fusion and formation of the heart tube; (4) caudal growth of the tube; (5) looping; (6) valve and chamber formation (Männer, 2000; Yelon, 2001). The heart loops naturally to the right, although a percentage of wild-type embryos, typically around 10 percent naturally, loop toward the left. There are mutants that randomize the direction of the looping, make it appear mostly to the left, or show no looping at all.

We can put together a model of the early formation of the bilateral lateral plate mesoderm, which can guide the direction of the looping. First, we assume a threshold-dependent, switchlike behavior for the dynamics of the left and the right epithelial sheets of cardiac precursor cells that originate at the lateral plate mesoderm. This function will have the form of a Hill equation, already used for modeling the Notch interactions. The Hill curve has a switchlike behavior when the Hill coefficient, a measure of cooperation, is sufficiently high (Hunding and Engelhardt,

1995). Using the Hill equations, each forming epithelium can be characterized independently:

$$T_R = \frac{j^h}{p^h + j^h}; \tag{8.7}$$

$$T_L = \frac{j^h}{p^h + j^h}, \tag{8.8}$$

where T is the tightness of the precardiac right and left tissue; j is the number or strength of cell-to-cell junctions in the epithelial sheet; h is the exponent that measures cooperation; and p is the point at which half of the possible value for T is reached.

The key idea of using the Hill equation for this kind of early heart formation modeling is that there has to be a cooperative effect among cell-to-cell junctions in the mesoderm that forms the heart. Thus, above a certain threshold the tissue forms an epithelium, whereas below this threshold, the tissue remains as loose mesenchyme. The cooperative effect depends on the value of h: the greater h is, the more abrupt the conversion between mesenchyme and epithelium. This predicts that as the density of junctions increases within the tissue, it will become easier to form a compact tissue that eventually behaves like an epithelium.

The formation of a tube is dependant upon the previous formation of the epithelium. Hence, it is restricted to a region of equations (8.7) and (8.8) in which both precardiac tissues form an epithelium. Bending will be determined by the difference between left and right numbers (or strength) of the junctions. When the left epithelium is "tighter" than the right one, bending occurs to the right. Conversely, left looping would occur when the right epithelium has more junctional density than the left. Tubes with no looping would occur when left and right epithelia are equally tight.

This model provides a quantitative basis for the looping phenomenon in the heart tube. Only one variable is relevant: the density of cell-to-cell junctions responsible for the formation of left and right precardial epithelia. Different morphologies can be predicted, and even the amount of bending could be calculated in a more complex model incorporating both the viscoelastic properties of the heart tissue and the forces exerted by the extracellular matrix. In addition, the model provides guidance to carry out further experimentation by analyzing the misexpression of genes that affect the formation of epithelia. There is already some evidence suggesting that these gene products may be related to the direction of the looping.

For example, N-cadherin, a calcium-dependent transmembrane protein that mediates cell-to-cell adherens junctions has been found to be involved in the randomization of the heart looping (Garcia-Castro et al., 2000). This protein is expressed in the left lateral plate mesoderm, where the left heart field epithelium is forming. Upon treating developing chick embryos with anti-N-cadherin, García-Castro et al. found randomization of heart looping. In the model presented here, this can be explained by the difference in tightness for each epithelium. Other proteins must also be involved in forming and maintaining the heart field epithelium in the right and left lateral plate mesoderm.

In zebra fish it has been found that the adhesion protein DM-GRASP is involved in left-right asymmetries of the heart and gut (Schilling et al., 1999). Mercola (2003) has reviewed the involvement of adhesion molecules such as N-cadherins and claudins in left-right specification; he also has noted the relationship between epithelial abnormalities and left-right inversion, as demonstrated by the thickening of the epithelium that surrounds the node in *inv* mutant mice.

An immediate prediction of the model is that by blocking the action of these other proteins, the epithelia will not be tight enough, and consequently the heart will not form a tube in the midline. That is equivalent to being in the region of the Hill curve that is not able to reach the threshold needed in density of junctions. In addition, by varying the density of proteins involved in the formation of each epithelium, or by modulating the amount of extracelllular calcium necessary for cadherin-cadherin interaction, it should be possible to direct the left and right looping of the heart (figure 8.5).

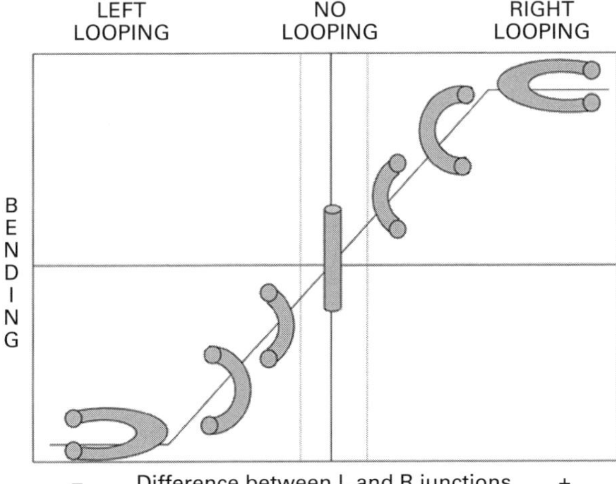

Figure 8.5
A modeling framework for adhesiveness. Using a Hill equation, as in the perinodal network, a model in which tissue changes from mesenchyme to epithelium that depends upon the total strength of cell-to-cell adhesion can be made. In this plot a morphospace of cardiac tubes that follows bending from the left to the right is shown. In the center of the plot, the difference between junctions in the left and right epithelium is zero; hence the tube is straight. (After Rasskin-Gutman and Izpisúa-Belmonte, 2004.)

The model can be seen as a cooperative action among mesoderm cells to form the mesenchyme-epithelium transition. One possibility is that cadherins might act as a "zipper," in which existing junctions facilitate the formation of new junctions. Evidence for such a behavior in cadherins has been provided (Shapiro et al., 1995), but the actual mechanism of cadherin-cadherin union remains controversial (Tepass et al., 2000). Another possibility is that each cell as a whole acts as a junction unit; in this case the cooperative phenomenon occurs as more cells are tightly united, facilitating the union of more cells. Yet another possibility is that the density of cadherins needs to be higher than a certain threshold in order to achieve tight junctions and form an epithelium. Varying this threshold (perhaps as a result of the presence of other adhesion molecules, different concentrations of extracellullar calcium, or specific differences in the catenin-actin intracellular anchorage of the cadherins) would vary the tightness of the epithelium.

Conclusion

Spherical elephants abound in science. They provide a source of conceptual ideas offering a set of specific predictions about natural phenomena that help to deepen our understanding of the world. In developmental biology this is obvious, both historically and in the light of present research agendas. Given the rich historical background of modeling in developmental biology, we can ask the following rhetorical question: Why has modeling not caught on in mainstream developmental biology? The answer is manifold. First, biologists are always very aware of the complexity of their living material and experimental setups, and any simplification seems to be a move in the wrong direction. Thus, mathematical modeling appears to be too far removed form "reality" and, hence, insubstantial. Second, the modeling tools and approaches themselves are a big challenge for the usually not numerically inclined experimental biologist. This is difficult to overcome, although there are now initiatives in academic curricula around the world to educate a growing number of young experimental biologists in the interdisciplinary crossroads of physics, mathematics, and computer science. Third, published models in the literature usually have been created by theoreticians who lack a solid experimental background and at times ignore the complex tasks involved in understanding the biological problem at hand. In other words, theoreticians very often create spherical elephants that are too rounded and that lack a handle to which the experimentalist can hold on.

These three problems in the transfer of information between experimentalists and theoreticians are difficult to solve. A question that the skeptical biologists usually raise is What have we learned from this model that we did not know before? Sometimes this is not easy to answer satisfactorily. The theoretical work may offer an almost trivial answer as a conclusion to a very elaborated model. The gain in this case comes from a better understanding of modeling itself, which can be handy later on for a similar, or sometimes totally unrelated, problem.

The theoretician might claim that the conclusion is "deep" and "far-reaching," but for the experimentalist this is a synonym for "evident." Another common question posed by the biologist is Why did we need a model to infer this relationship? This one is a matter of taste. Granted, sooner or later, without a model an experiment will offer evidence to substantiate a previously unknown relationship. However, an informed modeling approach can make this happen sooner rather than later, at times unveiling totally unpredicted relationships. That is the case, for instance, with statistical models of correlation from data obtained by microarray assays.

A point that needs to be made is that modeling also permeates the work of the experimentalist, although he or she may not be aware of it. For example, reaction kinetics, based on the law of mass action, is only a theoretical model that has been empirically deduced from experimental evidence. However, once the scientific community has accepted its universality, the validity of the equations is not put into question. Most software packages that are currently available for the lay biologist are based on reaction kinetics, and in order to use them efficiently, a thorough understanding of how enzymes and proteins react is desirable. However, there is no need to understand the details of the differential equations that characterize the involved reactions.

Which kinds of problems are tractable from a modeling perspective? And more important, which ones will not offer the developmental biologist "evident" answers, but instead interesting insights? The first question is an easy one, providing that one is sufficiently open-minded and willing to concede that some usefulness might come from modeling biological systems. Then the short answer is "everything." From molecular networks involved in a developmental pathway to the physical forces involved in organogenesis, at all levels of biological organization, modeling can do well for the experimentalist. Undoubtedly, many patterns and processes are used repeatedly during development in a variety of ways. Mathematical modeling provides a way to conceptualize these elements of biological design.

At a molecular level, the analysis of regulatory networks yields a finite number of dynamical relationships that occur over and over: oscillations, positive and negative feedback regulatory loops, molecular cooperation, and so. These are building blocks of the signaling pathways that are amenable to modeling using graph theory or differential equations. These are by far the regulatory elements that have been most analyzed and modeled, and they constitute the current effort of a very popular field that is now coalescing under the catchy name of "systems biology." Differential equations are the natural mathematical tool to explore the dynamics of molecular interactions, while graph theory (which is the theory of how elements interact within a network; see chapter 4, this volume) is making its way at a more conceptual level. The former can yield very accurate quantitative predictions; in contrast, graph theory mostly identifies patterns of interaction among the elements of the network.

To be sure, modeling sometimes is too much of an oversimplification of a very complex problem. An example is the gravitational law, which takes into account only the distance and the masses of two bodies in space to account for an empirical observation about their trajectories. Yet, the predictive power of Newton's formula overshadows its terrific simplification of the problem at hand. Equivalently, a similar predictive "developmental law" should also be based on simple data and undoubtedly will raise the suspicions of the die-hard experimentalist. The gravitational law is useful for practical things, such as ballistics or flying. In this case, the immediate application of a theoretical formula brings benefits right away. But what kind of applicability would a developmental law have? What sort of useful predictions would be drawn from it?

The answers to these important questions relate to the level of organization tackled by the model. At the molecular level, network modeling can predict new relationships among known genes or proteins (the segment polarity model of Odell and collaborators, von Dassow et al., 2000). At the cell behavior level, a cellular automaton can predict the aggregation patterns of cells following a chemoattractant (the Maree and Hogeweg model of *Dictyostelium discoideum* morphogenesis; Maree and Hogeweg, 2001). Finally, at the morphological level, geometry and forces can account for basic processes during embryo development such as invagination (Oster and collaborators' model of sea urchin gastrulation; Davidson et al., 1995).

But what could the experimentalist do with this information? Here we have the most important problem of data transfer between theory and experiment. The first task for the experimentalist is to test theoretical models with careful experimentation, which is the healthiest two-way relationship that can be asked for. But to do so, the theory must be attractive enough to catch the attention of an experimental lab. Also, the tests must be feasible; in other words, the theory must be testable by experimental means. And not only that, but the experiments required should fall within the normal protocols and procedures used in common labs.

In this chapter we have seen three instances of "spherical elephants" that tie together several aspects of left-right asymmetry in the early developing embryo: the differential expression of Nodal in the left lateral plate mesoderm of the chick embryo; the dynamics of Nodal flow due to cilia nonplanar beating in the mouse embryo; and the establishment of the preferential right-sided heart tube looping in the chick embryo.

Left-right early specification in the vertebrate body plan is a rich field of inquiry into basic processes during embryo development. A question that often arises, refers to why one side and not the other. Any molecular argument in this regard will fail the proximate-ultimate cause test, for no causal explanation can be said to be the most fundamental one from which everything originates. We need to work on more modest questions, particularly the mechanisms that generate specific outputs that are related to the establishment of the left-right axis. That is where the power of mathematical modeling with heuristics and predictions resides.

The models we have shown here highlight the adequacy of the theoretical point of view in a bursting field such as contemporary developmental biology. However, the prevalence of today's experimental take on science, and especially in our field, tends to darken the true importance of modeling and simulation approaches in favor of the accumulation of "hard evidence." There is a consensus that this amazing amount of data will be deciphered in a sudden, emergent explosion of *ideas* by self-organizing, academic, and corporate means. But without theory, data means nothing.

References

Blake J (1974) Hydrodynamic calculations on the movements of cilia and flagella. I. Paramecium. J Theor Biol 45: 183–203.

Brown NA, Wolpert L (1990) The development of handedness in left-right asymmetry. Development 109: 1–9.

Bosenberg MW, Massague J (1993) Juxtacrine cell signaling molecules. Curr Opin Cell Biol 5: 832–838.

Buceta J, Ibañes M, Rasskin-Gutman D, Okada Y, Hirokawa N, Izpisúa-Belmonte JC (2005) Nodal cilia dynamics and the specification of the left/right axis in early vertebrate embryo development. Biophys J 89: 2199–2209.

Capdevila J, Vogan KJ, Tabin CJ, Izpisua Belmonte JC (2000) Mechanisms of left-right determination in vertebrates. Cell 101: 9–21.

Cartwright JH, Piro O, Tuval I (2004) Fluid-dynamical basis of the embryonic development of left-right asymmetry in vertebrates. Proc Natl Acad Sci USA 101: 7234–7239.

Collier JR, Monk NA, Maini PK, Lewis JH (1996) Pattern formation by lateral inhibition with feedback: A mathematical model of Delta-Notch intercellular signalling. J Theor Biol 183: 429–446.

Crick F (1970) Diffusion in embryogenesis. Nature 225: 420–422.

Davidson LA, Koehl MA, Keller R, Oster GF (1995) How do sea urchins invaginate? Using biomechanics to distinguish between mechanisms of primary invagination. Development 121(7): 2005–2018.

Edelman GM (1988) Topobiology: An Introduction to Molecular Embryology. New York: Basic Books.

Essner JJ, Amack JD, Nyholm MK, Harris EB, Yost HJ (2005) Kupffer's vesicle is a ciliated organ of asymmetry in the zebrafish embryo that initiates left-right development of the brain, heart and gut. Development 132: 1247–1260.

Fox-Keller E (2002) Making Sense of Life: Explaining Biological Development Models, Metaphors, and Machines. Cambridge, Mass.: Harvard University Press.

Garcia-Castro MI, Vielmetter E, Bronner-Fraser M (2000) N-cadherin, a cell adhesion molecule involved in establishment of embryonic left-right asymmetry. Science 288(5468): 1047–1051.

Gierer A, Meinhardt H (1972). A theory of biological pattern formation. Kybernetik 12: 30–39.

Graham JH, Freeman DC, Emlen JM (1993) Antisymmetry, directional asymmetry, and dynamic morphogenesis. Genetica 89: 121–137.

Hunding A, Engelhardt R (1995) Early biological morphogenesis and nonlinear dynamics. J Theor Biol 173: 401–413.

Jouve C, Iimura T, Pourquie O (2002) Onset of the segmentation clock in the chick embryo: Evidence for oscillations in the somite precursors in the primitive streak. Development 129: 1107–1117.

Kawakami Y, Raya A, Raya RM, Rodriguez-Esteban C, Izpisúa-Belmonte JC (2005) Retinoic acid signalling links left-right asymmetric patterning and bilaterally symmetric somitogenesis in the zebrafish embryo. Nature 435: 165–171.

Kramer-Zucker AG, Olale F, Haycraft CJ, Yoder BK, Schier AF, Drummond IA (2005) Cilia-driven fluid flow in the zebra fish pronephros, brain and Kupffer's vesicle is required for normal organogenesis. Development 132: 1907–1921.

Krebs LT, Iwai N, Nonaka S, Welsh IC, Lan Y, Jiang R, Saijoh Y, O'Brien TP, Hamada H, Gridley T (2003) Notch signaling regulates left–right asymmetry determination by inducing Nodal expression. Genes Dev 17: 1207–1212.

Landau LD, Lifshitz EM (1959) Hydrodynamics. New York: Pergamon.

Levin M, Mercola M (1998) Gap junctions are involved in the early generation of left-right assymetry. Dev Biol 203: 90–105.

Männer J (2000) Cardiac looping in the chick embryo: A morphological review with special reference to terminological and biomechanical aspects of the looping process. Anat Rec 259(3): 248–262.

Maree AF, Hogeweg P (2001) How amoeboids self-organize into a fruiting body: Multicellular coordination in Dictyostelium discoideum. Proc Natl Acad Sci USA 98(7): 3879–3883.

Marszalek JR, Ruiz-Lozano P, Roberts E, Chien KR, Goldstein LS (1999) *Situs inversus* and embryonic ciliary morphogenesis defects in mouse mutants lacking the KIF3A subunit of kinesin-II. Proc Natl Acad Sci USA 96: 5043–5048.

Maxey MR, Riley JJ (1983) Equation of motion for a small rigid sphere in a nonuniform flow. Phys Fluids 26: 883–889.

Mercola M (2003) Left-right asymmetry: Nodal points. J Cell Sci 116: 3251–3257.

Nonaka S, Shiratori H, Saijoh Y, Hamada H (2002) Determination of left-right patterning of the mouse embryo by artificial nodal flow. Nature 418: 96–99.

Nonaka S, Tanaka Y, Okada Y, Takeda S, Harada A, Kanai Y, Kido M, Hirokawa N (1998) Randomization of left-right asymmetry due to loss of nodal cilia generating leftward flow of extraembryonic fluid in mice lacking KIF3B motor protein. Cell 95: 829–837.

Okada Y, Nonaka S, Tanaka Y, Saijoh Y, Hamada H, Hirokawa N (1999) Abnormal nodal flow precedes *situs inversus* in *iv* and *inv* mice. Mol Cell 4: 459–468.

Okada Y, Takeda S, Tanaka Y, Izpisúa-Belmonte JC, Hirokawa N (2005) Mechanism of Nodal flow: A conserved symmetry breaking event in left-right axis determination. Cell 121: 633–644.

Oppenheimer J (1967) Essays in the History of Embryology and Biology. Cambridge, Mass.: MIT Press.

Owen MR, Sherratt JA, Wearing HJ (2000) Lateral induction by juxtacrine signaling is a new mechanism for pattern formation. Dev Biol 217: 54–61.

Panin VM, Papayannopoulos V, Wilson R, Irvine KD (1997) Fringe modulates Notch–ligand interactions. Nature 387: 908–912.

Purcell EM (1977) Life at low Reynolds number. Amer J Phys 45: 3–11.

Rasskin-Gutman D, Buscalioni AD (2001) Theoretical morphology of the Archosaur (Reptilia: Diapsida) pelvic girdle. Paleobiology 27(1): 59–78.

Rasskin-Gutman D, Izpisúa-Belmonte JC (2004) Theoretical morphology of developmental asymmetries. BioEssays 26(4): 405–412.

Raya A, Izpisúa-Belmonte JC (2004) Unveiling the establishment of left-right asymmetry in the chick embryo. Mech Dev 121(9): 1043–1054.

Raya A, Kawakami Y, Rodriguez-Esteban C, Buscher D, Koth CM, Itoh T, Morita M, Raya RM, Dubova I, Bessa JG, de la Pompa JL, Izpisúa-Belmonte JC (2003) Notch activity induces Nodal expression and mediates the establishment of left-right asymmetry in vertebrate embryos. Genes Dev 17(10): 1213–1218.

Raya A, Kawakami Y, Rodriguez-Esteban C, Ibanes M, Rasskin-Gutman D, Rodriguez-Leon J, Buscher D, Feijo JA, Izpisúa-Belmonte JC (2004) Notch activity acts as a sensor for extracellular calcium during vertebrate left-right determination. Nature 427(6970): 121–128.

Schilling TF, Concordet JP, Ingham PW (1999) Regulation of left-right asymmetries in the zebrafish by *Shh* and *BMP4*. Dev Biol 210(2): 277–287.

Segel L (1980) Mathematical models in molecular and cellular biology. Cambridge: Cambridge University Press.

Shapiro L, Fannon AM, Kwong PD, Thompson A, Lehmann MS, Grubel G, Legrand JF, Als-Nielsen J, Colman DR, Hendrickson WA (1995) Structural basis of cell-cell adhesion by cadherins. Nature 374(6520): 327–337.

Shi S, Stanley P (2003) Protein O-fucosyltransferase 1 is an essential component of Notch signaling pathways. Proc Natl Acad Sci USA 100: 5234–5239.

Sleigh MA (1974) Cilia and Flagella. New York: Academic Press.

Takeda S, Yonekawa Y, Tanaka Y, Okada Y, Nonaka S, Hirokawa N (1999) Left-right asymmetry and kinesin superfamily protein KIF3A: New insights in determination of laterality and mesoderm induction by KIF3A−/− mice analysis. J Cell Biol 145: 825–836.

Tepass U, Truong K, Godt D, Ikura M, Peifer M (2000) Cadherins in embryonic and neural morphogenesis. Nat Rev Mol Cell Biol 1(2): 91–100.

Thompson D'AW (1942) On Growth and Form. Cambridge: Cambridge University Press.

Turing AM (1952) The chemical basis of morphogenesis. Phil Trans R Soc Lond B237: 37–72.

von Dassow G, Meir E, Munro EM, Odell GM (2000) The segment polarity network is a robust developmental module. Nature 406(6792): 188–192.

Wakabayashi K, Yagi Y, Kamiya R (1997) Ca2+-dependent waveform conversion in the flagellar axoneme of *Chlamydomonas* mutants lacking the central-pair/radial spoke system. Cell Motil Cytoskel 38: 22–28.

Weiss P (1939) Principles of Development. New York: Holt.

Wheatley D (2004) Primary cilia: New perspectives. Cell Biol Int 28: 75–77.

Wilkins AS (2002) The Evolution of Developmental Pathways. Sunderland Mass.: Sinauer.

Wimsatt WC (1987) False models as means to truer theories. In: Neutral Models in Biology (Nitecki M, Hoffman A, eds), 23–55. London: Oxford University Press.

Wolpert L (1969) Positional information and spatial pattern of cellular differentiation. J Theor Biol 25: 1–47.

Woolley DM, Vernon GG (2001) A study of helical and planar waves on sea urchin sperm flagella, with a theory of how they are generated. J Exp Biol 204: 1333–1345.

Yelon D (2001) Cardiac patterning and morphogenesis in zebrafish. Dev Dyn 222(4): 552–563.

9 Modeling Wave Propagation, Chemotaxis, Cell Adhesion, and Cell Sorting: Examples with *Dictyostelium*, Using a 3-D Cell-Based Model

Eirikur Palsson

The focus of this book is on modeling in biology, and the various chapters present different modeling approaches used on very diverse biological systems, ranging from dendritic morphology (chapter 6), neuronal growth and shape (chapter 10), and autism therapy (chapter 13) to plant evolution (chapter 14). From the models and systems discussed in each chapter, we can see many of the benefits of mathematical modeling in biology. In general, mathematical modeling in biology can provide us with a basic framework and help us understand complex systems. It does this by predicting the behavior of many "units" based on the rules of behavior for a single "unit." We can also use models to make predictions for some particular systems, get quantitative results, and then suggest experiments to verify those predictions.

Usually, when designing the model, one must make a number of assumptions and simplify things; these simplifications can reveal the most important processes underlying the observed behavior, and also can put constraints of what kinds of interactions are possible. In developmental biology one of the important objectives is gaining an understanding of the processes that govern the transformation of a single cell to the nonsymmetrical multicellular adult form. For example, the transformation from a fertilized egg to a newborn human, or from a *Dictyostelium* spore to a fully developed fruiting body. Here mathematical models become necessary to explain the phenomena of symmetry-breaking and how the genes control the process that leads from a zygote to a complete organism.

A very important use of a model, which is often ignored, is to predict new phenomena and to understand the subtleties of the biological system by putting them into the right context: for instance, explaining why, for biological and evolutionary reasons, a system may have certain features. Forgetting about this aspect is what often causes a conceptual gap between modelers and biologists. A biologist usually wants to know what the model can teach about the system he or she is working on. If the model is presented as just an animation of the biological system, the biologist may just respond, "So what? What use does this model have for me?" This problem of presenting a model from a mathematical point of view has, in my experience, turned many biologists away from appreciating the benefits of models in biology. But exposing biologists to examples where mathematical models are applied successfully should help change that attitude.

Wave propagation in many biological systems and cell movements in multicellular organisms are phenomena in biology where modeling can help answer questions that would be difficult to answer otherwise. The need for developing a model to study these phenomena comes from the fact that a large number of "individual units" are involved, and that makes understanding the complex interactions without a model very difficult. For example, in chemical waves these units are molecules, whereas in cell movements the natural unit is a single cell. Often, to make things more complex, wave propagation and signaling are directly involved in coordinating cell movements, which in turn affects the wave propagation.

In this chapter I will discuss some of the benefits of mathematical models of biological systems. I will focus on the developmental process in *Dictyostelium discoideum* (*Dd* hereafter), using a three-dimensional cell-based model to explore a number of features and questions that have come up. This model is based on individual cells, so it lets us investigate the effect of cell individuality, something that can't be explored using continuum-based models. The questions I am looking at are how the observed cAMP wave patterns, predominately spiral and target patterns, are generated and how these waves, in combination with chemotaxis, cell adhesion, and stiffness, orchestrate the movement of cells observed during *Dd* development. Many of the questions explored here have many things in common with other biological systems, and therefore a similar methodological approach can be used on them.

Wave Propagation and Cell Movements

Wave propagation is common in a number of biological systems—for example, waves of excitability in the heart (Davidenko et al., 1992; Jalife and Antzelevitch, 1979), calcium waves in a fertilized frog egg (Jaffe, 1995), waves of insect outbreak (Ludwig et al., 1978), wave spread of diseases such as rabies and the plague, and cAMP waves in *Dictyostelium discoideum*. The dynamics in many of these systems can be classified as that of excitable media, and a general understanding of excitable media gives insight into wave propagation in all these systems. Systems that are excitable can, under certain conditions, exhibit spiral wave propagation rather than planar traveling waves. A spiral wave is self-generating, whereas a planar or circular wave needs to be reinitiated. Spiral waves have been observed in *Dictyostelium* (Alcantara and Monk, 1974), in cardiac tissues (Davidenko et al., 1992), and in spruce budworm outbreak in the Alps (Ludwig et al., 1978).

There are a number of examples where cell movements play a vital part throughout the life spans of organisms. Bacteria and other single-cell organisms find food and avoid repellents by chemotaxis, and in animals, coordinated movements of cells and tissues occur throughout embryogenesis. In multicellular systems the relative movement of cells can lead to sorting out of different cell types and formation of specific structures and patterns, such as during gastrulation and in wound healing (Alberts et al., 1994), cancer cell invasion of tissues (Takeichi, 1993), the development of *Dictyostelium* (Bonner, 1967), and limb bud regeneration (Gilbert, 1997). In these multicellular systems, the collective cell motion can be quite different from the motion of isolated individual cells. Since the combined effect of cell-cell interactions and the production and propagation of chemotactic signals is often very complex, it is not always obvious which underlying mechanism is responsible for the observed patterns and movements.

I developed a three-dimensional cell-based model to explore these questions. Its purpose was to achieve visualization of cell movements in 3-D and gain an understanding of how simple cell-cell interactions, signaling, and adhesion lead to the observed complex cell movements.

I used *Dictyostelium discoideum* as a model system because understanding the collective motion both in the early individual cell stage and in the later multicellular stages sheds light on how cells in other developing systems coordinate their collective motion via short-range adhesion with other cells and long-range chemotactic signals. The cellular slime mold *Dd* has an interesting life cycle (figure 9.1). It begins with free-ranging amoebas that feed on bacteria and multiply by cell division (Bonner, 1967; Konijn et al., 1967). When the food source has been depleted, a few of the starving cells begin emitting a single cyclic adenosine-3′,5′-monophosphate (cAMP) pulse that is relayed from cell to cell, spreading outward as a circular wave. At the same time, cells move chemotactically toward the cAMP source. Early during starvation, the system is called weakly excitable because new circles emerge, only to die out. As time goes on, the excitability increases, circular waves begin to propagate, and those waves occasionally interact with new pulses, breaking up and forming spirals (Palsson and Cox, 1997).

These cAMP wave patterns can form in a field of randomly distributed cells because of the properties of the cAMP signaling system. When secreted cAMP levels rise above a certain threshold, cAMP binds to its receptor, stimulating internal cAMP production. Most of the cAMP is secreted, and external levels rise, which in turn stimulates increased production by binding to more receptors, creating a positive feedback loop. The system soon becomes adapted, however, and cAMP production halts. External cAMP levels then drop because of diffusion and the activity of the cAMP-degrading enzyme phosphodiesterase (PDE) (Wu et al., 1995), and cells slowly reenter a regime where they become sensitive to cAMP once again. This production and relay of cAMP pulses by the excitable cells, coupled with chemotactic movement toward the cAMP signaling center (a pacemaker region or the center of a spiral), facilitates the organization of large territories (Alcantara and Monk, 1974).

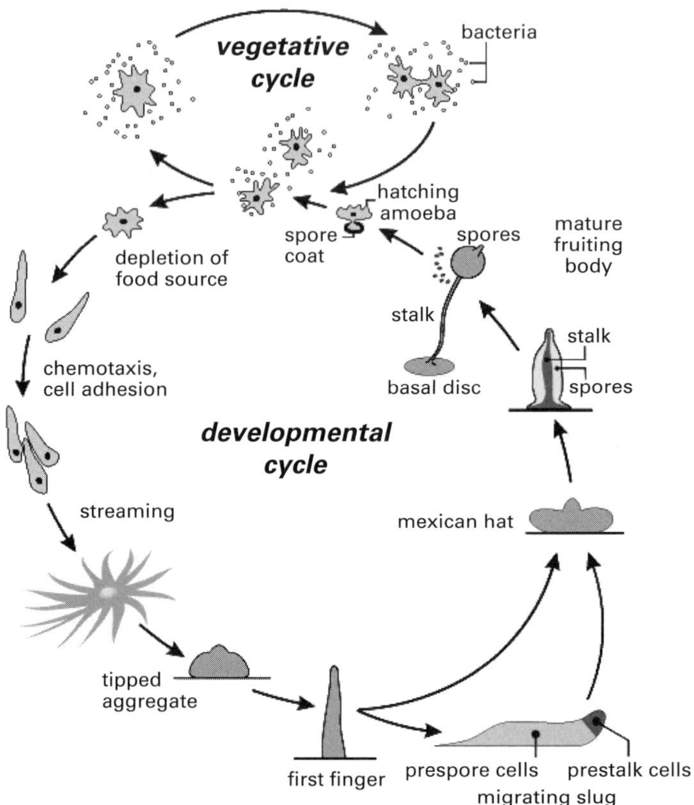

Figure 9.1
Dictyostelium discoideum life cycle. (Courtesy of Edward Cox, Princeton University.)

In early aggregation the cells move autonomously, but in late aggregation they form connected streams that flow toward the aggregation centers. The migration of cells toward the signaling center leads to the formation of a mound of cells that moves upward and eventually topples over to form a cigar-shaped mass called the slug (Bonner, 1967), containing up to 10^5 cells. cAMP waves are thought to be responsible for organizing the transformation of the mound into a slug, the migration of the slug over the substratum, and its culmination in a fruiting body (Siegert et al., 1992).

Models

The building blocks of my model are individual cells. To make it feasible to model cell motion in systems comprising thousands of cells, we stipulated that all cells are ellipsoids, and thus restricted the admissible deformations to those that only change the relative lengths of the semi-axes but conserve the cell volume. These computational cells have characteristics and responses that correspond to properties observed in real cells. These characteristics include the stiffness of the cell, cell adhesion, locomotive force generation, and response to environmental cues. The response of a cell depends on its internal parameter state, and on the information it receives from its external environment, which includes neighbor cells, the extracellular matrix, and chemical signals. Each cell is moved and deformed according to the equations of motion and deformation, based on all the forces acting on it, including adhesive and chemotactic forces from other cells.

The detailed aspects of cell movement have been discussed elsewhere (Elson et al., 1999; Sheetz et al., 1999; Small, 1989), so here I summarize the important aspects that most cells share. In suspension a cell is spherical; when resting on a surface it can be slightly flattened; and when it moves, it elongates in the direction of movement (Taylor et al., 1982). When subjected to a force, a cell initially resists deformation, an elastic response, but under sustained force the actin network in the cytoskeleton can break and re-form, which produces a viscous response (Chien et al., 1984; Evans, 1985a). When a *Dd* amoeboid cell moves, either randomly or in response to a chemotactic signal, it sends out pseudopods, one of which eventually dominates, and attaches it to the surroundings or to a neighbor cell (Wessels et al., 1994, 1998). This determines the direction of motion, as the cell realigns its axis toward the pseudopod and the rest of the cell body is "pulled" toward the attached pseudopod. Once the direction is established, the cell becomes polarized and is more likely to continue moving in the established direction.

The extensions of pseudopods and the retraction of the rest of the cell body require active force generation, and the force necessary for retraction is applied at the site of pseudopod attachment. When the cell is moving on a surface, the pseudopod attachment will be onto the surface, and the applied force is transmitted directly to the surface. However, when the cell is inside a multicellular aggregate, it must attach the pseudopod to another cell. It is important to realize that when a cell attaches its pseudopod to another cell and pulls itself forward, the other cell experiences an equal force in the opposite direction, and if that cell is not firmly attached, it is pulled backward; for a good discussion see Odell and Bonner (1986). Therefore, a complete model of cell movements must balance all the forces. Neglecting to balance the forces has been the main fault of many previous models. Also, since *Dd* cells can only pull toward an attachment—they can't push off—any upward movement of the cells can be achieved only by crawling up other cells. And these other cells must be semirigid so that they can support and transmit the force down to the surface.

There have been several approaches to modeling the collective motion of cells in multicellular systems: (1) models designed to simulate cell-sorting (Childress and

Percus, 1981; Mochizuki et al., 1996; Sulsky et al., 1984; Umeda and Inouye, 1999); (2) models of epithelial morphogenesis (Odell et al., 1981; Weliky and Oster, 1990); (3) one class of models using a cellular automata approach, often a Potts model (Glazier and Graner, 1993; Graner and Glazier, 1992; Maree and Hogeweg, 2002; Savill and Hogeweg, 1997), to represent the energy between the cells, where the cells exchange position with higher probabilities if that exchange is energetically favorable. Simulations of such models show patterns that resemble experimental observations, but limited insight is gained from them because they don't directly incorporate the details of the physical forces of cell-cell and cell-substrate interactions, nor do they balance the forces. Also, for instance, they fail to explain the mechanics of how the cells exchange positions.

Others have used continuum approaches, such as those of Odell and Bonner (1986) and Vasiev et al. (1997) to model the mound formation or slug movement in Dd. Here the slug is modeled as a viscous fluid, and the cells respond to cAMP chemotactically, as they do during the aggregation phase. While this approach can lead to solutions that apparently reflect aggregation, it is phenomenological, and no procedure is given to connect the forces postulated with experimentally measurable quantities such as the force exerted by a single cell. A limitation of these continuum-based models is that they do not take into account elasticity and size of the cells; cell individuality is lost. An individual-based model by Bretschneider et al. (1999) treats cells as semi-hard adhesive spheres and gives some insight into cell movement, but the cells do not deform, so there is very little cell sorting or exchange of positions.

One major problem with most of these models is that they fail to properly balance the active locomotive forces generated by the cells, or even ignore those forces. Some models have an active force that propels the cell, but no discussion of where that force is applied. Also, most of these models are two-dimensional, and therefore often do not adequately describe movement in multicellular systems.

The Design of the *Dictyostelium* Model

The basic units in the model, as mentioned above, are individual ellipsoidal cells, each of which is characterized by its location and orientation within the system, its state of stress, its internal dynamics, and the forces it exerts in response to the local microenvironment. If this is known for each cell, the movement of all cells, and hence of an aggregate of cells, can be calculated. A detailed description of the model can be found in Palsson (2001, 2007). Below is a summary of the features of the model.

The physical location and orientation of a cell are given by its center of mass coordinates and by the orientation and length of the principal axes, which we denote by the vectors **a**, **b**, **c**. We assume that cells are deformable ellipsoids, conserving the total volume under deformation, with axes of length a, b, and c. Each axis contains a

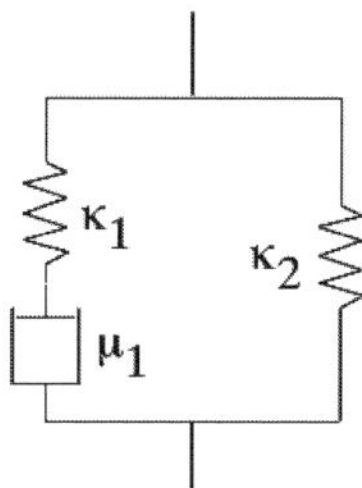

Figure 9.2
The combination of springs and dashpots that is used to model the viscoelastic properties of the ellipsoid axes. Here m1 is the viscous constant and κ_1 and κ_2 are the spring constants for Hookean springs.

nonlinear spring in parallel with a spring and a viscous element in series (figure 9.2). Following Chien et al. (1984) and Skalak et al. (1984), κ_1 and κ_2 are the spring constants and μ_1 is the viscocity of the element. This allows the cell to respond elastically to a brief force application and to deform in a viscous manner under prolonged force, then slowly relax back to a spherical shape when the force is removed. Because the cell must conserve volume, the elongation and compression of the three axes are not independent; they occur under the constraint that the volume remain constant. This gives us the following equations for the deformation of the i axis of a cell for a given force on that axis, $\mathbf{F}_i$.

$$\frac{dr_i}{dt} = \frac{\kappa_1(F_i - ff)}{\mu_1(\kappa_1 + \kappa_2)} + \frac{\frac{dF_i}{dt}}{(\kappa_1 + \kappa_2)} - r_i \frac{\kappa_1 \kappa_2}{\mu_1(\kappa_1 + \kappa_2)},$$

$$r_a r_b r_c = r_{a0} r_{b0} r_{c0} = V_{ellipse} \Big/ \left(\frac{4}{3}\pi\right).$$

Here r_i stands for the length of the a, b, or c axis in units of 10 μm, and ff is a modifying force that is calculated from the volume constraint by solving the two equations simultaneously (solution of a third-order polynomial for ff). The cell conserves volume, so when one axis is compressed, the other axes must be stretched, but due to the elasticity of the cell a counteracting force is generated, which is the basis for the modifying force, ff.

Orientation

In vivo, the first step in cell movement is to establish a dominant pseudopod and realign its anterior-posterior axis in its chosen direction (polarization). The selection of the dominant pseudopod is a stochastic process, strongly biased toward the gradient of a chemotactic signal, $\nabla\gamma$, and/or the direction of the cell's previous orientation. In the model the orientation of the cell, which is the direction it moves in, is determined as follows. At specific time intervals the cell determines if the gradient of the chemotactic signal is above some threshold. If it is, a new direction, r_{new}, is picked from a random Gaussian distribution around the direction of the gradient; otherwise, the new direction is picked randomly with a bias toward the previous direction. Here intervals of about 2 minutes where used, similar to the time of pseudopod formation in Dd (Wessels et al., 1994, 1998). After the cell has chosen a new direction and while the signal remains above threshold, the cell becomes polarized and does not easily change its orientation.

Forces

The forces acting on a cell are of three types: active, passive, and viscous drag. The active forces come from the locomotive forces that the cell applies or is exposed to by other cells. The passive forces arise from adhesive and elastic interactions between cells, and the viscous drag forces come from the viscous drag of the surrounding fluid and the drag associated with sliding past neighbor cells. The active force is generated when a cell moves either randomly or in response to a chemotactic signal. When a cell moves, it sends out a pseudopod in the direction of movement, attaches it, and applies a force to either another cell or the surface and, in effect, pulls the main body toward the pseudopod (Taylor et al., 1982).

In the model we assume that when cell i actively tries to move, it attaches a pseudopod to the substrate, if it is close enough, or to that neighbor cell j in front of it which has the smallest angle between the vector $\mathbf{r}_{ij}$, connecting the cell centers of i and j, and the direction of movement (figure 9.3a). The active force is always directed along the cell's anterior-posterior axis, which is the a-axis of the ellipsoid. When a cell attaches a pseudopod to the substrate on which it is moving, the active locomotive force is applied directly to the substrate, and the force on the moving cell is the reaction force that pulls the cell forward. However, when the pseudopod attaches to a neighbor cell, the active force pulls the neighbor cell toward the first cell, and the opposing reaction force pulls the first cell forward. Thus, the active force enters into the equation of motion for the second cell and the reaction force enters into the equation of motion for the first cell. These internal forces will cancel and will not produce any net force on the whole system unless some of the active force can be transmitted to the substrate through a network of connected cells, one or more of which is in contact with the substrate. This force balance is missing in all fluid-type models and the cellular Potts models. The magnitude of force generated for movement depends on several factors: what type of cell it is, the average local chemical concentration, and whether the cell is moving randomly or toward a chemical gradient.

When the cAMP concentration exceeds a threshold, the cell applies a chemotactic force in the direction of the cAMP gradient. If the cAMP concentration is below threshold, the cell will instead apply a force in a randomly chosen direction. When

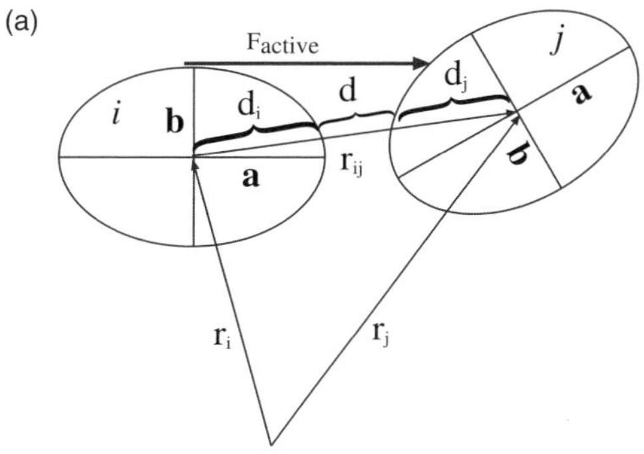

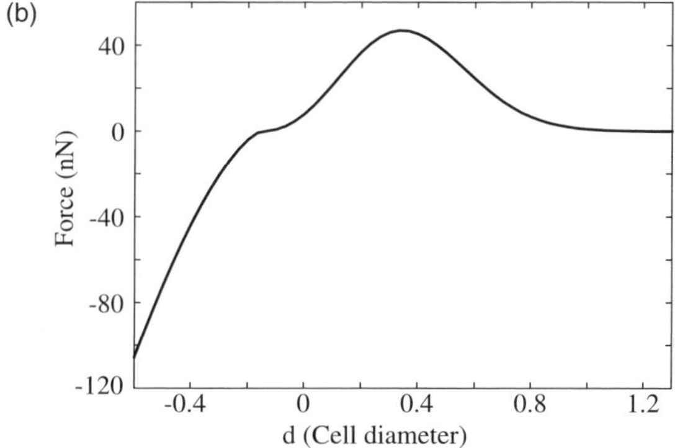

Figure 9.3
(a) Determination of d, the distance between the surfaces of two cells. F_{active} is the tactic force that the cell applies. (b) The force acting on a cell depends on the distance d between two cells.

the cell moves randomly, it does not migrate as much as when it is moving in response to a chemotactic signal. I model this observation by making the random force smaller than the chemotactic force. The active random force was set to be about 20 percent of the chemotactic force. The passive forces acting on a cell come from adhesive and elastic interactions with neighboring cells and are based on qualitative assumptions about the force, similar to the work by Evans (1985a, 1985b) and Zhu et al. (1994). The magnitude of the adhesion force between two cells depends on their proximity, since this determines the membrane area that two cells have in common, and thus how many adhesion molecules can bind. There is also a repelling force that arises from a cell's resistance to deformation and counteracts the adhesive force when cells get too close.

Estimates of the magnitude of the adhesion force and repulsive force between cells are given in Forgacs et al. (1998) and Phillips and Steinberg (1978). The passive force is always parallel to the $\mathbf{r}_{ij}$ direction ($\mathbf{r}_{ij}$ is the vector connecting the centers of the two

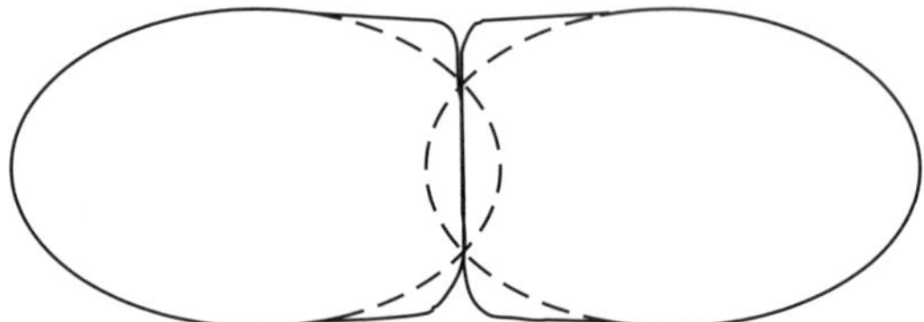

Figure 9.4
Two overlapping ellipsoids are used to represent close contact between neighboring cells. Here d is negative. The dotted lines indicate the ellipsoid shape.

cells). Figure 9.3b shows how the passive force varies with the distance d between the surfaces of the two ellipsoidal cells shown in figure 9.3a. Electron micrographs of multicellular tissues, for example, of a Dd slug (Fuchs et al., 1993), show that there is little free space between cells: they deform so as to fill up most available space and, as a result, have a larger contact area than if they were restricted to ellipsoidal shapes, as in the model. To compensate for this, the distance between the surfaces of the two ellipsoidal cells, d, does not represent the actual distance between two adhering cell membranes, but instead provides an estimate of how much surface area the cells have in common. In the model, the equilibrium state where the adhesive and repelling forces balance, occurs for $d = min_{dist}$ (figure 9.3b), which is slightly less than zero.

This compensates for our constraint that the cells are ellipsoids because cells do not overlap each other in real life, but instead deform into the vacant space (see figure 9.4). We choose the equilibrium value of min_{dist} to be such that the volume of a stacked cube of these cells would be close to the sum of the individual cell volumes, giving $min_{dist} \sim -0.1 \times \mathrm{r}_{cell}$ (r_{cell} is the radius of a typical cell). I used the following relation for the force between two cells:

$$F = \begin{cases} F_{comp}\chi(-x)^{3/2} & x < 0 \\ -\alpha\chi\{(x - x_0)\exp(-\lambda(x + x_0)^2) - v_0 \exp(-\lambda x^2)\} & x \geq 0 \end{cases}$$

$$\chi = \frac{r_{cell}}{2}\left(\frac{1}{d_i} + \frac{1}{d_j}\right), \quad x = \left(\frac{d}{r_{cell}} - b\right), \quad xr_{\vec{ij}} = (2\lambda)^{-1/2}, \quad b = min_{dist},$$

$$\mathbf{F} = F\frac{r_{\vec{ij}}}{\|r_{\vec{ij}}\|}.$$

Here, α is the adhesion strength, F_{comp} is the strength of the repelling force, and x_0 and v_0 are matching constants. In figure 9.3b the exponent is $\lambda = 7$. When the ellipsoids are deformed, their respective orientations, whether their longest axes are parallel or perpendicular, changes the common surface area between the two cells, and thus their adhesion strength. I take this into account by the value of χ in the above equations, where d_i and d_j are the distances, for ellipsoid i and j, respectively, to the surface from the cell center, along the vector $\mathbf{r}_{ij}$. The two types of forces listed above, the active and the passive, are "static" forces; they depend only on the separation of the cells and not on their velocities. The magnitude and direction of the total static force for cell i, $\mathbf{F}_i^{\text{stat}}$, is then found by summing all the above forces acting on the cell:

$$\mathbf{F}_i^{\text{stat}} = \mathbf{F}_{i(j/s)}^{act} + \sum_{j \in \mathcal{N}(i)} \mathbf{F}_{ij}^{pass} - \sum_{j \in \mathcal{N}(i)} \mathbf{F}_{ji}^{act}.$$

Here $\mathbf{F}_{i(j/s)}^{act}$ is the *reactive* force that comes from the *active* force that cell i applies to cell j or to the surface (s). $\mathbf{F}_{ij}^{pass}$ is the sum of the adhesive and elastic forces from neighbor cells on cell i, and $\mathbf{F}_{ij}^{act}$ is the active forces that neighbor cells exert on cell i.

The third type of force, the viscous drag force $\mathbf{F}^D$, depends on the velocity of the cells. One contribution comes from the viscosity of the surrounding fluid (usually water) and the viscosity due to cell-surface interactions, and is proportional to the speed of the cell relative to the laboratory frame. For a cell moving alone, this is the only contribution to the drag force. However, when the cell is moving in a multicellular system, there is another contribution to the drag force that comes from the viscosity between cells. We assume this force is directly proportional to the cells' relative velocity and their common surface area. Included in this cell-cell viscosity is the time-dependent force associated with the breaking of the cell-cell adhesion bond. Thus, in a multicellular system, $\mathbf{F}^D$ is a composite force, and depends on both the speed of the cell relative to the laboratory frame and on the relative velocity of the cell in question and its neighbor cells:

$$\mathbf{F}_i^D = \mu_s \frac{A_{is}}{A} \mathbf{v}_i + \mu_c \sum_{j \in \mathcal{N}(i)} \frac{A_{ij}}{A} (\mathbf{v}_i - \mathbf{v}_j),$$

$$A_i^T = \sum_{j \in \mathcal{N}(i)} \frac{A_{ij}}{A}.$$

Here A_{ij} is the common surface area between cell i and cell j, and A_{is} is the surface area of cell i that is in contact with the external fluid or surface and is not shared with any other cell. μ_s and μ_c are viscosity coefficients, $\mathcal{N}(i)$ denotes all the neighbors of i, and A is the surface area of a spherical cell of $(4\pi\mathrm{R}^2)$. μ_s was estimated from the measured velocity of a crawling amoeba and the force that it generates when it moves. Since the Reynolds number, $\Re$, of a moving cell 10 µm long and moving at a speed of 10 µm/min in water is very low, $\Re \sim 10^{-6}$, we can ignore the effect of inertia in the equation of motion. This in effect means that $\mathbf{F}^{\text{stat}}$ is balanced by the drag force, $\mathbf{F}^D$. Equating $\mathbf{F}^{\text{stat}}$ and $\mathbf{F}^D$ from the above equation gives us the equation of motion:

$$\frac{d\mathbf{x}_i}{dt} = \frac{\mathbf{F}^{\text{stat}} + \mu_c \sum_{j \in \mathcal{N}(i)} \frac{A_{ij}}{A} \mathbf{v}_j}{\mu_s \frac{A_{is}}{A} + \mu_c A_i^T}.$$

To describe the evolution of cAMP during aggregation, I modified a model of the cAMP signal transduction and relay system that had been used previously to study spiral wave formation during *Dd* aggregation (Martiel and Goldbeter, 1987; Palsson and Cox, 1996). The model was extended to three dimensions, and now the reaction-diffusion equations that govern the detection and production of cAMP are solved within each cell. The extracellular cAMP concentration is stored on a regular 3-D grid, and the diffusion and degradation of extracellular cAMP are calculated on this grid. The cells move within this grid but are not restricted to grid cubes, and therefore cAMP is interpolated from the grid to each cell. This is done by assigning the weight of each grid cube to be proportional to the fraction of the cell volume inside that grid:

$$[cAMP]_{cell\text{-}p} = \sum_{i=l-1}^{i=l+1} \sum_{j=m-1}^{j=m+1} \sum_{k=n-1}^{k=n+1} \frac{V_{cell\text{-}p}^{ijk}}{V_{cell\text{-}p}} [cAMP]_{ijk}.$$

Here $V_{\text{cell-p}}^{\text{ijk}}$ is the fraction of the volume of cell p inside grid cube i, j, k and $V_{\text{cell-p}}$ is the total cell volume. $[\text{cAMP}]_{\text{cell-p}}$ is the local external cAMP concentration at cell p and $[\text{cAMP}]_{\text{ijk}}$ is the cAMP concentration in grid cubes i, j, and k; l, m, and n are the x, y, and z locations of the cell, respectively. The production of cAMP by each cell is based on the dynamics of its internal signaling system in response to $[\text{cAMP}]_{\text{cell-p}}$. The secreted cAMP is then distributed back to the grid to provide the cAMP distribution for the next step:

$$[cAMP]_{ijk} = [cAMP]_{ijk} + \sum_{p \in \mathcal{N}(p)} \frac{V_{cell}^{ijk}}{V_{cell}} cAMP_{cell\text{-}p}^{secreted}.$$

A previous version of the model used a different model for the signaling system (Palsson and Othmer, 2000), and some of the results presented here are from that version. The results were qualitatively similar regardless of which model of the cAMP signaling system was used.

To determine the chemical gradient at each cell, it is assumed that a cell can sense a chemical gradient across its body. The chemical concentration is stored on a computational 3-D grid which overlies the cell positions. The cell compares the chemical concentration in all the grids adjacent to the grid cube that the cell belongs to. In three dimensions this amounts to twenty-six grid cubes around the cell, and from that information the chemical gradient, $\nabla\gamma$, for a cell is determined. The progression of the model is as follows.

Table 9.1
List of experimentally measured parameters

Parmeter	Typical range in model	Experimental justification
Active force, $\mathbf{F}_{act}$	$2–10 \times 10^{-3}$ dyne	$3–8 \times 10^{-3}$ dyne leukocytes (Usami et al., 1992) Cells in Dd slug (Inouye and Takeuchi, 1980)
Cell stiffness, κ	$0.3–1.0 \times 10^{-3}$ dyne/mm	$0.6–1.0 \times 10^{-3}$ dyne/mm (Bray, 1992) red blood cells—fibroblasts
Normalized viscosity, μ_s	$8–20 \times 10^{-3}$ dyne sec/mm	Estimated from active force and 20 μm/min cell velocity (Alcantara and Monk, 1974)
Cell adhesion, a	$0–10 \times 10^{-3}$ dyne	Estimated from surface tension measurements

1. For each cell
   a. Determine if the cAMP concentration in the cell's vicinity is above threshold. If it is, find the direction of the local cAMP gradient and orient the cell in that direction.
   b. Locate all cells that are within a minimum distance from each cell; these are the neighbor cells.
   c. Find all the forces that act on the cell, $\mathbf{F}^{stat}$, from each of the neighbor cells located in b. These are the passive and active forces that deform the axes of the ellipsoid, and move the cell according to the equation of motion.
2. Update the cAMP concentration.
3. Repeat 1 and 2 for every time step.

Parameter Values

In the model, the parameter values were chosen to lie within the experimentally observed ranges (see table 9.1). Since these parameters are based on experimental measurements, changes made in the model parameters reflect a specific biological mechanism or property. Therefore, the model allows us to explore the effect these mechanisms have on the slug movement and how sensitive the system is to variations in the parameters.

Results

The validity of the model was first tested by doing simulations that were then directly compared against experiments already done. I will briefly discuss these experiments here; a more detailed description can be found in Palsson (2001). The first step was modeling the behavior of a multicellular aggregate in the absence of chemotaxis. In these simulations, cells were compressed between two parallel plates, and the force

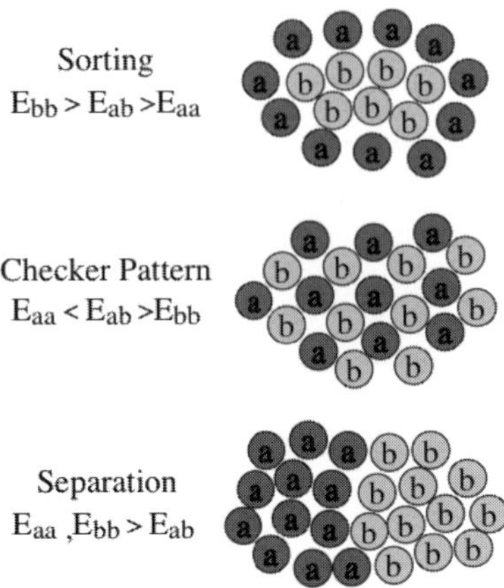

Figure 9.5
Sorting of two different cell types for different cell adhesions. E_{aa}, E_{bb}, and E_{ab} are the adhesion energies between cells of type a-a, b-b, and a-b, respectively.

exerted on the plates was determined. These results showed that the simulated multicellular aggregate had the characteristics of a viscoelastic fluid, and the surface tension was within the range of previous measurements of embryonic tissue by Steinberg and coworkers (Forgacs et al., 1998; Foty et al., 1996), using a similar configuration. The next step was to simulate cell-sorting due to differences in adhesion between the two cell types.

The sorting of cells due to differences in adhesion was first suggested by Steinberg (1975), and is referred to as the differential adhesion hypothesis (DAH). The DAH states that two cell types with different adhesion strengths, or differences in energy of adhesion, sort out in such a fashion that the cells with stronger adhesion sort together in the center, completely enveloped by the cells with weaker adhesion. The reasoning for this was based on thermodynamic considerations stating that in thermodynamic equilibrium, the potential energy is minimized.

Figure 9.5 shows the equilibrium configurations for cell-specific adhesion of two cell types, A and B. Steinberg's group verified the DAH by mixing cells that had different adhesion strengths, and showed that invariably at equilibrium the cells with stronger adhesion sorted to the center (Foty et al., 1996). I simulated this sorting of two cell types in an aggregate on which no external forces were imposed. Here the motive force arises from differences in adhesive forces between the two cell types and from the random force responsible for the random cell movement. I explored how changing the magnitude of the random motion, the compressibility, and the adhesion of the cells affected the rate and completeness of the cell sorting. Figure 9.6 shows a typical cell-sorting simulation. Here the light cells are two times more adhesive than

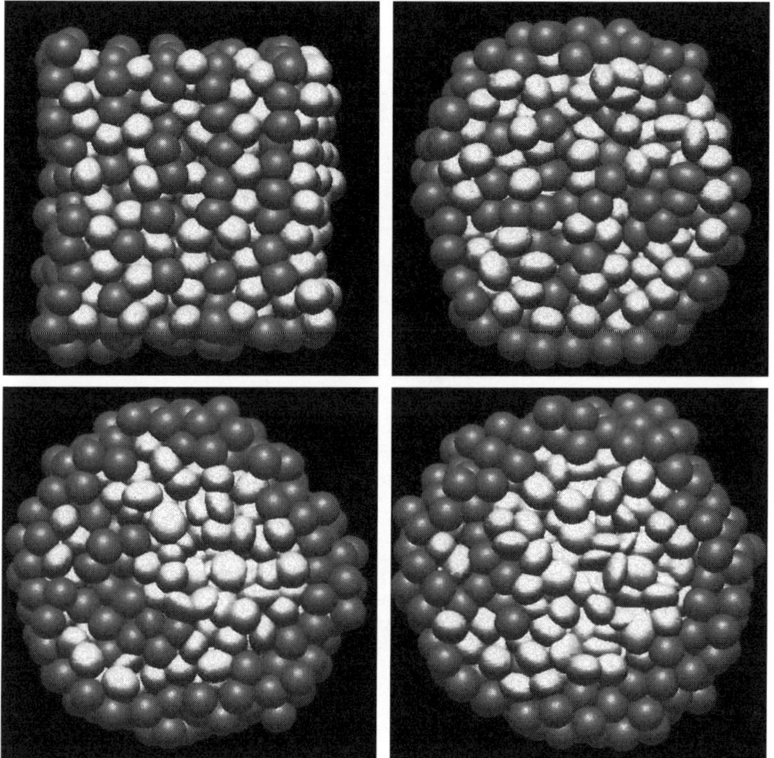

Figure 9.6
A cross section through the middle of the aggregate at four different time points: 10 min, 120 min, 280 min, and 320 min. Note that the light cells, which are more adhesive, sort to the middle and push the less adhesive dark cells to the outside.

the dark cells. As can be seen, the light cells move into the center of the aggregate, at the same time pushing the dark and less adhesive cells to the surface.

I also studied the effect that random cell movements, cell stiffness, and adhesion have on cell sorting. The results of these simulations are shown in figures 9.7 and 9.8. The sorting parameter, s, is found by dividing the number of neighbor cells of the same type by the total number of neighbor cells. Increasing the random motion increases the rate of sorting (figure 9.7). The time it takes to sort completely depends strongly on the magnitude of the random force. When there is no random force, there is hardly any sorting (solid line). Full sorting takes 20 minutes, 70 minutes, and 160 minutes when the random force is 30 nN, 20 nN, and 10 nN, respectively. Note that the sorting is more complete when the random force is not too high. This is because when the random movements are large, the cells move in and out of the minimum energy state rather than remaining fully sorted.

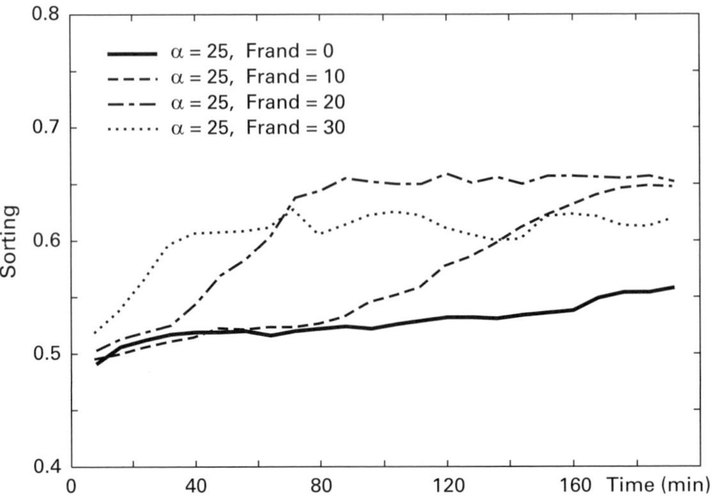

Figure 9.7
The sorting for two types of cells with a total of 400 cells having different random force. Solid line $F_{rand} = 0$ nN; dashed $F_{rand} = 10$ nN; dash-dot $F_{rand} = 20$ nN; dots $F_{rand} = 30$ nN. Maximum sorting occurs at 0.69. Here the adhesion strength is $\alpha = 25$. In figures 9.7 and 9.8, the adhesion energies for cell 1 and 2 are $E_{2\text{-}2} = 0.5 \times E_{1\text{-}1}$ and $E_{1\text{-}2} = 0.75 \times E_{1\text{-}1}$, and the spring constant is $\kappa = 200$ nN/10 μm, unless specified otherwise.

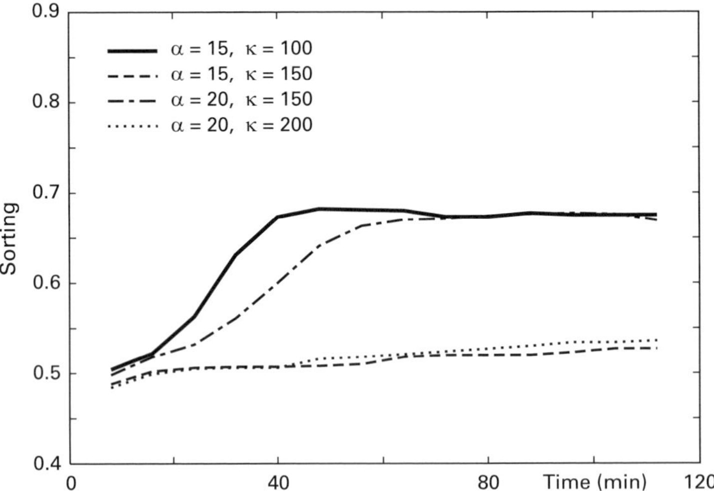

Figure 9.8
Changes in the spring constant, k, affect cell sorting. Low κ means easily deformable cells. Note that cells sort easily when the $\kappa = 100$ nN/10 μm is low and $a = 15$ (solid line), but the sorting is severely hampered when $\kappa = 150$ nN/10 μm (dashed line). Increasing the adhesion to $\alpha = 20$ counteracts the effect of increasing the compression constant (dashed-dotted line). If $\kappa = 200$ nN/10 μm, again there is very little sorting (dotted line). Random force is 15 nN.

Figure 9.9
Dark field photo of wild type DH1 (right) and mutant PDI ko #27 (left). Density $r = 5.5{}^{*}105\ \text{cm}^2$. Taken 8 hours and 25 minutes after plating.

Changing the stiffness of the cells affects the sorting quite dramatically (figure 9.8). When the cells are easily deformable with low spring constant ($\kappa = 100$ nN/10 μm), complete sorting occurs rapidly, since cells can easily move past each other. As the cells become stiffer ($\kappa = 150$ nN/10 μm), they do not sort unless the adhesion strength and/or the random cell movement is increased. The initial configuration, whether it is random or a regular 3-D checkerboard pattern, has very little effect on the time of sorting. The sorting occurs just slightly faster when the initial configuration is random.

This close agreement between the results of computational experiments and laboratory findings suggests that the model reproduces the essential features of multicellular systems, and therefore can be used to investigate other aspects of cell movement in tissue-like aggregates.

Simulations of Aggregation and Slug Movement in *Dictyostelium Discoideum*

Figure 9.9 shows the observed aggregation patterns for a *Dictyostelium* strain. The wild type is on the right and forms large aggregation territories. A phosphodiesterase inhibitor (PDI) knockout mutant is on the left, and here the territories are much smaller (Palsson et al., 1997). In the following simulations a few cells in the center, shown as darker cells, are assigned a higher number of cAMP receptors and are thus more excitable, effectively making them pacemakers. These pacemaker cells periodically secrete a cAMP pulse, initiating outward-propagating circular cAMP waves

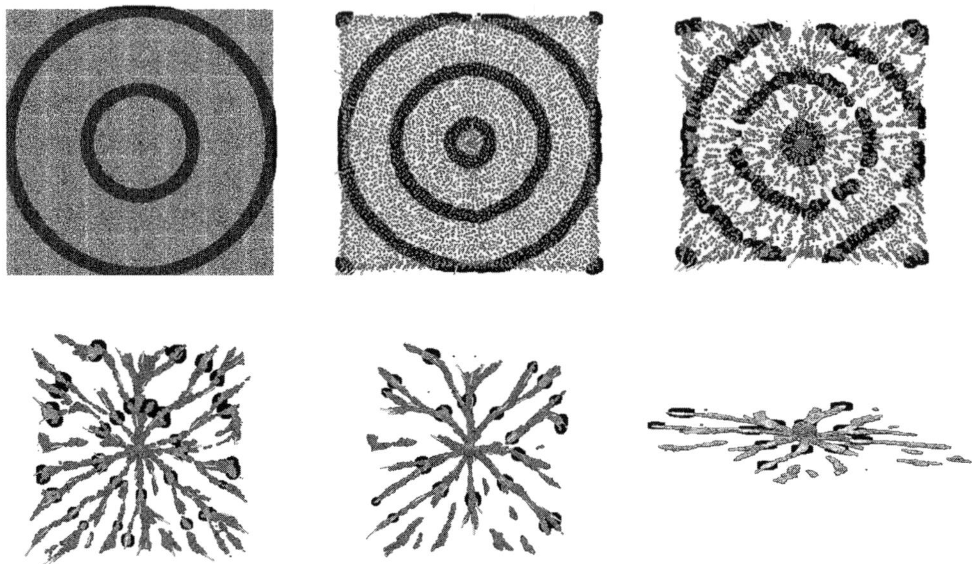

Figure 9.10
Aggregation of *Dictyostelium* cells in response to cAMP signaling from (slightly darker) pacemaker cells that are located in the center. Here the times for the frames are 25, 250, 500, 1000, 1500, and 2000 minutes, respectively. The dark bands correspond to high cAMP concentrations. The first five frames show the aggregation looking down on the plate, and the last frame shows the aggregation territory at an angle. The time for aggregation is longer than observed, since here the computational cells move at a speed five times less than experimental cells. The total number of cells is 40,000. The field of each figure is 4 mm × 4 mm.

that trigger relay of the cAMP signal and chemotaxis in the remainder of the aggregation territory. A spiral that generates rotating cAMP waves would also work as a signaling center (Palsson and Cox, 1996). However, since pacemakers are easier to initiate and both types have been observed (Alcantara and Monk, 1974; Gross et al., 1976), I chose to use pacemakers.

Figure 9.10 shows the time series of a simulation of the aggregation of 40,000 cells. The animation of the computational results (see http://www.sfu.ca/~epalsson/altenberg/tempdir22_movie12) shows both the evolution of streams during aggregation and how the pacemaker cells are lifted by the inward motion of the other cells. Figure 9.11 shows how the circular waves initiated from the pacemaker cells can sometimes break up and form spirals. These spirals are self-sustaining; the tip continuously rotates around a core and reinitiates the wave. The spirals do not need pacemaker cells for survival. The result is the formation of many smaller aggregation territories, as can be clearly seen in the last frame of figure 9.11, which is a common phenomenon in *Dd* aggregation.

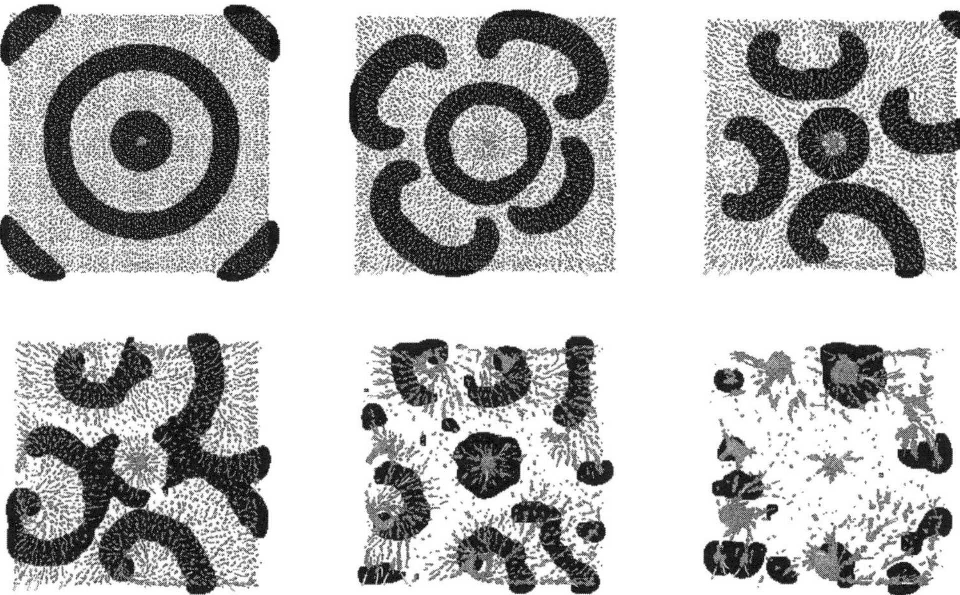

Figure 9.11
Aggregation of *Dictyostelium* cells in response to cAMP signaling from slightly darker pacemaker cells that are located in the center. We can clearly see the formation of several spirals that end up forming new aggregation territories. Here the times for the frames are 60, 110, 160, 250, 375, and 500 minutes, respectively. The dark bands correspond to high cAMP concentrations. The total number of cells is 10,000. The field of each figure is 2 mm $\times$ 2 mm.

Dictyostelium cells have both a membrane-bound phosphodiesterase (PDE_m) and a secreted phosphodiesterase (PDE_s), both of which degrade cAMP and have a strong effect on the excitability of the system. If the activity of PDE is too high, cAMP is degraded rapidly and no cAMP waves can propagate. However, if PDE activity is low, all the cells become pacemakers. At even lower activity levels the system gets saturated with cAMP and no waves can propagate.

I explored the differences in the effects that membrane-bound and secreted PDE have on the aggregation. Figure 9.12 shows the results from two simulations where the ratio between PDE_m and PDE_s was changed. In the upper half the PDE_m/PDE_s ratio is high (≈ 8) and aggregation proceeds normally. In the lower half the PDE_m/PDE_s ratio is much lower (≈ 1), and aggregation is interrupted later on, when the cell density has increased. This happens because at higher densities many other cells become pacemakers and begin to signal and initiate new territories. An explanation for this is that as the density increases, the excitability increases because the production of cAMP is proportional to the number of cells but the activity of PDE_s per unit

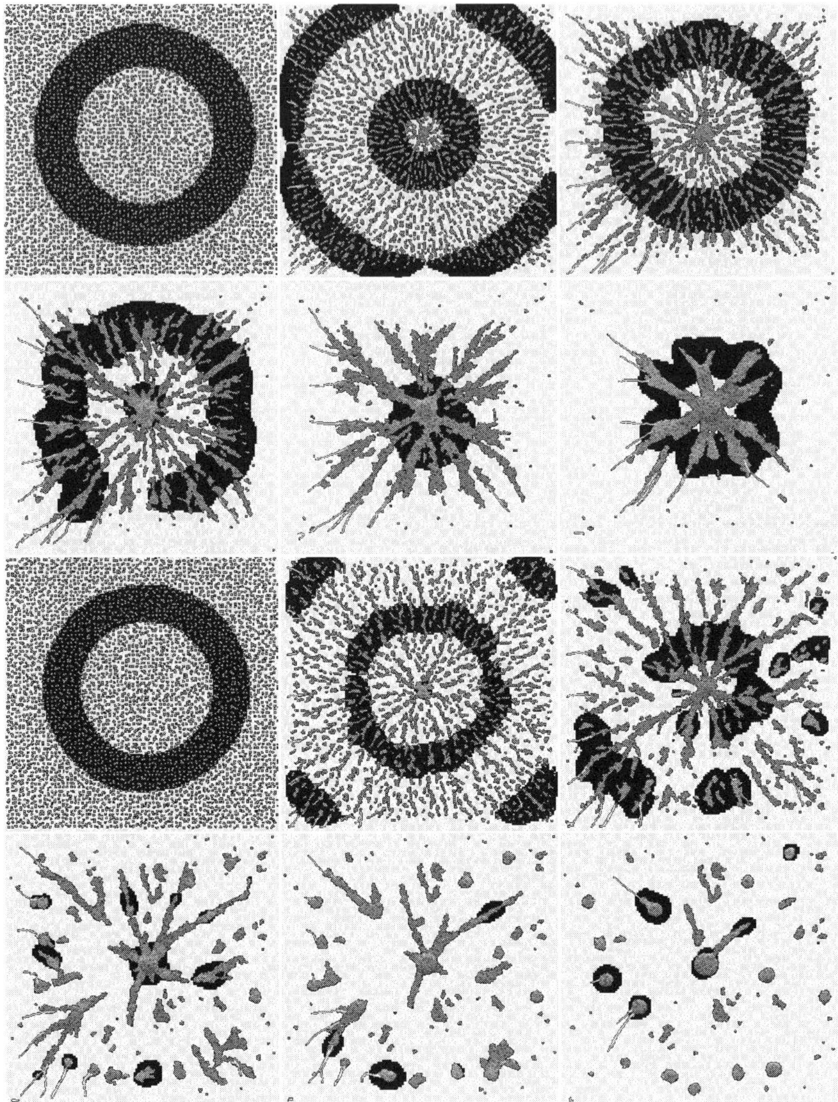

Figure 9.12
Aggregation of *Dictyostelium* cells for low and high PDE_m/PDE_s ratios. cAMP signaling is from slightly darker pacemaker cells that are located in the center. In the upper half the PDE_m/PDE_s ratio is high (≈ 8), as in figures 9.10 and 9.11, and aggregation proceeds normally. In the lower half the PDE_m/PDE_s ratio is much lower (≈ 1), and aggregation is interrupted later on, when the density has increased, since many new signaling centers form. The times for the frames are 20, 100, 200, 300, 400, and 600 minutes, respectively. The dark bands correspond to high cAMP concentrations. The total number of cells is 4,900. The field of each figure is 1.4 mm × 1.4 mm.

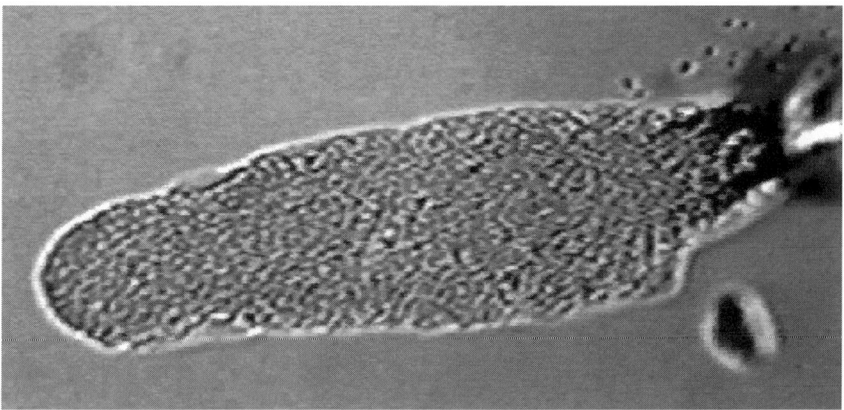

Figure 9.13
A 2-D Dd slug with about 2000 cells moving between the boundary of oil and a glass plate. The slug is about 500 μm long. (Courtesy of John Bonner.)

volume remains the same, so many new cells become pacemakers. This increase in excitability is reduced when the PDE_m/PDE_s ratio is high because, as the cell density increases, the activity of PDE_m per unit volume increases as well, since the PDE_m is associated with the cells. Therefore, new pacemaker cells are not formed and the aggregation is not disrupted.

I also studied cell movement and cell-sorting in a 2-D slug to simulate experiments by Bonner (1998). Bonner created an essentially flat *Dd* slug by forcing slugs to move between a glass plate and a layer of mineral oil (figure 9.13). His results show that many of the characteristic patterns of movement that are observed in 3-D aggregates, such as rotary cell motion, also occur in 2-D. To simulate this, the cells were constrained between two parallel plates separated by 1.5 cell diameters, thus effectively forcing the slug to be about only one to two cell layers thick. The cell interactions with the plates were the same above and below, which differed from the experimental conditions. However, the results show that this difference is apparently not very significant. Initially the cells were arranged in a regular square between the two plates, and a few pacemaker cells (dark cells) were placed on the midline at the right edge. As before, these cells initiated cAMP waves that were relayed throughout the slug, which began to elongate and move forward. In the movie of this simulation one can clearly see the cells move forward in response to the cAMP wave emanating from the pacemakers, and this forward motion pushes the pacemaker cells forward, giving rise to a net forward movement for the slug.

In the simulation shown in figure 9.14, a new group of pacemakers was grafted onto one side of the slug after fifty minutes of migration. These new pacemakers began to send out cAMP signals, and thus compete with the pacemakers at the anterior

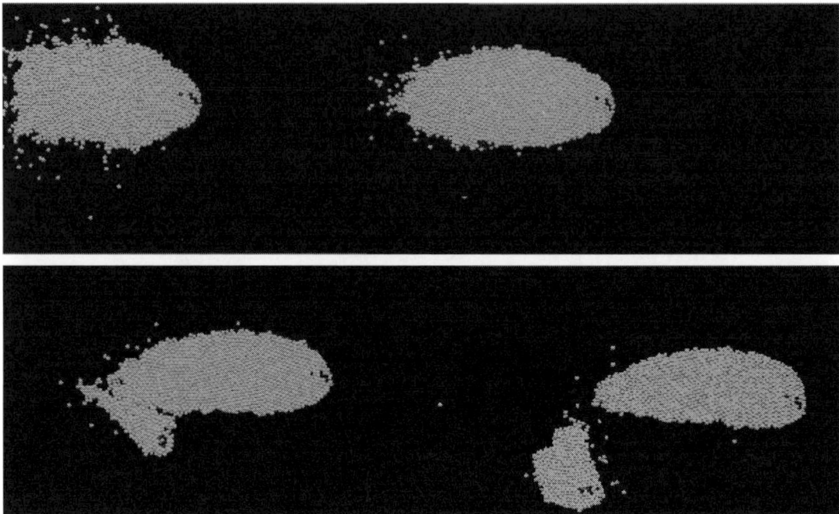

Figure 9.14
The movement of a *Dictyostelium* slug with 2500 cells in response to a few pacemaker cells at the front (red cells). The frames were taken after 60, 80, 100, and 120 minutes, respectively. At time = 50 minutes I initiated another pacemaker at the side of the slug.

end. Eventually the signals from the grafted pacemakers dominated the signals from the anterior pacemaker in a region surrounding the grafted pacemakers, with the result that the slug split into two smaller slugs, each controlled by its own pacemaker (Palsson, 2001). Such splitting is common in grafting experiments using 3-D slugs, but recently it has also been observed in 2-D slugs (Bonner, 1998). If the cell-cell adhesion is too strong, the slug does not split. Instead, after some period of tug-of-war, one of the pacemaker centers eventually entrains the other, and the rest of the slug moves toward that center (results not shown).

Figure 9.15 shows a typical slug and the corresponding cAMP wave. The cAMP wave shown in dark bands propagating down the slug is initiated by the darker pacemaker cells at the front; the rest of the cells move toward the origin of the wave.

In early aggregation the more powerful pre-stalk cells move about 75 percent faster, and quickly pass the pre-spore cells. To explore sorting of pre-stalk and prespore cells during the Dd slug stage, I did simulations of a thicker slug composed of those two cell types. The active force is 75 percent larger for the pre-stalk cells. Initially these cell types were uniformly distributed in the slug. I discovered that even though the more powerful cells have a tendency to move toward the front, most of them had not sorted to the front at the end of four hours. In these simulations the adhesive forces were identical for both cell types, suggesting that when the cells inter-

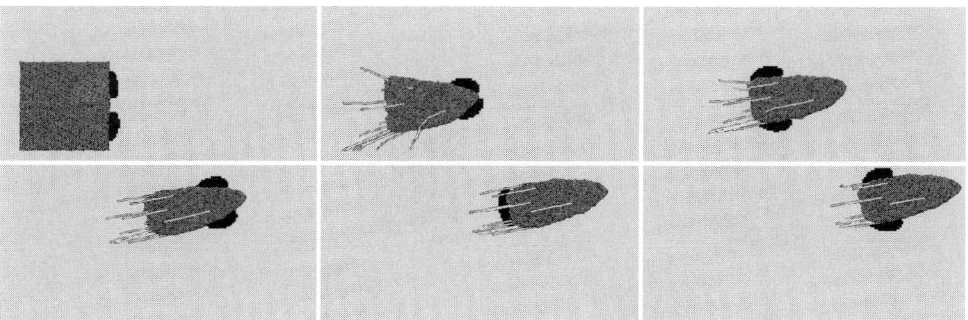

Figure 9.15
Slug movement in response to cAMP waves initiated from the slightly darker pacemakers at the front. The dark bands show the cAMP wave (high levels of cAMP). The frames are 150 minutes apart.

act strongly, as they do in the slug, simply exerting more force is not enough to produce rapid sorting. Making the cells more deformable improves the sorting, but not dramatically.

A mechanism that might speed up the cell-sorting would be a column of more active cells moving toward the front, in which case the less active cells wouldn't be as much in the way. I decided to explore this possibility. When cells adhere more strongly to cells of the same type (i.e., cell-specific cell adhesion), the different cell populations tend to group together inside the aggregate. It has been shown that in a disaggregated *Dd* slug, pre-spore and pre-stalk cells preferentially adhere to themselves (Takeuchi et al., 1988; Tasaka and Takeuchi, 1979). Based on this knowledge, I set out to determine if the sorting out of pre-spore and pre-stalk cells was enhanced when the cells have cell-specific cell adhesion.

Figure 9.16 shows a time series of the cell sorting in a slug. The chemotactic force for the gray cells (pre-stalk) is 75 percent larger than for the lighter cells (pre-spore). The pre-stalk cells are more adhesive than pre-spore cells, and they rapidly sort to the front, though a few scattered cells remain in the posterior part of the slug. This can be seen more clearly in the longitudinal sections of the slug on the right. In real slugs these cells have been found, and are called anterior-like cells (Sternfeld and David, 1982). Note that the pre-stalk cells form small groups that move as a unit to the front of the slug. This grouping of the pre-stalk cells helps them move past the pre-spore cells, since now the collective chemotactic force of the group is sufficient to allow them to do so. This is possible because the total chemotactic force of the group is proportional to the number of cells in the group, whereas the force required to deform and move past the cells ahead is proportional to the cross section of the group. Therefore, the easiest way for stronger cells to move to the front is in a narrow stream up the center, because here the cells do not need to deform much and the

Figure 9.16
The sorting of pre-stalk (gray) cells and pre-spore (lighter) cells in a *Dictyostelium* slug for different cell adhesion. The figures on the right are longitudinal sections of the slug figure on the left that show the lower half. The dark cells at the front are pacemakers (visible only in the cross-sectional view) periodically emitting cAMP. The gray cells have a chemotactic force that is 75 percent greater than the green cells. The times for the frames are 125, 188, 250, 375, and 500 minutes as you move down the picture. Adhesion energy: $E_{light,light} = 1.0$, $E_{light,gray} = 1.3$, $E_{gray,gray} = 1.6$. Sorting of pre-stalk cells to the front is facilitated when the adhesion between pre-spore and pre-stalk cells differs.

force required to move past the slower cells comes mostly from the drag from sliding past them. This could explain why streams of cells are often observed moving up the center of the slug (Odell and Bonner, 1986).

Conclusions

The objective of this chapter was to demonstrate that models provide an invaluable tool for exploring how cells in multicellular systems interact to produce collective cell movements. These cell movements arise from the behavior of individual cells in response to the chemical and mechanical environment they experience. Because of the complex cell-cell interactions, the collective motion is often quite different from the motion of isolated individuals, as evidenced by tissue movements during embryogenesis (Gilbert, 1997) or in the slug stage of *Dictyostelium*. I introduced a model that simulated, in 3-D, the movement of cells in multicellular systems and enabled visualization of the results. I used this model to explore cell-sorting and movement, cAMP wave propagation in early aggregation, and slug movement in *Dd*. I showed a number of simulations that highlighted different phenomena in cell-cell communication and interaction, and I also made predictions and compared some of the results with experimental findings.

A comparison of my simulations with experimental observations will help determine the most important factors controlling the movement of cells. Many of these simulations were relatively simple to do with the model, whereas it would be quite difficult, although still feasible, to perform those same experiments in vivo. Thus the model facilitates exploring a number of different scenarios. When interesting results are found, such as the effects of membrane and secreted PDE, the next step is to set up in vivo experiments to check the predictions of the model. This is yet another example of the benefits of mathematical modeling in biology, which have been demonstrated in the other chapters in this book.

It was shown that the simulations of cell aggregates behave as viscoelastic fluids, in agreement with experimental results from embryonic tissues. But the cell aggregate still retains cell individuality, as demonstrated in the cell-sorting simulations. Since the cells have specified size, we can study the effects of cell size and cell deformability on cell sorting and motility, which is not possible with fluid-based models. The importance of cell individuality was demonstrated for both chemotactic and adhesion-driven cell sorting, since the stiffness, cell-cell adhesion strength, and random movement of the cells significantly affected the rate and completeness of cell sorting.

The model simulations showed that the observed collective, coordinated motion of cells in *Dd*, from aggregation to slug movements, follows directly from the behavior of individual cells; no additional assumptions or mechanisms are necessary to produce motion of a slug. Using the model, I have simulated large numbers of interacting cells, accounting for both the passive interactions, due to adhesion and cell stiffness, and the active, locomotive forces, combined with cell-cell signaling. When cells inside a multicellular system actively try to move, they can do so only by pulling on or pushing off neighbor cells. Since all the forces are carefully balanced in the model, I could use it to study the difference between movement where cells get traction from the boundary and movement where they get traction from other cells.

From the aggregation simulations one can see how the stream formation comes out naturally; all that is needed is to couple the cAMP secretion with the cells. Essentially this coupling causes the cAMP secretion to be higher where the cell density is higher, and since the cells aggregate in the direction of the cAMP gradient, the cells chemotact toward higher-density regions, further increasing the density. When the active locomotive force that the cells apply is increased, the cell speed increases as well, and this affects the overall pattern and cell movements. When cell speed is high (active locomotive force is large), cells tend to move straight toward the center rather than join into streams, so streams do not really develop.

A new discovery that came from the simulations is the importance of the ratio of membrane-bound phosphodiesterase (PDE_m) to secreted phosphodiesterase (PDE_s), which has not previously been explored by either modelers or experimentalists. I demonstrated that PDE_m is necessary to make the system less sensitive to the cell density changes that occur during aggregation. When the ratio of membrane-bound to secreted PDE becomes too low, spontaneous oscillating centers form when the cell density increases during aggregation. This disrupts the aggregation and leads to the formation of many aggregation centers, giving rise to smaller territories that might have an evolutionary disadvantage. Aggregation under the standard conditions proceeds normally even without any secreted PDE.

So why do the cells have external PDE? Possibly the secreted PDE is needed in early aggregation for low cell densities. Recall that cAMP waves can propagate only when the average global PDE activity is within a limited range (Palsson and Cox, 1997). If the activity is too high, the cAMP is degraded too fast and no waves propagate, and if the activity is too low, the cAMP concentration reaches a high steady state, which is not excitable, and thus, again, no waves propagate. In low densities, the local PDE activity (around each cell) must be high in order to achieve global PDE activity within the correct range. This prevents any cAMP pulse initiation around the cells, and no waves form. If the local PDE activity is lowered enough to allow cAMP pulse initiation, then the global PDE activity becomes too low and no cAMP waves can propagate. This problem could be rectified by a secreted PDE, thus raising the global PDE activity above some minimum level. A future use of the model will be to get a more detailed understanding of the interaction between the membrane-bound and secreted PDE.

Another important discovery and understanding comes from the cell-sorting simulations in the 2-D slug, where the pre-stalk cells have a 75 percent larger chemotactic force than the pre-spore cells. I demonstrated how important cell-specific adhesion is for the sorting time of the pre-stalk cell to the front of the slug. From these results I suggest that chemotaxis alone is not sufficient to achieve the observed sorting in *Dd* in a timely manner, but that combined with cell-specific adhesion, the pre-stalk cells sort to the front of the slug in a rapid manner, and distinct pre-spore/pre-stalk boundaries are formed, as observed in the *Dd* slug. These observations suggest why the pre-stalk and pre-spore cells exhibit cell-specific adhesion (Lam et al., 1981), since it may be necessary in order to achieve proper separation of the two cell types in the mound and the slug stages of *Dd*.

A specific focus is studying the cell movement and sorting of pre-spore and pre-stalk cells in the late mound and slug stages of *Dictyostelium*. We want to explore, in the full 3-D system, the interaction of cAMP signal propagation and cell movements, as well as what type of signal propagation can give rise to the observed cell movement (Rietdorf et al., 1996; Siegert and Weijer, 1991). In this system the movement of cells has been observed using confocal microscopy, but so far the spatial distribution and profiles of the chemotactic signals can't be measured inside the slug. So here the model can help us by showing what kind of signal profiles can give rise to the observed cell movements. Being able to visualize the movement and trajectories of individual cells can help us gain understanding of and insight into how cell movement and sorting transpire and the effect that external or internal factors have. This application of the model can be useful in basic research as well as for teaching purposes.

References

Alberts B, Bray D, Lewis J, Raff M, Roberts K, Watson JD (1994) Molecular Biology of the Cell (3d ed). New York: Garland.

Alcantara F, Monk M (1974) Signal propagation during aggregation in the slime mould *Dictyostelium discoideum*. J Gen Microbiol 85: 321–334.

Bonner JT (1967) The Cellular Slime Molds (2d ed, rev. and enl.). Princeton, N.J.: Princeton University Press.

Bonner JT (1998) A way of following individual cells in the migrating slugs of *Dictyostelium discoideum*. Proc Natl Acad Sci USA 95: 9355–9359.

Bray, D. 1992. Cell Movements. New York: Garland.

Bretschneider T, Vasiev B, Cornelis JW (1999) A model for *Dictyostelium* slug movement. J Theoret Biol 199(2): 125–136.

Chien S, Schmid-Schonbein GW, Sung KLP, Schmaizer EA, Skalak R (1984) Viscoelastic properties of leukocytes. In: White Cell Mechanics: Basic Science and Clinical Aspects, (Meiselman HJ, Lichtman MA, LaCelle PL, eds), 3–18. New York: A. R. Liss.

Childress S, Percus JK (1981) Nonlinear aspects of chemotaxis. Math Biosci 56: 217–237.

Davidenko JM, Pertsov AV, Salomonsz R, Baxter W, Jalife J (1992) Stationary and drifting spiral waves of excitation in isolated cardiac muscle. Nature 355(6358): 349–351.

Elson EL, Felder SF, Jay PY, Kolodney MS, Pasternak C (1999) Forces in cell locomotion. Biochem Soc Symp 65: 299–314.

Evans EA (1985a) Detailed mechanics of membrane-membrane adhesion and separation, I. Continuum of molecular cross-bridges. Biophys J 48: 175–183.

Evans EA (1985b) Detailed mechanics of membrane-membrane adhesion and separation, II. Discrete kinetically trapped molecular cross-bridges. Biophys J 48: 185–192.

Forgacs G, Foty RA, Shafrir Y, Steinberg MS (1998) Viscoelastic properties of living embryonic tissues: A quantitative study. Biophys J 74: 2227–2234.

Foty RA, Pfleger CM, Forgacs G, Steinberg MS (1996) Surface tensions of embryonic tissues predict their mutual envelopment behavior. Development 122: 1611–1620.

Fuchs M, Jones MK, Williams KL (1993) Characterisation of an epithelium-like layer of cells in the multicellular *Dictyostelium discoideum* slug. J Cell Sci 105: 243–253.

Gilbert SF (1997) Developmental Biology (5th ed). Sunderland, Mass.: Sinauer.

Glazier J, Graner F (1993) Simulations of the differential adhesion driven rearrangement of biological cells. Phys Rev E47: 2128–2154.

Graner F, Glazier J (1992) Simulations of biological cell sorting using a two-dimensional extended Potts model. Phys Rev Let 69: 2013–2016.

Gross JD, Peacey MJ, Trevan DJ (1976) Signal emission and signal propagation during early aggregation in *Dictyostelium discoideum*. J Cell Sci 22: 645–656.

Inouye K, Takeuchi I (1980) Motive force of the migrating pseudoplasmodium of the cellular slime mold *Dictostelium disoideum*. J Cell Sci 41: 53–64.

Jaffe LF (1995) Calcium waves and development. Ciba Found Symp 188: 4–12.

Jalife J, Antzelevitch C (1979) Phase resetting and annihilation of pacemaker activity in cardiac tissue. Science 206(4419): 695–697.

Konijn TM, van de Meene JGC, Bonner JT, Barkley DS (1967) The acrasin activity of adenosine-3′,5′-cyclic phosphate. Proc Natl Acad Sci USA 58: 1152–1154.

Lam TY, Pickering G, Geltosky J, Siu CH (1981) Differential cell cohesiveness expressed by prespore and prestalk cells of *Dictyostelium discoideum*. Differentiation 20: 22–28.

Ludwig D, Jones DD, Holling CS (1978) Qualitative analysis of insect outbreak systems: The spruce budworm and forest. J Anim Ecol 47: 315–332.

Maree AFM, Hogeweg P (2002) Modelling *Dictyostelium discoideum* morphogenesis: The culmination. Bull Math Biol 64: 327–353.

Martiel JL, Goldbeter A (1987) A model based on receptor desensitization for cyclic AMP signaling in *Dictyostelium* cells. Biophys J 52: 807–828.

Mochizuki A, Iwasa Y, Takeda Y (1996) A stochastic model for cell sorting and measuring cell-cell adhesion. J Theor Biol 179: 129–146.

Odell GM, Bonner J (1986) How the *dictyostelium dicoideum* grex crawls. Philos Trans Roy Soc Lond B312: 487–525.

Odell GM, Oster G, Alberch P, Burnside B (1981) The mechanical basis of morphogenesis. I. Epithelial folding and invagination. Dev Biol 85(2): 446–462.

Palsson E (2001) A three-dimensional model of cell movement in multicellular systems. Fut Gen Comput Sys 17: 835–852.

Palsson E (2006) Study of cell sorting and movement in multicellular systems due to chemotaxis and cell adhesion, using a three dimensional model of cell movement. Submitted to Biophys J.

Palsson E, Cox EC (1996) Origin and evolution of circular waves and spirals in *Dictyostelium discoideum* territories. Proc Natl Acad Sci USA 93: 1151–1155.

Palsson E, Cox EC (1997) On the origin of spiral waves in aggregating *Dictyostelium*. In: Dictyostelium: A Model System for Cell and Developmental Biology (Maeda Y, Inouye K, Takeuchi I, eds), 411–423. Tokyo: Universal Academy Press.

Palsson E, Lee KJ, Goldstein RE, Franke J, Kessin RH, Cox EC (1997) Selection for spiral waves in the social amoebae *Dictyostelium*. Proc Natl Acad Sci USA 94: 13719–13723.

Palsson E, Othmer H (2000) A model for individual and collective cell movement in *Dictyostelium discoideum*. Proc Natl Acad Sci USA 97: 10448–10453.

Phillips HM, Steinberg MS (1978) Embryonic tissues as elasticoviscous liquids. I. Rapid and slow shape changes in centrifuged cell aggregates. J Cell Sci 30: 1–20.

Rietdorf J, Siegert F, Weijer CJ (1996) Analysis of optical density wave propagation and cell movement during mound formation in *Dictyostelium discoideum*. Dev Biol 177: 427–438.

Savill N, Hogeweg P (1997) Modeling morphogenesis: From single cells to crawling slugs. J Theor Biol 184: 229–235.

Sheetz MP, Felsenfeld D, Galbraith CG, Choquet D (1999) Cell migration as a five-step cycle. Biochem Soc Symp 65: 233–243.

Siegert F, Steinbock O, Weijer CJ, Muller SC (1992) Three-dimensional autowaves control cell motion in *Dictyostelium* slugs. In: Spatio-Temporal Organization in Nonequilibrium Systems (Muller SC, Plesser T, eds), 241–243. Dortmund, Germany: Projekt Verlag.

Siegert F, Weijer CJ (1991) Analysis of optical density wave propagation and cell movement in the cellular slime mould *Dictyostelium discoideum*. Physica D49: 224–232.

Skalak R, Chien S, Schmid-Schönbein GB (1984) Viscoelastic deformation of white cells: Theory and analysis. In: White Cell Mechanics: Basic Science and Clinical Aspects (Meiselman HJ, Lictman MA, LaCelle PL, eds), p. 3–18. New York: A. R. Liss.

Small JV (1989) Microfilament-based motility in non-muscle cells. Curr Opin Cell Biol 1: 75–79.

Steinberg MS (1975) Adhesion-guided multicellular assembly: A commentary upon the postulates, real and imagined, of the differential adhesion hypothesis, with special attention to computer simulations of cell sorting. J Theor Biol 55: 431–443.

Sternfeld J, David CN (1982) Fate and regulation of anterior-like cells in *Dictyostelium* slugs. Dev Biol 93: 111–118.

Sulsky D, Childress S, Percus JK (1984) A model of cell sorting. J Theor Biol 106: 275–301.

Takeichi M (1993) Cadherins in cancer: Implications for invasion and metastasis. Curr Opin Cell Biol 5: 806–811.

Takeuchi I, Kakutani T, Tasaka M (1988) Cell behaviour during formation of prestalk/prespore pattern in submerged agglomerates of *Dictyostelium discoideum*. Dev Genet 9: 607–614.

Tasaka M, Takeuchi I (1979) Sorting out behaviour of disaggregated cells in the absence of morphogenesis in *Dictyostelium discoideum*. J Embryol Exp Morp 49: 89–102.

Taylor DL, Heiple J, Wang YL, Luna EJ, Tanasugarn L, Brier J, Swanson J, Fechheimer M, Amato P, Rockwell M, Daley G (1982) Cellular and molecular aspects of amoeboid movement. CSH Symp Quant Biol 46: 101–111.

Umeda T, Inouye K (1999) Theoretical model for morphogenesis and cell sorting in *Dictyostelium discoideum*. Physica D126: 189–200.

Usami S, Wung S-L, Skierczynski BA, Skalak R, Chien S (1992) Locomotion forces generated by a polymorphonuclear leukocyte. Biophys J Biophys Soc 63: 1663–1666.

Vasiev B, Siegert F, Weijer CJ (1997) A hydrodynamic model for *Dictyostelium discoideum* mound formation. J Theor Biol 184: 441–450.

Weliky M, Oster G (1990) The mechanical basis of cell rearrangement. Development 109: 373–386.

Wessels D, Vawter-Hugart H, Murray J, Soll DR (1994) Three-dimensional dynamics of pseudopod formation and the regulation of turning during the motility cycle of *Dictyostelium*. Cell Motil Cytoskel 27: 1–12.

Wessels D, Voss E, Bergen NV, Burns R, Stites J, Soll DR (1998) A computer-assisted system for reconstructing and interpreting the dynamic three-dimensional relationships of the outer surface, nucleus, and pseudopods of crawling cells. Cell Motil Cytoskel 41: 225–246.

Wu L, Franke J, Blanton RL, Podgorski GJ, Kessin RH (1995) The phosphodiesterase secreted by prestalk cells is necessary for *Dictyostelium* morphogenesis. Dev Biol 167: 1–8.

Zhu C, Williams TE, Delobel J, Xia D, Offerman MK (1994) A cell-cell adhesion model for the analysis of micropipette experiments. In: Cell Mechanics and Cellular Engineering (Mow VC, Guilak F, Tran-Son-Tay R, Hochmuth RM, eds), 160–180. New York: Springer.

10 Modeling Neuronal Growth and Shape

Jaap van Pelt and Harry B. M. Uylings

Neurons are the cells of the nervous system involved in the processing of information. With their highly branched axons and dendrites they exchange electrical signals through a dense network of synaptic connections. Incoming electrical signals at an individual neuron are integrated by the dendritic trees, and trigger this nerve cell to generate new electrical signals that are relayed via the axonal processes to other neurons. The information-processing characteristics of a neuron depend on this signal integration process, which is determined by the shapes of dendrites and their electrophysiological properties (e.g., Mason and Larkman, 1990). Dendrites have a limited extension in space, while axons may run over long distances. Neurons appear in the nervous system with an enormous diversity in shapes, which is thought to contribute to their functional specialization. An illustration of this diversity is given in figure 10.1, which shows dendritic shape differentiation in length scales and in the details of the branching patterns.

Neurons attain their shapes through a developmental process of elongation and branching of their axonal and dendritic extensions (collectively called neurites). Shape differentiation must therefore find its origin in the details of this developmental process. How shape characteristics of neurons reflect developmental information is nevertheless a question difficult to solve, not only because of the complexity of dendritic shapes but also because of the dynamic process of dendritic development. Theoretical and computational modeling approaches are required for the quantification of dendritic shapes and of the developmental process through which these shapes emerge.

Modeling in Biology

In order to gain understanding of nervous system development and function, experimental work needs to be complemented by theoretical analysis and computer simulation. Even for biological systems in which all the components are known,

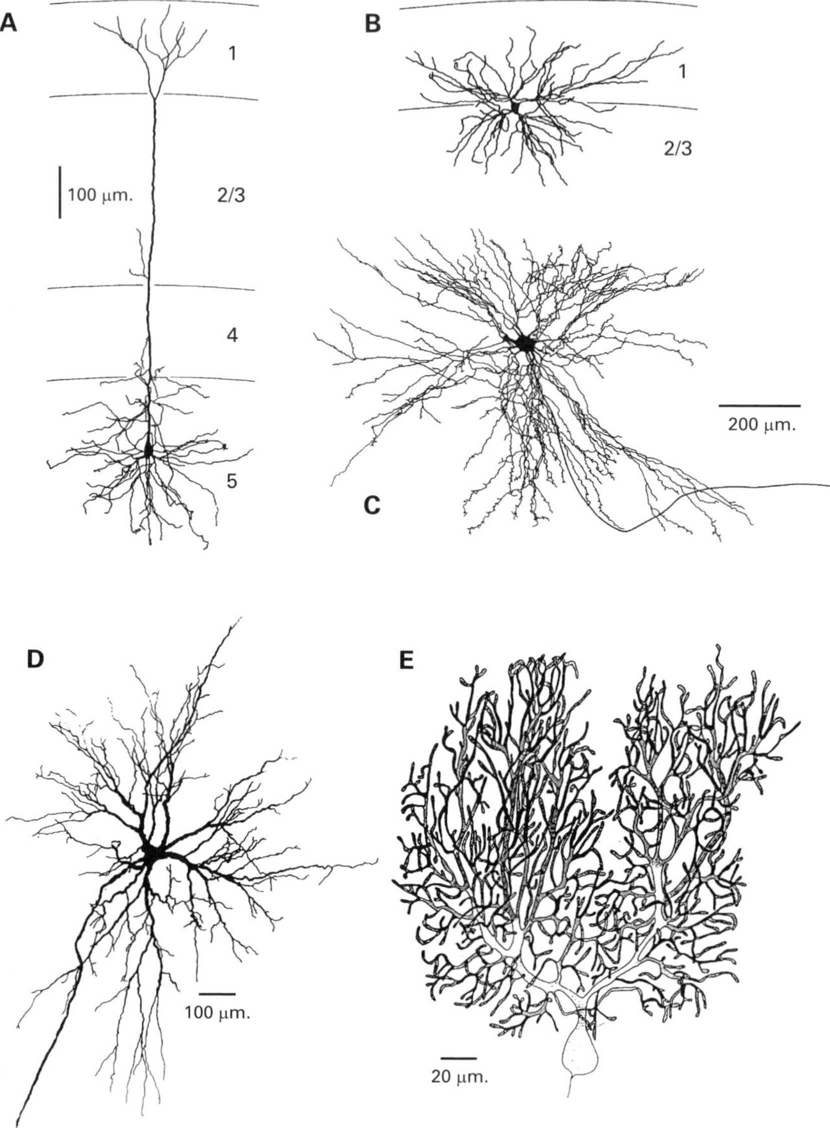

Figure 10.1
Neuronal branching patterns from different cell types in the nervous system. (A) rat cortical layer 5 pyramidal cells. (Adapted from Larkman and Mason, 1990.) (B) rat cortical layer 2/3 pyramidal cells. (Adapted from Larkman and Mason, 1990.) (C) newborn kitten α-motoneuron. (Adapted from Uhlfhake and Cullheim, 1988.) (D) cat deep superior colliculus neurons. (Adapted from Schierwagen and Grantyn, 1986; Schierwagen, 1986.) (E) rat cerebellum Purkinje cell. (Adapted from Berry and Bradley, 1976b.)

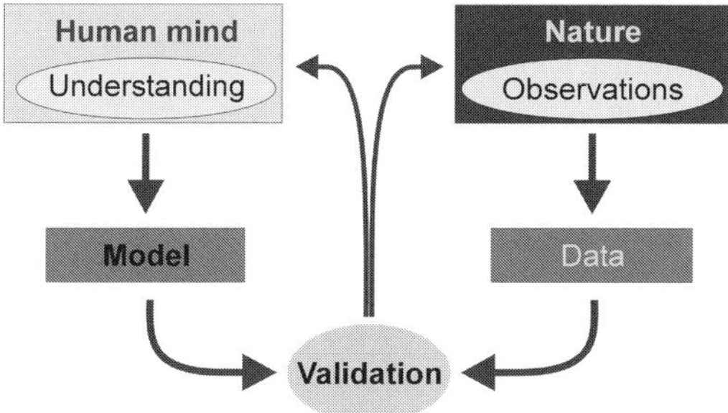

Figure 10.2
Role of a model as a tool to implement understanding from the human mind, to express hypotheses, and to validate predictions against experimental data originating from observations of nature.

computational models are necessary to explore and understand how the components interact to make the system work, and how phenomena at different levels of organization or description are linked.

A model can be considered as a theoretical/computational tool to quantitatively implement knowledge and hypotheses about a system and to enable the study of the properties of that system. Modeling approaches depend strongly on the questions being raised (see also chapter 2, this volume). Knowledge and hypotheses are products of the human mind. They originate from human understanding of nature but need models for quantitative implementation. Nature is explored through human observations and is represented by quantitative data. Model validation of the experimental data provides the basis for understanding the data, thereby bridging human mind and nature (figure 10.2).

Biological systems are complex at any level of their structural and functional organization, from the molecular up to the behavioral level. They consequently represent systems of (almost) infinite dimensionality. Building models of biological systems thus requires a well-thought-out approach to reduce this dimensionality to a meaningful and treatable level. Starting with a focus on selected mechanisms or phenomena, the level of dimensional reduction strongly depends on the particular research question, and may range from the most reductionist "conceptual" models to complex, but still computational treatable models. For instance, a typical method to reduce dimensionality is the statistical one in which the many degrees of freedom are lumped into a single probability function and the system is regarded as stochastic (Monte Carlo approach). Other modeling approaches may integrate all available knowledge, aiming at studying the fine structure of the system's behavior. Biophysical models aim at understanding a well-defined system through the underlying biophysical laws of nature.

Biological systems distinguish themselves from physical systems in that they show organization at all levels of spatial and temporal scales—apparent, for instance, in genetic regulation of protein synthesis at the molecular level and in feedback, homeostatic, and adaptive mechanisms throughout the organism. Of course, biological systems are subjected to (bio)physical constraints which, for instance, become apparent in growth cone behavior when conservation laws result in limiting resources and competitive phenomena during neurite outgrowth.

Biology of Neurite Outgrowth and Neuronal Morphogenesis

Neurite Outgrowth and Branching

Neurons grow out by a process of elongation and branching of their neuritic extensions. These processes are governed by the dynamic behavior of growth cones, highly motile structures at the tips of outgrowing neurites. Neurite elongation requires the polymerization of tubulin into cytoskeletal microtubules. Branching occurs when a growth cone, including its actin cytoskeletal meshwork, splits in two parts while partitioning the number of microtubules, and each daughter growth cone continues to grow on its own (e.g., Black, 1994; Kater et al., 1994; Kater and Rehder, 1995; Letourneau et al. 1994). Growth cones may also retract by microtubule depolymerization, or may redirect their orientations. The dynamic behavior of a growth cone results from the integrated outcome of many intracellular mechanisms and interactions with its local environment. These include the exploratory interactions of its filopodia with a variety of extracellular (e.g., guidance, chemorepellant, chemoattractant) molecules in the local environment (e.g. Kater et al., 1994; McAllister, 2002; Whitford et al., 2002), receptor-mediated transmembrane signaling to cytoplasmic regulatory systems (e.g., Letourneau et al., 1994), intracellular regulatory genetic and molecular signaling pathways (e.g., Song and Poo, 2001), and electrical activity (e.g., Cline, 1999; Ramakers et al., 1998, 2001; Zhang and Poo, 2001).

Microtubule Cytoskeleton

In addition to regulatory mechanisms, dendritic outgrowth is also subjected to constraints limiting modulatory responsiveness. For instance, neurite elongation proceeds by growing mitrotubules, which requires the production, transport, and polymerization of tubulin cytoskeletal elements. The production rate of cytoskeletal elements will set an upper limit to the averaged total length increase of the dendrite. In addition, the division of flow of cytoskeletal elements at a bifurcation will modulate the elongation rate of the daughter branches. Earlier model studies of microtubule polymerization (neurite elongation) in relation to tubulin production and transport demonstrated how limited supply conditions may lead to competition between growth cones for tubulin, resulting in alternating advance and immobilization (Van Ooyen, 2001). Empirical evidence for such competitive behavior has been obtained from time-lapse studies of outgrowing neurons in tissue culture by Ramakers (see Costa et al., 2002).

Synapse Stabilization

Limiting resources may also lead to competitive phenomena between axons, where target-derived neurotrophins are required for the maintenance and stabilization of synaptic axonal endings on target dendrites. Modeling studies have shown that the precise manner in which neurotrophins regulate the growth of axons determines which patterns of target innervation can develop (Van Ooyen and Willshaw, 1999). Taking a spatial dimension into account, these authors show that the distance between axonal endings on the target dendritic tree mitigates competition and permits the coexistence of axons (Van Ooyen and Willshaw, 2000). These examples illustrate the complexity of the outgrowth process in terms of involved molecules, interactions, signaling pathways, and constraints.

The integrated action and the details of all these intracellular and extracellular mechanisms will finally form the basis of dendritic morphological characteristics, of morphological differentiation between cell types, and of the diversity between neurons of the same type (e.g., Acebes and Ferrus, 2000).

Modeling Neuronal Shape

The complexity of dendritic shape has for a long time been a challenge to computational modelers to extract regularities or common features and to synthesize realistic structures. Present studies show high levels of sophistication in the realism of reconstruction of dendritic trees, including the 3-D orientation of branches and environmental influences (e.g., Ascoli et al., 2001; Ascoli 2002a, 2002b; chapter 6 in this volume). An analytical approach was used by Kliemann (1987), Carriquiry et al. (1991), and Uemura et al. (1995), who considered the segments at a given centrifugal order as individuals of a generation which may give rise to individuals in a next generation (by producing a bifurcation point with two daughter segments). Mathematically, such a dendritic reconstruction could be described by a Galton-Watson process, based on empirically obtained splitting probabilities for defining per generation whether or not a segment will be a terminal or an intermediate one. Tamori (1993) used a set of six "fundamental parameters" to describe dendritic morphology, and analyzed branch angles using a principle of least effective volume. Growth models have been applied by Smit et al. (1972), Berry and Bradley (1976a), Van Pelt and Verwer (1983), Ireland et al. (1985), Horsfield et al. (1987), and Nowakowski et al. (1992). The growth model reviewed in this chapter is developed to answer the question of whether morphological details and variability observed in dendritic branching patterns can be understood quantitatively in terms of a minimal set of growth rules.

Parameterization of Dendritic Shape

A first step requires the parameterization of dendritic shape (i.e., the reduction of morphological complexity into a set of well-defined shape parameters). Dendritic shape can be quantified in terms of number of branch points, number of segments, length of segments, connectivity pattern of segments (topological structure), segment diameters, curvature of segments, and embedding in 3-D space. Experimentally reconstructed dendrites show characteristics and variability in all these measures (e.g., Ascoli et al., 2001; Uylings and Van Pelt, 2002; Scorcioni et al., 2004; chapter 6 in this volume) which require modeling studies to make specific choices on shape parameters to be investigated.

Assumptions and Structure of the Model

The second step is to develop the model structure on the basis of well-defined assumptions. In our approach, dendrites are described as binary rooted trees. Intermediate and terminal segments are distinguished and labeled with a centrifugal order indicating the number of segments on the path from the root to the segment. As discussed above, dendrites develop from initial protrusions from the cell body into branched and elongated structures through dynamic actions of growth cones. Although these actions may include forward migration and branching, but also retraction or complete disappearance, at a sufficiently coarse time scale the net outcome is still a growing dendrite with increasing length and number of branches, as is schematically illustrated in figure 10.3.

In the model description a phenomenological approach is used in which the intra- and extracellular mechanisms and processes themselves are not modeled but rather their final outcome. To this end, the detailed dynamic actions of outgrowing neurites are coarse-grained and modeled by sustained processes of elongation and branching. The multitude of mechanisms involved in growth cone behavior (see preceding section) justifies a stochastic description in which terminal segments elongate and branch with given probabilities per unit of time. These elongation and branching probability functions form the basic components of the model. The challenge is to find minimal but sufficient expressions for these functions to describe morphological characteristics and variability. When the search is successful, these functions reflect important biological properties in the development of neuronal morphology. Branching and elongation are assumed to be independent, making it possible to study the two processes separately.

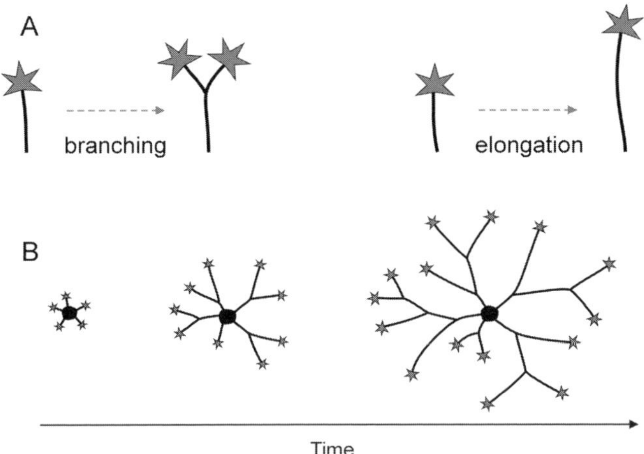

Figure 10.3
(A) Basic actions of a growth cone, splitting of a growth cone resulting in neurite branching, and migration of a growth cone, resulting in neuritic elongation. (B) Schematic illustration of the branching and elongation process mediated by growth cones. Initial protrusions from the cell body develop into mature dendritic trees through a process of growth cone elongation and branching.

Variation in Topological Tree Types

To study the variation in the connectivity patterns of segments, dendrites will be reduced to binary tree types, ignoring any metrical property, as is illustrated in figure 10.4. Only a finite number of tree types are possible for a given number of segments. A branching event in a given tree increases its number of branch points and segments by one. Dependent on the branching segment, the given tree transforms into one of the tree types in the next row in figure 10.4. Not all transitions are possible between tree types in successive rows. The dashed arrows in figure 10.4A denote all possible transitions when branching is allowed at any segment in a tree. When branching is restricted to terminal segments (as in the dendritic growth model), this will further reduce the number of possible transitions (figure 10.4B).

In the stochastic branching process, each segment in a tree has a given probability to be selected for branching, which will add to the probability of the transition of that tree to one of its successors (outgoing dashed arrows for a given tree type in figure 10.4). The figure illustrates that different paths starting from the single segment may end in a particular tree type. A given number of branching events will result in one of the trees in a row (with a number of terminal segments equal to 1 plus the number of branching events). Thus, the probability that a given tree type is obtained results from the probabilities of all the possible paths, and thus of all the transitions (growth steps) in these paths. Both the number of possible paths and the probability

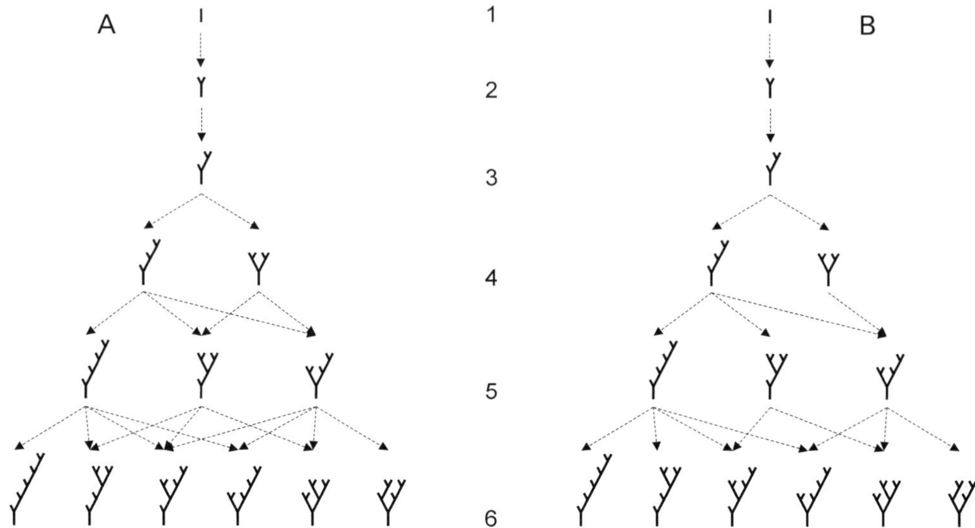

Figure 10.4
Possible transitions between topological tree types by branching (A) of any of the segments in the tree, or (B) of one of the terminal segments only.

of each growth step depend on the specific rules for branching of the individual segments in the branching tree. The eventual probability distribution of tree types can be obtained by enumerating all the growth paths and their associated probabilities.

In a series of studies these probabilities have been calculated for a number of branching rules: (1) random segmental branching when each segment has the same probability to be selected for branching; (2) random terminal branching when only terminal segments can be selected for branching with equal probabilities; and (3) a branching scheme where the probability of branching is modulated by the position of the segment in the tree (e.g., Van Pelt and Verwer, 1986). These studies have demonstrated how the topological variation between tree types depends on the specific rules for branching of the individual segments (e.g., figure 10.5).

In an extensive study of experimental reconstructed motoneurons Dityatev et al. (1995) showed that the model, as simple as it is, successfully described the topological variability in a large variety of motoneuronal dendrites. Additionally, the analysis of reconstructed dendritic branching patterns has shown that their topological variation is consistent with the terminal branching mode, which is in agreement with the process of neurite outgrowth where growth cones at the neuritic tips are the sites of branching. The analysis of tree-type frequencies in terms of the mode of branching can be applied to other natural branching patterns as well, such as river systems. Where the branching process in river systems is more difficult to investigate, given

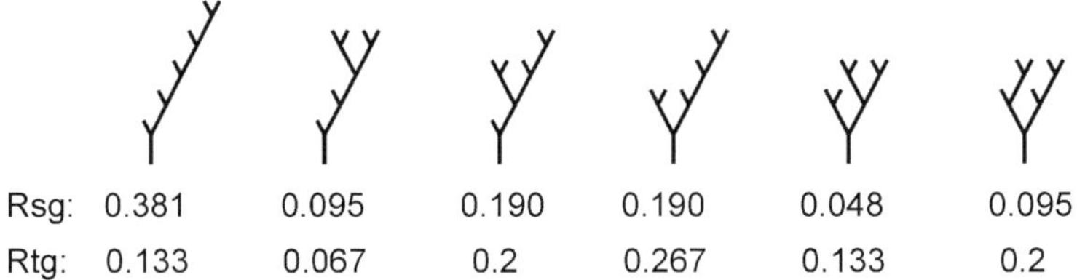

Figure 10.5
Probabilities of tree types grown with the random segmental growth (rsg) and the random terminal growth (rtg) modes of branching. (Van Pelt and Verwer, 1983.)

the time scale of development, the topological analysis suggested a process of terminal branching with a preference for the distal parts of the river as an alternative to the most studied random segmental growth mode (Van Pelt et al., 1989). This finding supported the headward network growth hypothesis adopted by many geomorphologists.

Variation in the Number of Terminal Segments

The probability distribution of topological tree types has been obtained for a growth process with a given number of branching events. The occurrence of a branching event, however, is stochastic, implying that after a given time interval, trees have been obtained with a varying number of terminal segments. For the calculation of the terminal segment number distribution in a population of trees, we will ignore the topological differentiation and label a tree only by its terminal segment number. This process requires a description in time with branching probabilities of segments defined as functions of time. In addition, we assume that the branching probability per unit of time of a terminal segment may depend on the momentary number of terminal segments in the growing tree. The stochastic nature of the branching process implies that a varying number of branching events occurs in a growing tree in a given time interval. This is illustrated in figure 10.6, which shows that trees with different numbers of terminal segments at time t may give rise to trees with n terminal segments at time $t + \Delta t$, depending on the number of branching events in time period Δt. All these possibilities contribute to the probability of trees with n terminal segments at time t.

For the calculation of the distribution, one needs to include all the possible growth sequences in a certain period of time. To this end Van Pelt and Uylings (2002) used the recurrent equation

$$P_{tree}(n, t + \Delta t) = \sum_{j=0}^{n/2} P_{tree}(n - j, t) \cdot \binom{n - j}{j} \cdot [\bar{p}_s(t \,|\, n - j)\, \Delta t]^j \cdot [(1 - \bar{p}_s(t \,|\, n - j))\, \Delta t]^{n-2j}, \tag{10.1}$$

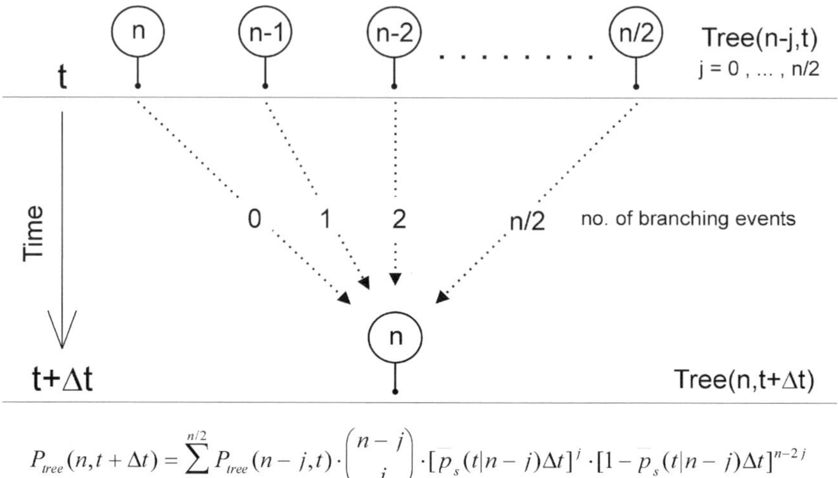

$$P_{tree}(n,t+\Delta t) = \sum_{j=0}^{n/2} P_{tree}(n-j,t) \cdot \binom{n-j}{j} \cdot [\bar{p}_s(t|n-j)\Delta t]^j \cdot [1-\bar{p}_s(t|n-j)\Delta t]^{n-2j}$$

Figure 10.6
Trees with n terminal segments may originate from trees of different sizes, depending on the number of branching events per time interval.

with $P_{tree}(n, t)$ being the probability of having a tree with n terminal segments at time t, and $\bar{p}_s(t \mid n)$ denoting the probability of branching of a terminal segment, averaged over all n terminal segments in the tree. A tree with n terminal segments at time $t + \Delta t$ emerges when j branching events occur in time interval $[t, t + \Delta t]$ in a tree with $n - j$ terminal segments. The recurrent equation expresses the probabilities of all these possible contributions from $j = 0, \dots, n/2$. The last two terms express the probability that in a tree with $n - j$ terminal segments, a number j of them will branch and the remaining $n - 2j$ terminal segments will not. The combinatorial coefficient $\binom{n-j}{j}$ expresses the number of possible ways of selecting j terminal segments from the existing $n - j$ ones. Note that the probability distribution of trees with varying terminal segment numbers is normalized at each time point t, and thus obeys $\sum_{n=1}^{n_{\max}} P_{tree}(n, t) = 1$, $\forall t$ with $n_{\max}$ denoting the maximal possible terminal segment number at time t.

Finally, the branching probabilities of segments per unit of time need to be defined in order to solve the recurrent equation. Here, we make some explicit assumptions. All the tips of a dendritic tree are assumed to participate in the branching process. Each tip is assumed to branch with a certain probability per unit of time, which may change during the outgrowth process. The time-dependent branching probability of a tip is assumed to consist of a time-dependent baseline branching rate function $D(t)$, a term dependent on the total number of tips n^{-E} with a free parameter E, and a term dependent on the position of the tip $2^{-S\gamma}$, with γ denoting the number of segments on the path from the root to the segment (centrifugal order), and S indicating a free parameter (this dependency was mentioned in the section on topological variation). The choice for implementing n-dependency by the term n^{-E} was made for its simplicity, its modulatory power and attractive behavior at $E = 0$, and also at $E = 1$ when, with $1/n$, the branching probability becomes inversely related to the number of segments. Then, the total branching rate in a growing dendrite is independent of its increasing number of terminal segments.

The choice for implementing position dependency by the term $2^{-S\gamma}$ was inspired by the binary nature of the branching structure, its modulatory power, and the attractive behavior at $S = 0$, and also at $S = 1$, when $2^{-\gamma}$ illustrates the distribution of some conserved quantity over the segments at different centrifugal orders. These choices for the model structure have been made in order to investigate whether the position of a segment in the tree or the total number of segments influences the branching probability of the individual segments.

$$p_s(t \mid n, \gamma) = D(t) n^{-E} 2^{-S\gamma} / C(t) \tag{10.2}$$

The position-dependent term $2^{-S\gamma}$ implements a particular branching scheme for the segments in a tree, defining the topological variation as is discussed and illustrated in the preceding section and in figure 10.4. For $S = 0$ the branching scheme becomes the random terminal mode of growth. The term $C(t)$ normalizes, at each point in (developmental) time, the position-dependency of all tips. Then the term $2^{-S\gamma}/C(t)$ contributes only to the topological variation and not to the growth rate of the tree (the increase in number of tips in the tree per unit of time). Thus, the mean branching probability of a terminal segment, averaged over all terminal segments, is given by

$$\bar{p}_s(t \mid n) = \frac{1}{n} \sum_{i=1}^{n} D(t) n^{-E} 2^{-S\gamma_i} / C(t) = D(t) n^{-E}. \tag{10.3}$$

Inserting this expression into equation (10.1) yields

$$P_{tree}(n, t + \Delta t) = \sum_{j=0}^{n/2} P_{tree}(n - j, t) \cdot \binom{n - j}{j} \cdot [D(t)(n - j)^{-E} \Delta t]^{j} \cdot [(1 - D(t)(n - j)^{-E} \Delta t)]^{n-2j}. \tag{10.4}$$

Introducing the function $B(t) = \int D(t)\, dt$, with $dB(t) = D(t).dt$, and taking $\Delta B(t) = (dB(t)/dt)\Delta t = D(t).\Delta t$, we obtain

$$P_{tree}(n, t + \Delta t) = \sum_{j=0}^{n/2} P_{tree}(n - j, t) \cdot \binom{n - j}{j} \cdot [\Delta B(t)(n - j)^{-E}]^{j} \cdot [(1 - \Delta B(t)(n - j)^{-E})]^{n-2j}. \tag{10.5}$$

Each iteration thus depends only on the value $\Delta B(t)$. For a given time period T and a given time step size Δt, we have $N = T/\Delta t$ iterations, with the function $B(t)$ incrementing with $\sum^N \Delta B(t) = B(T)$. For sufficiently small time steps Δt, the outcome of the iterations should not depend on the actual step size Δt, but only on the sum of steps $\Delta B(t)$, and thus on the value $B(T)$. This implies that the terminal segment number distribution at time $t = T$ is determined only by the function value $B(T)$, and not by the detailed shape of the function $D(t)$. The recursion can now be simplified by taking a number of N_c constant steps ΔB_c, such that $\sum^{N_c} \Delta B_c = B_c$ as

$$P_{tree}(n, t + \Delta t) = \sum_{j=0}^{n/2} P_{tree}(n - j, t) \cdot \binom{n - j}{j} \cdot [\Delta B_c(n - j)^{-E}]^j \cdot [(1 - \Delta B_c(n - j)^{-E})]^{n-2j}, \quad (10.6)$$

with a time period associated with this number of iterations equal to $T_{B_c} = \sum^{N_c} \Delta t$.

An example of the family of distributions obtained for $B = 6$ and for several values of the parameter E is given in figure 10.7.

The curves in figure 10.7 show the effect of the parameter E on the shape of the terminal segment number distribution. The value $E = 0$ results in a monotonically decreasing function (with a long tail exceeding the size of the figure), while for positive values of E a modal shape is obtained with decreasing mean and standard deviation for increasing values of E. Actually, a direct relation exists between the parameter pair (B, E) and the mean and standard deviation of the terminal segment number distributions, allowing a mapping of (B, E)-space onto (mean, *SD*)-space (Van Pelt et al., 1997). Empirical distributions of the terminal segment number do

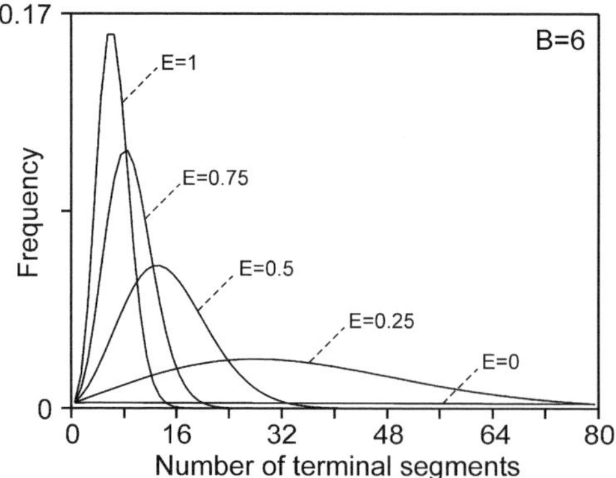

Figure 10.7
Frequency distributions of the terminal segment number of trees obtained for the parameter value $B = 6$ and for different values of the parameter E. Note that the mean and the width of the curves (SD) decrease with increasing values of E. For $E = 0$, the distribution has a monotonically decreasing shape.

not show exponential shapes. They all have a clear modal shape, indicating that a zero value for parameter E is not realistic; E should be positive. For positive values of E the branching probability of a terminal segment decreases with an increasing number of tips (eq. 10.2), suggesting some competitive mechanism in the branching process.

Empirical distributions for the terminal segment number have been analyzed for a number of different cell types. All these empirical distributions could be matched with model distributions by optimizing the values for the parameters B and E (Van Pelt et al., 1997). This is a remarkable result for a model that is based on only a few basic assumptions of the branching process. A scatter plot of the optimized parameter values in (B, E)-space showed a strong clustering per cell type, indicating that these values provide a significant characterization of cell types on the basis of the shape of their branching patterns (Van Ooyen and Van Pelt, 2002).

Growth Curves of the Number of Terminal Segments

During outgrowth the terminal segment number distribution in a population of dendrites will change with time, reflecting in its mean and standard deviation the increase in the number of terminal segments. At each point in time the shape of the distribution is determined by the parameter values B and E, with B also a function of time $(B(t) = \int D(t)\,dt)$. The growth of a population of dendrites thus can be calculated when the time course of the function $D(t)$ is known. $D(t)$, however, is unknown, and represents all factors influencing the growth process that are not covered by the dependence on the number of terminal segments. The function $D(t)$ thus represents an essential determinant of the outgrowth process.

Van Pelt and Uylings (2002) calculated dendritic growth curves for different choices of the function $D(t)$. It was shown that a constant function $D(t)$ resulted in an exponentially increasing growth curve, while an exponentially decreasing function $D(t) = ce^{-t/\tau}$ resulted in an asymptotic shape for the growth curve. Experimental data on the growth of the mean number of terminal segments in populations of dendrites at different time points during development did not show an exponential growth pattern. In fact, the experimental data were in good agreement with the growth curve predicted by the exponentially decreasing function $D(t)$. These results indicate that the baseline branching rate function $D(t)$, representing the basic drive for branching of a neuron, is not constant but rapidly decreases from the start of outgrowth (figure 10.8).

From the recursion in equation (10.1), an expression can be derived for the time pattern of the mean number of terminal segments per tree $\bar{n}(t)$, averaged over the population of trees, as $d\bar{n}(t)/dt = D(t)\overline{n^{1-E}(t)}$ (publication in preparation). In previous studies we have investigated the growth equation $dn(t)/dt = D(t)n(t)^{1-E}$ for a virtual tree with a real variable $n(t)$ for its number of terminal segments. In this equation the discrete process of adding integer numbers of terminal segments per growth

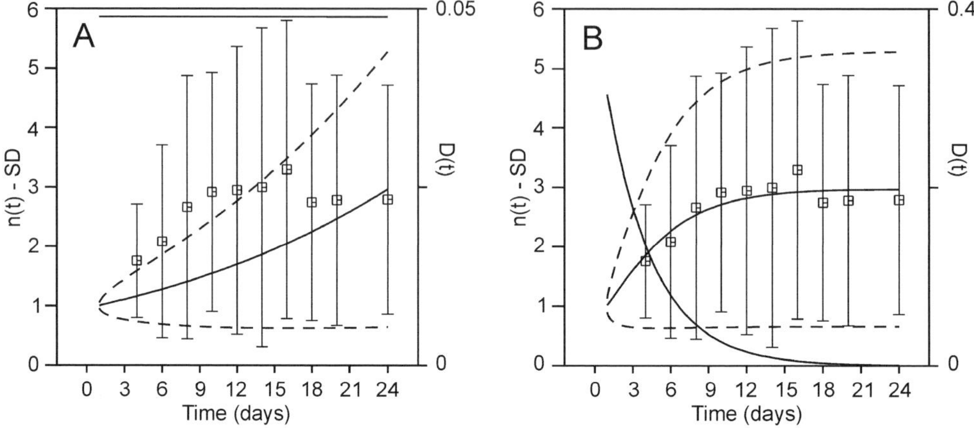

Figure 10.8
Growth curves of the population mean and standard deviation (SD) of the terminal segment number in dendritic trees during their outgrowth. (Van Pelt and Uylings, 2002.) The data points and SD bars originate from reconstructed dendrites of rat cortical multipolar nonpyramidal cells at several time points during their development. (Parnavelas and Uylings, 1980; Uylings et al., 1980.) The model growth curves (continuous curve for the mean, and dashed lines for the SD values) are calculated for parameter values $E = 0.05$ and $B = 1.12$. For panel (A) a constant baseline branching rate function is assumed (see horizontal line at top of panel), and for panel (B) an exponentially decreasing baseline branching rate function $D(t) = ce^{-t/\tau}$ with $\tau = 3.7d$ is assumed. With the latter choice an excellent agreement is obtained between model predictions and experimental data in the time course of the mean and of the SD values. Left ordinate scales refer to the number of terminal segments. Right ordinate scales refer to the baseline branching rate function.

step has been changed into a continuous process of adding infinitesimal values to a virtual number of terminal segments. The growth pattern of the virtual tree coincides with the population mean for the parameter value $E = 0$ or $E = 1$.

The accuracy with which the growth function of the virtual tree approximates the population mean growth function for other values of the parameter E is presently under study. The analytical treatability of the virtual tree growth function made it possible to quantitatively investigate the effect of different baseline branching rate functions on the virtual tree growth patterns (Van Pelt and Uylings, 2002, 2003). These studies showed that even decreasing power functions for the baseline branching rate were not able to suppress the growth rate of the tree to zero values. Indeed, exponential functions were able to compensate for the proliferation of branching segments and force the tree growth rate to zero values, thus terminating the growth phase.

Neurite Elongation

The lengths of segments in a dendritic tree are determined by both the branching rates and the elongation rates of neurites. The distribution of segment lengths in a

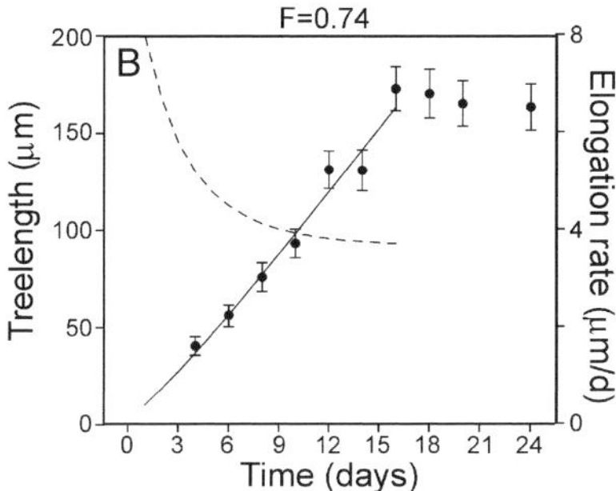

Figure 10.9
Mean and SEM (data points) of the total length of dendritic trees of rat cortical multipolar nonpyramidal cells at several time points during development. (Parnavelas and Uylings, 1980; Uylings et al., 1980.) The tree length growth function (solid line) was obtained by using the optimized parameter values for the branching process and by fitting the tree growth length function to the data points, resulting in optimized parameter values $L_0 = 10$ μm, $F = 0.73$ and $v_0 = 8.1$ μm/day. (Van Pelt and Uylings, 2003.) The corresponding elongation rate function $v(t) = v_0 n(t)^{-F}$ is shown as a dashed line.

tree, and its developmental changes with time, is thus the outcome of a complex, integrated process of branching and elongation rates during development. So far we have dealt with the branching process, and we were able to derive quantitative details in terms of baseline branching rate functions and size dependencies from experimental dendritic reconstructions. This approach thus also holds promise for the challenge to derive quantitative details of the elongation process, using these model-based approaches.

In a recent study, a first attempt was made to estimate elongation rates from an experimental data set of rat multipolar nonpyramidal neurons (Van Pelt and Uylings, 2003). In particular, the question was addressed of whether elongation rates were constant during the growth of a dendrite or would reflect competitive effects because of the proliferation of elongating neurites. The elongation rate is assumed to depend on the number of terminal segments $v(t) = v_0 n(t)^{-F}$, with v_0 denoting the elongation rate of the initial segment at the start of the branching process, and with a dependence on the number of terminal segments modulated by the exponent F.

Using the virtual tree growth function for this segment number, an estimate could be formulated for the total tree length growth function,

$$L(t) = L_0 + v_0 \int_{t_0}^{t} [1 + EB_\infty(1 - e^{-(s-t_0)/\tau})]^{(1-F)/E}\, ds$$

(Van Pelt and Uylings, 2003), with L_0 denoting the initial length of daughter branches after a branching event. Comparison with the observed growth function suggested a small competitive effect in the elongation rate (figure 10.9). These rat multipolar nonpyramidal dendritic trees are, however, small—on average, about

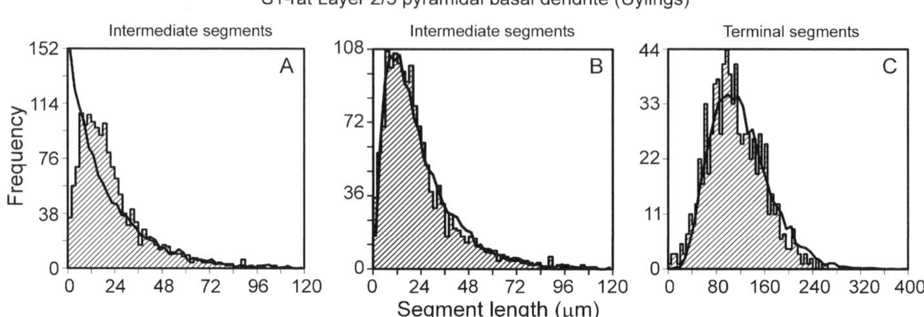

Figure 10.10
Segment lengths' distributions in S_1-rat cortical layer 2/3 pyramidal cell basal dendrites (Uylings et al., 1994) of intermediate segments (A and B, plotted with different scales) and terminal segments (C). Model predictions were obtained for optimized parameter values without (A) and with (B, C) initial lengths of daughter segments after a branching event (Van Pelt et al., 2001).

three terminal segments per dendrite. Developmental data sets of larger trees are therefore required to show more clearly any possible competitive effect on the elongation rate.

The model-based analysis of segment length distributions has also focused attention on the elongation process during and following a branching event. When a branching event is a point process in time, and the daughter branches immediately continue their outgrowth, one expects a monotonic decreasing length distribution with the highest probabilities for the shortest segments (figure 10.10A). Observed segment length distributions, however, have a modal shape in contrast to these expectations. To answer this question, it is important to realize that in the model the branching process is reduced to a point process both in time and in space. A biological growth cone, however, has a spatial extent while its branching proceeds during a certain period in time. To solve this problem, we have assumed that after a branching event in the model, the daughter branches already appear with a certain initial length (figure 10.11), resulting in a model length distribution closely matching the observed one (figure 10.10B). Alternatively, Nowakowski et al. (1992) assumed in their model description a transient reduction in the branching probability. Both approaches resulted in a good description of the intermediate segment length distribution.

Terminal segments are generally longer than intermediate ones, as shown in figure 10.10B and 10.10C. This may occur, for instance, when dendritic development proceeds in two different phases, the first one showing that neurites branch and elongate, followed by a second phase with neuritic elongation only, as is the case in rat pyramidal cell basal dendrites. Assuming different elongation rates during both phases,

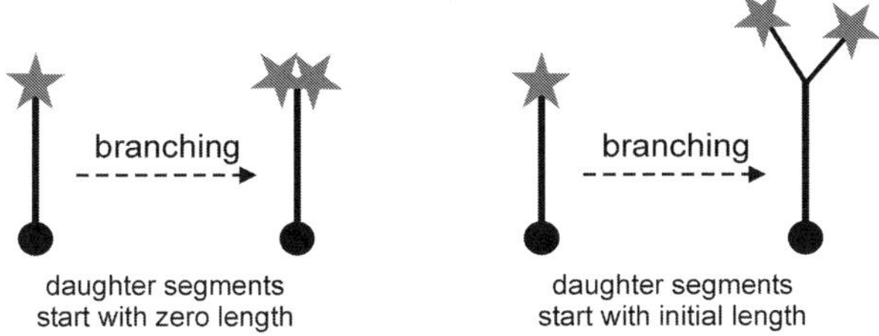

Figure 10.11
Schematic view of the initial length assumption of daughter segments after a branching event to comply with the noninfinitesimal character of branching events in space-time.

the observed segment length distributions could be reproduced accurately for these cell types. Even with constant elongation rates during dendritic development, changing branching rates will influence segment length distributions in dendritic trees in a complex, nontrivial way. For studying these questions, a good description of the branching process is required. Therefore, we are now at the beginning of quantitative exploration of the elongation rates, with the distributions of observed dendritic shape parameters as a rich source of information.

Variability in the Shape of Dendritic Trees

An impression of the shape properties and variability is given in figure 10.12, which illustrates a population of model-generated dendrograms reflecting the properties of cat deep layer superior colliculus neurons (Van Pelt et al., 2001). The examples show the variation in the number of terminal segments per tree and in the length of both intermediate and terminal segments. Segment diameters have been assigned according to the following rule. Terminal segments are first assigned a mean diameter of 1.05 μm, followed by the calculation of the diameter of a parent segment d_p from the diameters of its daughters d_1 and d_2, using the branch power rule $d_p^e = d_1^e + d_2^e$, with the branch power e randomly sampled from a normal distribution with mean = 1.47 and SD = 0.3 (Schierwagen and Grantyn, 1986).

Conclusion

Mathematical modeling contributes significantly to the understanding and the quantitative description of biological structures and processes. This has been illustrated in this chapter for the shape and variability of dendritic branching patterns. The modeling approach has demonstrated that (1) randomness in elongation and branching is sufficient to explain the variability in the ultimate shapes; (2) the branching probability depends on the momentary number of terminal segments in the growing tree; and

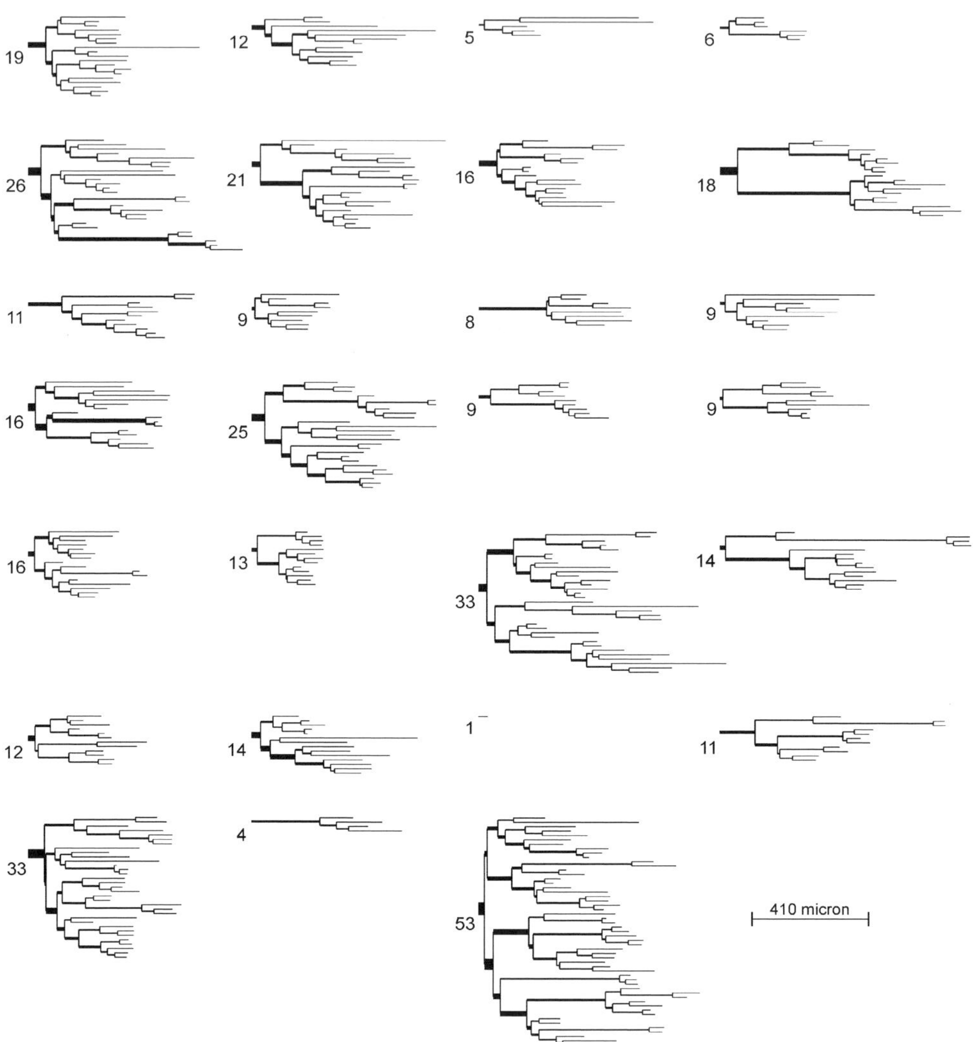

Figure 10.12
Illustration of morphological variability in a set of model dendrites generated using parameter values optimized for dendrites of cat superior colliculus neurons. (Schierwagen and Grantyn, 1986; Van Pelt et al., 2001.) Variability is shown in the number, the length, and the connectivity pattern of the segments. Terminal segment diameters have been assigned a mean value of 1.05 μm. Parent segments have been assigned a diameter using a power law relation with the daughter segment diameters, with a branch power randomly drawn from a normal distribution (see text). The number of terminal segments of each tree is indicated at the root of the tree.

(3) the branching probability depends on an intrinsic component (baseline branching rate function) that rapidly (as strongly as exponentially) decreases with time of development. Ongoing studies of segment length distributions are expected to clarify whether neurite elongation rates are also subjected to competitive influences due to the proliferation of the number of segments in the growing tree. For these modeling studies dendritic shape complexity was reduced to skeleton trees maintaining segment number, segment length, and segment connectivity (topology).

The conclusions from these modeling studies need to be followed by experimental investigations of the cell biological mechanisms underlying the baseline branching rate and the dependencies on the number of terminal segments. Clearly, the modeling approach was essential to unravel the influence of the branching and elongation processes on the segment length structure in dendritic trees.

References

Acebes A, Ferrus A (2000) Cellular and molecular features of axon collaterals and dendrites. Trends Neurosci 23: 557–565.

Ascoli GA (2002a) Neuroanatomical algorithms for dendritic modeling. Network 13: 247–260.

Ascoli GA (ed) (2002b) Computational Neuroanatomy: Principles and Methods. Totowa, N.J.: Humana Press.

Ascoli GA, Krichmar JL, Scorcioni R, Nasuto SJ, Senft SL (2001) Computer generation and quantitative morphometric analysis of virtual neurons. Anat Embryol 204: 283–301.

Berry M, Bradley PM (1976a) The application of network analysis to the study of branching patterns of large dendritic fields. Brain Res 109: 111–132.

Berry M, Bradley P (1976b) The growth of the dendritic trees of Purkinje cells in the cerebellum of the rat. Brain Res 112: 1–35.

Black MM (1994) Microtubule transport and assembly cooperate to generate the microtubule array of growing axons. In: The Self-Organizing Brain: From Growth Cones to Functional Networks (Van Pelt J, Corner MA, Uylings HBM, Lopes da Silva FH, eds), 61–77. Amsterdam: Elsevier.

Carriquiry AL, Ireland WP, Kliemann W, Uemura E (1991) Statistical evaluations of dendritic growth models. Bull Math Biol 53: 579–589.

Cline HT (1999) Development of dendrites. In: Dendrites (Stuart G, Spruston N, Häusser M, eds), 35–56. Oxford: Oxford University Press.

Costa LF, Manoel ETM, Faucereau F, Chelly J, Van Pelt J, Ramakers G (2002) A shape analysis framework for neuromorphomery. Network 13: 283–310.

Dityatev AE, Chmykhova NM, Studer L, Karamian OA, Kozhanov VM, Clamann HP (1995) Comparison of the topology and growth rules of motoneuronal dendrites. J Comp Neurol 363: 505–516.

Horsfield K, Woldenberg MJ, Bowes CL (1987) Sequential and synchronous growth models related to vertex analysis and branching ratios. Bull Math Biol 49: 413–430.

Ireland W, Heidel J, Uemura E (1985) A mathematical model for the growth of dendritic trees. Neurosci Lett 54: 243–249.

Kater SB, Davenport RW, Guthrie PB (1994) Filopodia as detectors of environmental cues: Signal integration through changes in growth cone calcium levels. In: The Self-Organizing Brain: From Growth Cones to Functional Networks (Van Pelt J, Corner MA, Uylings HBM, Lopes da Silva FH eds), 49–60. Amsterdam: Elsevier.

Kater SB, Rehder V (1995) The sensory-motor role of growth cone filopodia. Curr Opin Neurobiol 5: 68–74.

Kliemann WA (1987) Stochastic dynamic model for the characterization of the geometrical structure of dendritic processes. Bull Math Biol 49: 135–152.

Larkman A, Mason A (1990) Correlations between morphology and electrophysiology of pyramidal neurons in slices of rat visual cortex. I. Establishment of cell classes. J Neurosci 10: 1407–1414.

Letourneau PC, Snow DM, Gomez TM (1994) Growth cone motility: Substratum-bound molecules, cytoplasmic [Ca^{2+}] and Ca^{2+}-regulated proteins. In: The Self-Organizing Brain: From Growth Cones to Functional Networks (Van Pelt J, Corner MA, Uylings HBM, Lopes da Silva FH, eds), 35–48. Amsterdam: Elsevier.

Mason A, Larkman A (1990) Correlations between morphology and electrophysiology of pyramidal neurons in slices of rat visual cortex. II. Electrophysiology. J Neurosci 10: 1415–1428.

McAllister AK (2002) Conserved cues for axon and dendritic growth in the developing cortex. Neuron 33: 2–4.

Nowakowski RS, Hayes NL, Egger MD (1992) Competitive interactions during dendritic growth: A simple stochastic growth algorithm. Brain Res 576: 152.

Parnavelas JG, Uylings HBM (1980) The growth of non-pyramidal neurons in the visual cortex of the rat: A morphometric study. Brain Res 193: 373–382.

Ramakers GJA, Avci B, Van Hulten P, Van Ooyen A, Van Pelt J, Pool CW, Lequin MB (2001) The role of calcium signaling in early axonal and dendritic morphogenesis of rat cerebral cortex neurons under non-stimulated growth conditions. Dev Brain Res 126: 163–172.

Ramakers GJA, Winter J, Hoogland TM, Lequin MB, Van Hulten P, Van Pelt J, Pool CW (1998) Depolarization stimulates lamellipodia formation and axonal but not dendritic branching in cultured rat cerebral cortex neurons. Dev Brain Res 108: 205–216.

Schierwagen A (1986) Segmental cable modeling of electrotonic transfer properties of deep superior colliculus neurons in the cat. J Hirnforsch 27: 679–690.

Schierwagen A, Grantyn R (1986) Quantitative morphological analysis of deep superior colliculus neurons stained intracellularly with HRP in the cat. J Hirnforsch 27: 611–623.

Scorcioni R, Lazarewicz MT, Ascoli GA (2004) Quantitative morphometry of hippocampal pyramidal cells: Differences between anatomical classes and reconstructing laboratories. J Comp Neurol 473: 177–193.

Smit GJ, Uylings HBM, Veldmaat-Wansink L (1972) The branching pattern in dendrites of cortical neurons. Acta Morphol Neerl-Scand 9: 253–274.

Song HJ, Poo MM (2001) The cell biology of neuronal navigation. Nat Cell Biol 3: E81–E88.

Tamori Y (1993) Theory of dendritic morphology. Phys Rev E148: 3124–3129.

Uemura E, Carriquiry A, Kliemann W, Goodwin J (1995) Mathematical modeling of dendritic growth in vitro. Brain Res 671: 187–194.

Uhlfhake B, Cullheim S (1988) Postnatal development of cat hind limb motoneurons. II: In vivo morphology of dendritic growth cones and the maturation of dendrite morphology. J Comp Neurol 278: 88–102.

Uylings HBM (2000) Development of the cerebral cortex in rodents and man. Eur J Morphol 38: 309–312.

Uylings HBM, Parnavelas JG, Walg H, Veltman WAM (1980) The morphometry of the branching pattern of developing non-pyramidal neurons in the visual cortex of rats. Mikroskopie 37: 220–224.

Uylings HBM, Van Pelt J (2002) Measures for quantifying dendritic arborizations. Network 13: 397–414.

Van Ooyen A (2001) Competition in the development of nerve connections: A review of models. Network 12: R1–R47.

Van Ooyen A, Van Pelt J (2002) Competition in neuronal morphogenesis and the development of nerve connections. In: Computational Neuroanatomy: Principles and Methods (Ascoli G, ed), 219–244. Totowa, N.J.: Humana Press.

Van Ooyen A, Willshaw DJ (1999) Competition for neurotrophic factor in the development of nerve connections. Proc Roy Soc Lond B266: 883–892.

Van Ooyen A, Willshaw DJ (2000) Development of nerve connections under the control of neurotrophic factors: Parallels with consumer-resource systems in population biology. J Theor Biol 206: 195–210.

Van Pelt J, Dityatev AE, Uylings HBM (1997) Natural variability in the number of dendritic segments: Model-based inferences about branching during neurite outgrowth. J Comp Neurol 387: 325–340.

Van Pelt J, Schierwagen A, Uylings HBM (2001) Modeling dendritic morphological complexity of deep layer cat superior colliculus neurons. Neurocomputing 38–40: 403–408.

Van Pelt J, Uylings HBM (2002) Branching rates and growth functions in the outgrowth of dendritic branching patterns. Network 13: 261–281.

Van Pelt J, Uylings HBM (2003) Growth functions in dendritic outgrowth. Brain and Mind 4: 51–65.

Van Pelt J, Van Ooyen A, Uylings HBM (2001) Modeling dendritic geometry and the development of nerve connections. In: Computational Neuroscience: Realistic Modeling for Experimentalists (with CD-ROM) (De Schutter E ed; CD-ROM by Cannon, RC), 179–208. Boca Raton, Fla.: CRC Press.

Van Pelt J, Verwer RWH (1983) The exact probabilities of branching patterns under segmental and terminal growth hypotheses. Bull Math Biol 45: 269–285.

Van Pelt J, Verwer RWH (1986) Topological properties of binary trees grown with order-dependent branching probabilities. Bull Math Biol 48: 197–211.

Van Pelt J, Woldenberg MJ, Verwer RWH (1989) Two generalized topological models of stream network growth. J Geol 97: 281–299.

Whitford KL, Marillat V, Stein EW, Goodman CS, Tessier-Lavigne M, Chedotal A, Ghosh A (2002) Regulation of cortical dendrite development by slit-robo interactions. Neuron 33: 47–61.

Zhang LI, Poo MM (2001) Electrical activity and development of neural circuits. Nat Neurosci 4 (suppl): 1207–1214.

V BEHAVIOR

Animal and human behaviors have properties of intrinsically complex biological systems that require subtle strategies of decomposition, recomposition, and systems analysis. Models assist these approaches in many of the ways described in other sections of this volume (heuristic, analytical, etc.), but in the case of behavior, models have come to play a distinct new role as reified active devices (robots) that provide an independent source of behavioral data. All three chapters of this section reflect the increasing use of robots in the study of both the bottom-up generation of *in silico* behaviors and of biologically based behavior. This marks the beginning of a new era of modeling agent-based behaviors and their interactions with other artificial or biological agents, as pioneered by artificial life proponents such as Luc Steels and others.

The major advantage of modeling approaches in the behavioral sciences is the capacity to study systems' properties by treating many components simultaneously. In particular there is a requirement to relate the different levels of organization that underlie biological behaviors, such as chemicophysical, molecular, developmental, and constructional. Representations of these multiple components of behavioral systems is a basic prerequisite for their modeling. Upon this dynamical interaction, patterns can be implemented that lead to functional, analytical, explanatory, and predictive models of biological behavior. On the other hand, the multitude of biological components will never be covered by a single model, and therefore robot-based and robot interaction-based modeling, in which all underlying components are known, can concentrate more directly on the features of emergent behaviors that are not easily addressed by biological models.

Jeffrey Schank and Thomas Koehnle (chapter 11) provide an entry to this perspective in their chapter on complex biobehavioral systems. They treat the issue of biological modeling from a multimodel perspective in which models are representations that serve as tools for understanding. They emphasize that a proper understanding of the role of models in biological research requires an appreciation of their limitations as well as a knowledge of their phylogeny. Whereas certain models lead to dead ends, others are modified and elaborated, and can be understood only in their historical context (i.e., the phylogeny of a model itself has an important heuristic role). Schank and Koehnle characterize nine dimensions of models that can evolve independently but have necessary trade-off relations among them. They show how this approach can assist in the appraisal of new conceptual and empirical work and help establish the relation with artificial agent models. The empirical system they present is the study of sensorimotor development in rat pups, and they briefly introduce a robotic approach to model essential physical parameters of such pups.

Chrystopher Nehaniv et al. (chapter 12) introduce information theory as a novel approach to modeling biological phenomena and then apply it to the evolution of sensory channels and information-processing systems in natural and artificial organizations. Their goal is to arrive at what they call "generalized biology," a highly abstract ("coordinate-free") biology of principles rather than mechanisms that includes such fields as astrobiology and artificial life. Within generalized biology, modeling plays a central role in that models serve as mediators between mechanisms and general theories. Using the information concept as an example, Nehaniv et al. discuss several conceptual problems and develop a behavioral modeling framework based on the Sony AIBO robot. This allows them to embed theoretical ideas within a concrete model robot and to explore the visual sensory maps of the robot in a variety of different contexts. Central to this approach is the concept of perception-action loops. Further embedding their ideas about information flow within an evolutionary perspective, Nehaniv et al. demonstrate that maximizing information flow within the model can lead to rather well structured representations of the world presented to the model.

Their contribution raises important points about the relationship between phenomena, models, and theories in biology. These distinctions can be blurred, as in the case of robots and other artificial life creations, since they can simultaneously be seen as a model, an instantiation of a theory, and a phenomenon in its own right.

Iain Werry and Kerstin Dautenhahn (chapter 13) demonstrate a practical application of using robot models in assisting therapy of behavioral disorders in humans. They concentrate on human-robot interaction in autistic children. It is shown how mobile robots are able to engage these children in various kinds of interaction. Moreover, their work emphasizes the emergent, proactive, and embodied nature of behavioral interactions. The robot's actions that connect with the human behavior can lead to complex patterns that cannot be reduced to the robot's programmed behavior. Also, these complex patterns require active participation of both interactors (i.e., they do not occur as prespecified behaviors). This approach raises the question of to what extent a robot represents a "model" of a human being and whether it is a more or less adequate participant in its interaction with others. Beyond the treatment of autism, the experimental setup presented by Werry and Dautenhahn can serve as a model for using robots in medical treatments of other behavioral disorders and learning disabilities, and in rehabilitation.

11 Modeling Complex Biobehavioral Systems

Jeffrey C. Schank and Thomas J. Koehnle

Why Behavior Matters

Genomics, proteomics, neuroscience, and the study of transgenic organisms are among the hottest and most heavily funded areas of the life sciences. Although each of these areas is relevant to explaining behavior, a fundamental problem concerns how these and other areas of research can be integrated to create detailed and dynamic explanations and predictions of behavior. After all, what we really want to know is how genes, proteins, and neurons are structurally and dynamically organized to produce an organism that can interact with its environment. One of the major challenges for behavioral research is devising multimodel and multilevel modeling strategies for understanding the complex processes that produce behavior.

The behaviors of animals are similar to the behaviors of any physical system—they are sequences of physical events. This suggests that animal behavior can be investigated in ways similar to the study of other physical systems. However, animal behaviors differ from many other physical events in two important ways: (1) they are generated by complex interactions between processes acting at and across multiple levels of organization; and (2) animals are active agents in the generation of their own behavior. Unlike the behaviors of systems that are the subject of physics, we are hard-pressed to find simple physical laws that explain the behaviors of animals. Indeed, to quote Barnett:

> Since environments vary, the details of individual experience in the wild state are far from uniform even within a small population. Hence, as far as behavior is concerned, the diversity already present at fertilization is magnified and made more complex by the acquired adjustments of each individual. If cannon balls or guided missiles or planets had personalities of their own, and their behavior changed progressively in response to repeated stimuli, the subject of physics would present additional difficulties. Physicists would be obliged to make observations on vast numbers of individual objects and to construct case histories of each: they would even perhaps be able to predict the behavior of individual bodies (from the previous observations of those bodies), but they would still be faced with the problem of grouping the individuals in classes concerning which more general predictions could be made. This is the situation with which students of complex behavior have to cope. (Barnett, 1975:166)

However, complexity in the study of animal behavior need not be inimical to the development of adequate methods and theories to explain and predict what animals do. We may not be able pursue general, universal laws of behavior, as behaviorists once sought to do, but we can systematically explain behavior by building multilevel models (see chapter 8 in this volume for further discussion of multilevel models).

Historically, there have been two general approaches to building models: top-down and bottom-up. Top-down research strategies bear the closest resemblance to the physical sciences. General laws or principles are formulated and specific predictions are mathematically or logically deduced from them. A model of a system is then derived from these principles. This is an appealing approach because it is inherently systematic, yielding testable predictions (e.g., in philosophy of science, a classic example is Hemple, 1966; and in biology, Lotka, 1956). Top-down approaches are problematic for the study of behavior because we lack the general laws and first principles that would allow us to systematically model the behavior of different organisms in different contexts.

Bottom-up, reductionist, decompositionalist, and isolationist research strategies (Wimsatt, 1997) start without a systematic theory of organisms and behavior. Instead, complex biological systems are conceptually and methodologically dissected into their constituents, and each component is studied in isolation. However, this approach faces at least three problematic and unanswered questions concerning their integration into theories of whole organisms. First, are the properties of the components the same in isolation as in the whole organism? Second, how can the pieces be put back together to form systematic models of the whole organism and behavior? Third, if organisms rearrange themselves, construct their environments (cf. Dewey, 1896; Lewontin, 1982; Bateson, 1988), and thus are causes of their own behavior, how can we understand their behavior from information about their parts? This dynamic aspect of living systems is a likely source of nonlinearity in behavior, and makes simple piecemeal reassembly of isolated systems components unlikely to work. Thus, the fundamental problem facing any strategy for understanding behavior is understanding the organization of parts at all biological levels.

If neither top-down nor bottom-up strategies are likely to succeed, is the study of behavior hopeless? We think not. Recently, systems strategies have emerged which treat all elements simultaneously as both causes and effects (e.g., Beer, 2003). Though the conceptual advantages of these approaches have been obvious, until recently the methodological and empirical tools capable of implementing them have been all but nonexistent. With these new tools, the problems of modeling the organization of parts can be increasingly addressed.

Integrative strategies are clearly desirable, but the devil is in the details. How do we integrate information from genomics, proteomics, and neurobiology obtained from bottom-up strategies with simple heuristics, optimal rules, game-theoretic, or other abstract approaches derived from top-down approaches to understand how organisms are organized to produce behavior? We are not yet at a stage where we can specify strategies to achieve integration, and perhaps such strategies are not possible. Instead, we think it likely that the integrative explanations of behavior we seek will emerge from a multimodel research orientation. We will argue that if models are used and recognized as important at all levels of behavioral research (including multiple models at the same level), understanding will emerge from their coordinated use.

Below, we illustrate the involvement of models at all levels of analysis in biobehavioral research. All models have limitations and are subject to modification or outright rejection. Multiple models can compensate for weaknesses and limitations in any one model. We believe that the goals of the biobehavioral sciences can be achieved only by an understanding of the role models play in various aspects of research (e.g., how they are coordinated and assessed; see also chapter 2 in this volume).

In the next three sections, we will introduce the notion of models as representations that function as tools for understanding. As representations and tools for understanding, models have important properties and limitations that can be described as the dimensions of models (see chapter 2 in this volume for further philosophical analysis of this perspective). The process of model-building is analogous in some respects to a biological phylogeny (that is, a phylogenetic tree may be a model of model-building, and the development a of phylogeny of models may be what we call a scientific theory in the biobehavioral sciences). We will show that not only is the phylogenetic development of models descriptive of successful models in science, but it also is essential to the model-building process. It comprises the modification of previous models as well as illustrating that all models carry with them prior conceptual and empirical validation and errors. Indeed, it provides a convenient and useful way to present the structure of theory and theoretical developments in the biological sciences.

Models

Models are representations used as aids for solving problems and as tools for understanding. As a representation, we assume a correspondence between some of the properties of the model and some of the properties of the system modeled (see Hesse, 1966; Giere, 1999 for similar views). The correspondence can be detailed and qualitative, and can comprise different kinds of properties. As an aid for problem-solving, models must be manipulable and analyzable so that we can do things with them and make inferences about them. We use models to study systems that are complex and little understood. However, unlike tools for constructing physical artifacts, we are never sure whether we are using the best model for the job. Indeed, we can never have a perfect model; we know by their very nature that models are limited (Wimsatt, 1987). As tools and representations, they may contain nonobvious flaws and biases. The sooner we discover them, the better off we are. From an evolutionary perspective, this suggests that most models will lead to dead ends, while others will be modified and elaborated to apply to different conditions and phenomena. Later, we will discuss how "phylogenies" of models emerge from modeling processes and how they, too, are tools for understanding and play critical roles in modeling.

It is also little recognized how extensively models are used in science. Typically, we think of models as mathematical or computational representations which are used to derive theoretical predictions or explanations of behavior. However, by viewing models as representations that are tools for understanding, we find them at more levels of analysis than is commonly recognized. An experiment is a model system (see chapter 2 in this volume for a discussion of pluralistic accounts of modeling strategies in science). Laboratory organisms are models with similarities and differences in the correspondences of their properties to their counterparts in the wild. Indeed, the U.S. National Institutes of Health are pushing for the development of knockout and transgenic animal models of human diseases and behavioral disorders. The apparatuses used to contain and manipulate organisms are models that hold constant some variables and allow the manipulation of others (e.g., see Alberts et al., 2004). Data are models and representations of the behavior of the organisms we study.

The models we use in biobehavioral research, even at different levels of analysis, are not independent of each other. We will show that data are also models laden with assumptions, and that we cannot assume strict epistemic priority of data over theoretical models. Models play different roles, but any model can play a role in the assessment and rejection of another. This makes the process of modeling biobehavioral systems iterative and more complex, but not necessarily intractable.

Models and the Dimensions for Appraising Them

Levins (1966), Webb (2001), and Leonelli (chapter 2 in this volume) outlined *dimensions of models* to consider the relative merits of different approaches to elaborating, analyzing, classifying, and testing models. Models have at least nine dimensions spanning epistemological, ontological, and practical issues (figure 11.1). For Levin (1966) and Webb (1999, 200), *medium* concerns the composition of a model (e.g., mathematical or robotic), *realism*, the fidelity of system properties represented in a model, *detail*, the number and complexity of mechanisms or properties represented in a model, *generality*, the variety of systems representable by a model, *match*, how well predictions and explanations derived from a model fit data, and *precision*, quan-

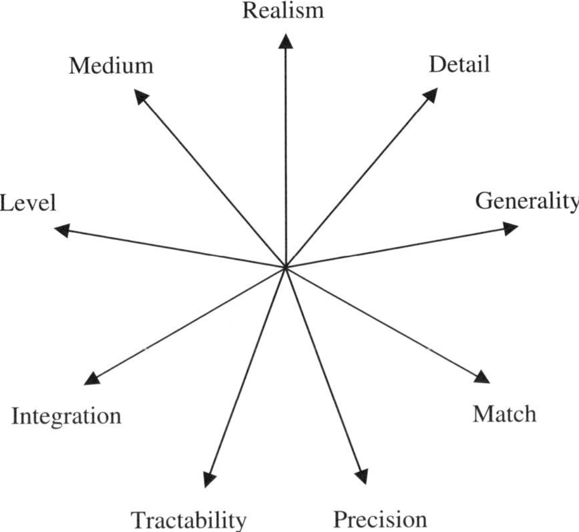

Figure 11.1
Dimensions of models. Theories, simulations, and experiments can be viewed as special kinds of representations of our beliefs about reality, or models. As more and more experiments are conducted, realism, precision, and match should increase. Evaluating models along all their relevant dimensions highlights their advantages and limitations. The utility of a given model is best judged by the progress it embodies, rather than by its formal adherence to statistical convention. (Adapted from Koehnle and Schank, 2003.)

titative accuracy of a model in predicting or explaining data. To these we (Koehnle and Schank, 2003) have added three more practical dimensions: *tractability* concerns how amenable a model is to analytical, computational, or experimental techniques of analysis; *integration* concerns how well a model can be integrated with other models or how a model contributes to integration of information across levels, which may represent different aspects or levels of a system; and *level* concerns the level(s) of organization of a system represented by a model.

Each model used in developing a theory or solving problems within that theory will vary in position along each of these nine dimensions. It is important to realize that these dimensions are not strictly independent. Models with high precision, detail, and realism might tend to be less tractable, on average, than models with more generality and less integration, although, as we shall see below, this is not knowable a priori.

We believe that thinking about biological models in terms of these dimensions is helpful for two reasons: (1) appraisals of new conceptual, empirical, and methodological works will be facilitated by using these dimensions; (2) recognizing the trade-offs between dimensions, in particular the practical dimensions, can help to anticipate future problems within a theory, and thereby mark off the frontier of a research domain. We will treat each of these in more depth.

The appraisal of new conceptual and empirical works will be facilitated by using a dimensional framework. Although there is a practical framework for evaluating and choosing between rival theories (e.g., Laudan, 1977, 1996), identifying and resolving new or anomalous problems *within* theories remains, at best, haphazard. For example, shall we count Pulliam's (1975) rejection of Schoener's (1971) assumption that optimal foragers seek to maximize energy intake as progressive? The recognition of nutritional constraints on organisms enhanced optimal foraging models by adding more *realism* and *detail* compared with earlier models developed under Schoener's (1971) assumptions about rate maximization, or the alternative assumption that animals should seek to minimize exposure to predators.

Though it seems intuitive that such increases in realism and detail should come at the cost of *generality*, whether or not advancing along a particular dimension will automatically carry penalties in other dimensions cannot be determined a priori. For example, in addition to their conceptual utility, nutrient constraint models were also progressive in the empirical domain, generating a great deal of insight into problems of food choices in herbivores (Belovsky, 1978; Duncan and Gordon, 1999), primates (Oftedal, 1991; Whiten et al., 1991), and birds (Thompson et al., 1987; Murphy and Pearcy, 1993). This advance in explaining foraging behavior in terms of nutrient needs has come at a cost: failure to explain situations where there are no such needs.

Nevertheless, the nutrient constraint model spawned greater *integration* and the recognition of the importance of additional *levels*, permitting synthesis of empirical studies at the level of foraging (Murphy and King, 1989; Murphy and Pearcy, 1993) and empirical studies at the level of individual behavior and neurobiology (Koehnle and Gietzen, 2005). This interplay between models of different types, or at different levels, points out the potential for finding and exploiting useful interactions under a dimensional framework.

In the actual practice of science, assessing progress is critical for both the short term and the long term. Ideally, all models would advance simultaneously along all dimensions, but advances along a particular dimension may constitute progress even if those advances entail losses along another model dimension. If, for example, a model adds detail but loses generality compared with its competitors and allies, should such a loss count against that model? We believe that the best way to answer this question is to relativize it to the research context in which the model is embedded. Certain fields (e.g., developmental systems theory) may place a premium on the level of realism or detail in describing the complex interactions among genes, cells, organs, and the animal's behavior. Within this research context, a loss of generality might well be perceived not merely as inconsequential but also as a positive attribute of a model. Evolutionary ecology, on the other hand, might place a premium on explaining the largest number of behaviors with the smallest number of general models.

Recognizing the trade-offs made along the nine model dimensions in constructing a particular model can help us to anticipate future problems within a research program, thereby marking off the frontier. As models become increasingly complex, they may become less tractable, but it may become possible to identify the empirical and methodological choke points that need attention (e.g., Beer, 2003; see "The Need for New Statistical Models," below). In addition, lateral shifts between different models may permit progress by routing around the practical obstacles. For example, if laboratory or field studies cannot be conducted with adequate controls, it may be possible to place greater reliance on simulation or robotic models (Schank, 2001b). However, introducing robotic models of behavior introduces considerable dimensional complexity because robotic models are really multimodel systems that vary on the dimension of medium, being both physical and computational (May et al., 2006).

For example, we have designed robots to model essential physical parameters of Norway rat pups (Schank et al., 2004; May et al., 2006). These robots have the same length:width ratio as rat pups, approximately 3:1, where the head constitutes one third of the length. Body shape was modeled with an aluminum skirt outfitted with fourteen micro/limit (touch) switches attached to brass strips to allow for a 360° tactile-sensory range. Because rat pups up to ten to twelve days of age move primarily by using their back legs (unpublished observations), the robots were equipped with rear-driven wheels with differential drive on a single chassis (figure 11.2).

There are at least two levels of modeling with robotics, each generating variations from basic designs. First, there are models of the cognitive architecture implemented in a robot's CPU. Many different architectures can be devised for relating sensory input to motor output (e.g., probabilistic, rule-based, mixed, and neural networks). We are currently exploring three classes of control systems, with a number of variants in each (Schank et al., 2004; May et al., 2006). Second, there is the physical body of the robot, in which many aspects of an organism can be represented in different robotic models, including sensor arrays and systems (e.g., tactile, thermal, vestibular, and visual) and aspects of body morphology, such as flexibility (figure 11.2).

Phylogenies of Models

Because models always involve idealizations and modeling is a constructive process, successful modeling always begins with simple models or even model schemata. Subsequent models often add elements or recombine elements from previous models. Thus, in an important sense, these basic models—if successful at generating new and more detailed models—belong high on the dimension of generality (Schank, 2001b).

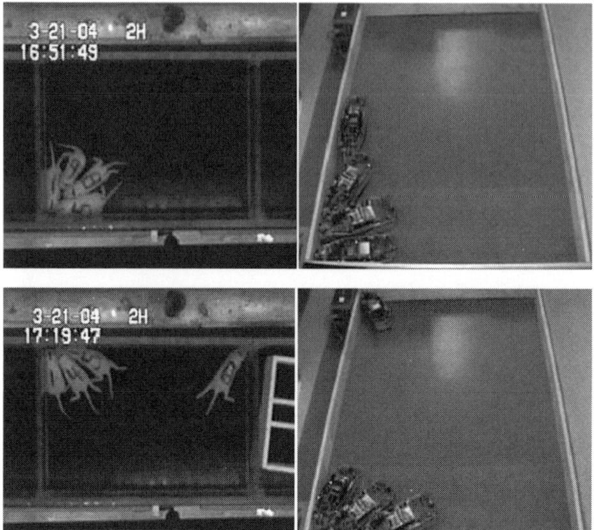

Figure 11.2
Groups of four rat pups and robots. Robots model rat pups both physically and computationally.

Building models from the starting point of basic models is a direct consequence of the fact that modelers do not know all the relevant factors or mechanisms involved in a biological system of interest. Building models of biological systems is often a piecemeal evolutionary process, which is an inevitable consequence of constructing models along the nine dimensions discussed above. At each step, the modeling process is influenced by a thicket of external and internal factors directly and indirectly affecting the evolution of models along any of the dimensions of modeling.

In some cases, the phylogenetic relationships among models can be specified by tracing the assumptions or limiting conditions that would make later models mathematically identical to earlier models. We will illustrate this kind of relationship with the basic Nicholson-Bailey predator-prey model. The basic Nicholson-Bailey model is

$$\begin{aligned} N_{t+1} &= \lambda N_t e^{-aP_t} \\ P_{t+1} &= N_t(1 - e^{-aP_t}), \end{aligned} \tag{11.1}$$

where N_t and P_t are prey and predator populations densities, respectively; λ is the prey growth rate; and a is the effect of predator density on prey density. In this simple and basic model, the growth of prey at each time interval is a function of the number of prey that survive and their growth rate, while the number of predators or parasitoids is directly related to the number of prey taken. The subsequent develop-

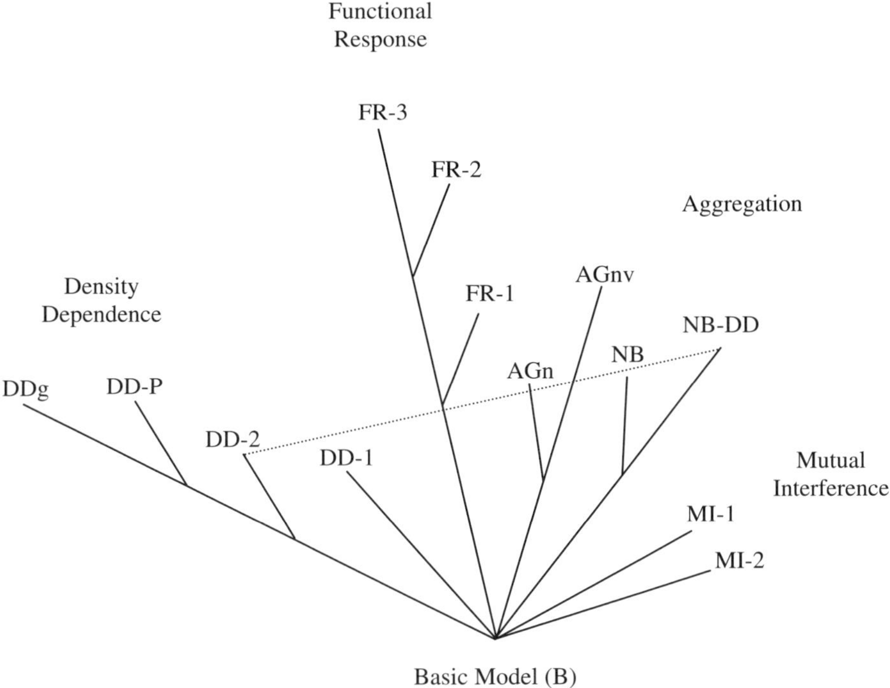

Figure 11.3
A very partial phylogeny of Nicholson-Bailey predator/parasitoid-prey models. Note that unlike biological phylogenies, a model can have more than one ancestral model and models are not necessarily chronologically related to each other.

ment of Nicholson-Bailey models has aimed at discovering mechanisms of population stability and modeling more realistic aspects of predator/parasitoid and prey systems (Hassell, 1978). Thus, the mathematical development of these models has been primarily along the dimension of realism constrained by the dimension of analytical tractability (though more recently, the latter constraint has been relaxed a bit with the development of easily accessible computers).

Using the technique of proving mathematical identity between models under limiting and special assumptions, a phylogeny of Nicholson-Bailey models can be constructed from the original basic model up to many current models. We have constructed a partial phylogeny of Nicholson-Bailey models from some of these models (see figures 11.3 and 11.4; these models are found in Hassell [1978], and the more recent papers by van Baalen and Sabelis [1999] and King and Hastings [2003]).

These types of model relationships in science are ubiquitous. For example, the binomial distribution can be viewed as the basic model distribution for much of analytical statistics, since both the Poisson and the normal distributions can be derived

Model	Reduction to	Model	Reduction to
Density Dependence		**Aggregation**	
$N_{t+1} = \lambda N_t^{1-b} e^{-aP_t}$ $P_{t+1} = N_t^{1-b}(1-e^{-aP_t})$ (DD-1)	$b = 0 \Rightarrow$ (B)	$N_{t+1} = \lambda N_t \sum_{i=1}^{n} \alpha_i e^{-a\beta_i P_t}$ $P_{t+1} = N_t (1 - \sum_{i=1}^{n} \alpha_i e^{-a\beta_i P_t})$ (AGn)	$\alpha_i = \beta_i = 1/n \Rightarrow$ (B)
$N_{t+1} = N_t e^{r(1-N_t/K)-aP_t}$ $P_{t+1} = N_t (1-e^{-aP_t})$ (DD-2)	$r = \ln\lambda$, $K \to \infty$ $\Rightarrow$ (B)	$N_{t+1} = N_t \sum_{i=1}^{n} \alpha_i \lambda_i e^{-a\beta_i P_t}$ $P_{t+1} = N_t (1 - \sum_{i=1}^{n} \alpha_i e^{-a\beta_i P_t})$ (AGnv)	λ_i is constant $\Rightarrow$ (AGn)
$N_{t+1} = N_t e^{r(1-N_t/K)-aP_t}$ $P_{t+1} = c[N_t (1-e^{-aP_t}) - \beta P_t]$ (DD-P)	$c = 1, \beta = 0 \Rightarrow$ (DD-2)	$N_{t+1} = \lambda N_t (1 + \frac{aP_t}{k})^{-k}$ $P_{t+1} = N_t [1 - (1 + \frac{aP_t}{k})^{-k}]$ (NB)	$k \to \infty \Rightarrow e^{aP_t} \Rightarrow$ (B)
$N_{ij,t+1} = N_{ij,t} e^{r(1-N_t/K)-aP_{IJ,t}}$ $P_{IJ,t+1} = N_{ij,t}(1-e^{-aP_{IJ,t}})$ (DDg-2) $N_t = N_{AA,t} + N_{Aa,t} + N_{aa,t}$	$i = j, I = J \Rightarrow$ (DD-2)	$N_{t+1} = N_t e^{r(1-N_t/K)} (1 + \frac{aP_t}{k})^{-k}$ $P_{t+1} = N_t [1 - (1 + \frac{aP_t}{k})^{-k}]$ (NB-DD)	$r = \ln\lambda$, $K \to \infty$ $\Rightarrow$ (NB)
Functional Response		**Mutual Interference**	
$N_{t+1} = \lambda N_t e^{-a'TP_t}$ $P_{t+1} = N_t (1-e^{-a'TP_t})$ (FR-1)	$T = 1 \Rightarrow$ (B)	$N_{t+1} = \lambda N_t e^{-QP_t^{1-m}}$ $P_{t+1} = N_t (1-e^{-QP_t^{1-m}})$ (MI-1)	$m = 0, Q = a \Rightarrow$ (B)
$N_{t+1} = \lambda N_t e^{-\frac{a'TP_t}{1+a'T_hN_t}}$ $P_{t+1} = N_t (1-e^{-\frac{a'TP_t}{1+a'T_hN_t}})$ (FR-2)	$T_h = 0 \Rightarrow$ (FR-1)	$N_{t+1} = \lambda N_t e^{-\frac{aTP_t}{1+bT_w(P_t-1)}}$ $P_{t+1} = N_t (1-e^{-\frac{aTP_t}{1+bT_w(P_t-1)}})$ (MI-2)	$bT_w = 0 \Rightarrow$ (B)
$N_{t+1} = \lambda N_t e^{-\frac{bN_tTP_t}{1+cN_t+bT_hN_t^2}}$ $P_{t+1} = N_t (1-e^{-\frac{bN_tTP_t}{1+cN_t+bT_hN_t^2}})$ (FR-3)	$c = 0, b = a' \Rightarrow$ (FR-2)		

DD-1: density dependence (Hassell, 1978, p. 23)
DD-2: density dependence (Hassell, 1978, p. 24)
DD-P: density dependence with predator rate of increase (Hassell, 1978, p. 116)
DDg: density dependence with single-locus genetic models (Flatt & Scheuring, 2004, p. 243)
FR-1, FR-2, FR-3: functional responses of predators to prey (Hassell, 1978, pp. 30-34)
AGn: prey aggregation in patches, explicit model (Hassell, 1978, p. 61)
AGnv: prey aggregation with prey fecundity varying with patches (van Baalen & Sabelis,1999, p. 71)
NB: negative binomial model of aggregation (Hassell, 1978, p. 74)
NB-DD: negative binomial model with density dependence (Hassell, 1978, p. 77)
MI-1: predator/parasitoid mutual interference (Hassell, 1978, p. 84)
MI-2: predator/parasitoid mutual interference (Hassell, 1978, p. 87)

Figure 11.4
The mathematical forms of the models represented in the phylogeny in figure 11.2, together with the assumptions under which a model is identical to its ancestor.

from the binomial under limiting assumptions of continuous mathematics (Stigler, 1986). Analytical statistics such as analysis of variance, linear regression, correlation, and factor analysis are all based on the basic schema of the general linear model. Thus, looking for mathematical identities based on special or limiting mathematical assumptions is an important way of reconstructing model phylogenies.

Indeed, the idea of thinking about the systematic development of models as phylogenies is not new. Stephens and Krebs (1986) present foraging theory as a "phylogeny" of models related to two basic patch and prey models based on changed assumptions. Stone (1996) has directly applied cladistic analysis to shell models starting with Mosely's 1838 geometric model. His analysis is even broader, incorporating assumptions of models, the nature of their parameters, and even whether they include computer graphics. While biological phylogenies may be constrained to such cladistic analyses, we believe that phylogenetic reconstruction of model relationships is more complicated because a given model may have multiple predecessors, as illustrated in figure 11.3.

Model phylogenies are by no means limited to mathematical or computational models. Animal models also fall neatly under this analysis. For example, almost all knockout and transgenic mice are created from five inbred strains (Bothe et al., 2004). Indeed, the limits of viewing model development through the lens of biological phylogenies are unclear. If models are seen as the product of an evolutionary process, there might be processes analogous to selection, drift, self-organization, and so on. Models might be targets of selection, since they have heritable variation and compete for limited resources: funding, data, and scientists to believe in them. There might even be a process analogous to sexual selection, as the explosion of game-theoretic models after the publication of Maynard Smith and Price's (1973) classic on animal conflict, or the more recent increased interest in models of female choice, seem to demonstrate. Models may have aesthetic as well as epistemological and practical dimensions.

More interesting, and perhaps alarming, than processes of selection in model development is the possible role of drift. A few models might, at random, survive cuts in funding or decreases in interest, and act as founders for future research. This would indicate a decoupling between explanatory power and model proliferation. Scientific diagrams such as Weismann diagrams, representing the relationship between soma and germ line, may indicate that drift plays a role in the evolution of models (Griesemer and Wimsatt, 1989). Such a process also seems to have taken place at the level of methods and apparatus in the behavioral sciences. Prior to Fisher's (1935) exposition of the analysis of variance, methods for data description and analysis were varied and complex. The null-hypothesis test, however, replaced this complexity with a narrow range of inferential statistical tools (see "The Need for New Statistical Models," below).

At the level of laboratory apparatus, we see similar reduction of variation over time, going from the complex housing environments used by Richter to study rat behavior (Moran and Schulkin, 2000) down to the Skinner box. Complicated apparatus eventually reappeared in the context of economic analysis of behavior (e.g., Cooper and Mason, 2001; Collier et al., 1972), demonstrating sequential loss and reinvention of complexity at the level of scientific equipment. Interestingly, the degree of complexity at each stage in the development of modern laboratory apparatuses was tightly correlated with the complexity of analytical methods and theoretical constructs, suggesting that practical considerations at least in part drive theoretical progress (see also Koehnle and Schank, 2003). Timberlake (2002) has argued that by "tuning" an apparatus to produce robust behavior, researchers can discover niche-related behaviors in the organisms they study.

We believe that the concept of model phylogenies is important to the model-building process for three reasons. First, they document the development of theories through model variants. Second, because variant models are built from previous models, they carry with them the successes of previous models and also all the hidden flaws. By recognizing that modeling processes generate model phylogenies, we may better be able to trace potential errors. Third, model phylogenies clarify what is meant by a general model. On the phylogenetic approach, a general model is one that gives rise to a large number of descendants that apply to a wide variety of phenomena and theoretical problems. Thus, general models are simple and often schematic.

This contrasts sharply with the view held by Orzack and Sober (1993), who view the most general models as the most detailed (i.e., containing all the mathematical relationships and parameters necessary to derive all other related models through simplifying and limiting assumptions). However, when this view is applied to actual models, such as the Nicholson-Bailey family of models, it leads to biologically absurd models (i.e., models that contain all parameters in descendant models) as the most general (Schank, 2001b).

Modeling Behavior

Perhaps the central problem for the study of behavior is how to connect processes that span multiple levels (i.e., genetic, neural, morphological, physical environment) and specify their organization (see chapter 8 in this volume for related discussion). How do we map the brain to behavior in a way that relates neural processing to the production of behavior, and not merely correlates of brain activation? How do we do this while recognizing that a brain is situated in a body, which can be in indefinitely many physical contexts? How do we build models in which animals actively reorganize themselves and their environments through their behavior, over their lives and over their evolution?

We advocate a multimodeling approach. This does not necessarily mean that each researcher must use multiple types of models to investigate behavior, but that the community of behavioral researchers should embrace the use of multiple types of models. Models that address weaknesses inherent in their allies and competitors can interact, resulting in better models and theories (Wimsatt, 1987).

Complex systems models might provide opportunities to combine data from both bottom-up and top-down approaches. Describing animal behavior in terms of coordinates in a phase space is not very intuitive, but it captures activity at all the dimensions relevant to the production of a given behavior. Moreover, systems models can also serve to visualize or implement top-down approaches to behavior. For example, in his model of a simple evolved agent, Beer (2003) used a phase space model to explain and describe the agent's behavior. The model itself captured neural, motor, cognitive, and environmental levels of organization. The agent seemed to apply a simple heuristic in the tasks it was asked to perform (Koehnle and Schank, 2003), demonstrating that the phase space was able to capture important features normally thought of as top-down approaches.

At this point, the major stumbling blocks preventing widespread application of systems models are the lack of adequate tools for the analysis of the models and the systems they represent (Beer, 2003; Nepomnyashchikh and Podgornyj, 2003), a lack of heuristics and tradecraft in knowing which levels or dimensions are relevant in a given model, and paucity or intractability of data available for use by modelers.

Agent-based probabilistic models are especially useful for modeling behaviors in social and group contexts when there are emergent features of behavior, but these models are notoriously difficult to analyze (Schank, 2001a). An agent-based modeling approach is often necessitated by having more information about the individuals than about groups, which cannot be put into a tractable mathematical model. Probabilistic models in particular are important, not merely because they reflect our uncertainty about the variables affecting individual behavior, but also because noise may be an essential feature of organisms (Coulson et al., 2004). Finally, we can use these models in combination with genetic algorithms and Monte Carlo simulations to perform theoretically based statistical analyses of data (Schank, 2001a; Schank and Alberts, 2000b).

The agent-based models we use represent behavioral events as probabilistic functions of time, internal states, and sensory inputs. The basic probabilistic representation is the change in activity state over time, which has the schematic form

$$p(A_t \mid I_{t-\Delta t}, s_1, \ldots, s_n, i_1, \ldots, i_m) \quad \text{and} \quad p(I_t \mid A_{t-\Delta t}, s_1, \ldots, s_n, i_1, \ldots, i_m) \tag{11.2}$$

where A is sensorimotor activity, I is inactivity, s represents time-dependent internal states, and i represents time-dependent sensory inputs. The equations in (11.2) provide a complete schematic form for transitions between active and inactive states.

Specific behavioral outputs are conditional on activity:

$$p(B_t \mid A_{t-\Delta t}, s_1, \ldots, s_n, i_1, \ldots, i_m). \tag{11.3}$$

Models are built from this representation scheme by specifying variables, parameters, and the mathematical functions relating variables and parameters to probabilities of behavioral events (Schank and Alberts, 2000b). From this basic model schema, a variety of probabilistic models of activity can be developed. For example, we can ignore internal states, and condition activity only on the prior activity state (Schank and Alberts, 1997; see probabilities in [11.4])

$$\begin{aligned} &p(A_t \mid I_{t-\Delta t}) \\ &p(I_t \mid A_{t-\Delta t}), \end{aligned} \tag{11.4}$$

and if we include the number of active and inactive pups that a given pup physically contacts and allow probabilities to change with time (Schank and Alberts, 2000a), we get equations (11.5)

$$\begin{aligned} p(A_t \mid I_{t-\Delta t}, N_A, N_I) &= P(A_{t=0}, \mid I_{t-\Delta t}, 0, 0) e^{t[a_{AI} N_A + b_{AI} N_I + c_{AI}] + c_{AI} N_A + d_{AI} N_I} \\ p(I_t \mid A_{t-\Delta t}, N_A, N_I) &= 1 - P(A_{t=0}, \mid I_{t-\Delta t}, 0, 0) e^{t[a_{AA} N_A + b_{AA} N_I + c_{AA}] + c_{AA} N_A + d_{AA} N_I}, \end{aligned} \tag{11.5}$$

which describe how transitional probabilities change as a function of time and the number of active and inactive pups a given pup is in contact with (Schank and Alberts, 2000) The set of equations in (11.5) can be further elaborated to include contact with different types of objects and stimuli.

Data

Data are usually considered sacred in science. Theories, models, and hypotheses may come and go, but data are treated as permanent and firm. Darwin (1882) recognized this view of data as a serious problem for science.

False facts are highly injurious to the progress of science, for they often endure long; but false views, if supported by some evidence, do little harm, for every one takes a salutary pleasure in proving their falseness: and when this is done, one path towards error is closed and the road to truth is often at the same time opened. (Darwin, 1882:606)

Theoretical models are far less problematic in this regard because we are much more willing to give them up or revise them in light of new data. This is because we are taught very early that data are epistemically superior to the theoretical models they are gathered to test. The assumptions behind this superiority have a long and tortuous history (see, e.g., Laudan, 1977; Overton, 1994), and are quite beyond the scope of this chapter. Suffice it to say that if we revise our assumptions just a bit, it is possible to interpret even raw data as a kind of representation, or model, of the systems we study.

Data are models because they have limitations, make assumptions, and, as with any models, may have obvious or hidden biases. Altmann (1974) discussed different techniques for recording data in which visual observations are transformed into symbolic output. Additional transformations may be made to generate summary statistics or to transform the data to fit the assumptions of a statistical model, allowing further conclusions to be derived. At each step, data (as models) must be checked to assess their accuracy as representations. For example, consider two independent observers watching an animal in an arena. Right away, we begin to appraise the "quality" of the data by asking about the methods used in recording the behaviors, and statistics such as interobserver reliability.

How do we get high interrater reliabilities in behavioral research? Specifying operational definitions of the behaviors to be observed and training observers to follow the operational definitions is the typical approach. Operational definitions are themselves models for how to symbolically extract information from the vast amount of sensory information we take in or the visual information we record on videotape. As with any model, operational definitions leave out information that may turn out to be relevant or may contain biases.

In our modeling studies of sensorimotor development of rat pups, we videotaped pups from directly above an arena (Schank and Alberts, 2000a, 2000b; Schank et al., 2004; May et al., 2006). Videotaped data are the initial representation of the organism in an experimental context and are the raw data from which we extract measurements. Videotaped raw data are too intractable to work with. Thus, we have to extract information from the videotapes. Taking into consideration several factors, including the reliability of humans in extracting features from videotapes (Schank and Alberts, 2000b), we sample a frame every *n* seconds, and record nose and base of tail coordinates (Schank et al., 2004; May et al., 2006). This creates a new data set for each experiment consisting of Cartesian coordinates for the tip of the snout and the base of the tail for each pup (figure 11.5).

Extracting measurements from the tip of the snout and the base of the tail of each pup greatly increases the tractability of the data, but at the cost of detail and realism.

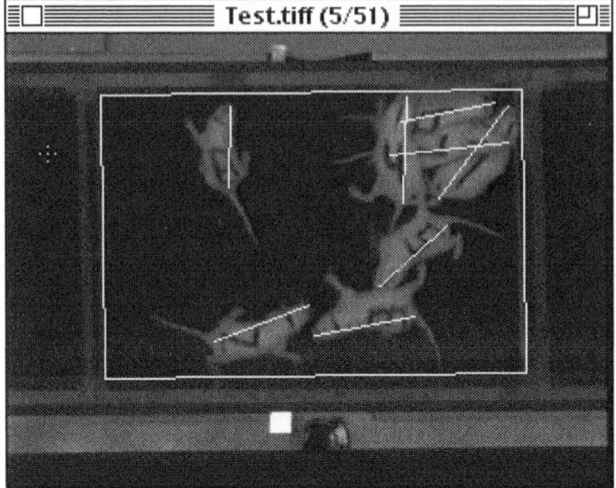

Figure 11.5
Scoring data by recording tip of snout and base of tail coordinates. The recording algorithms correct for the degree to which the arena is not aligned squarely on the videotape.

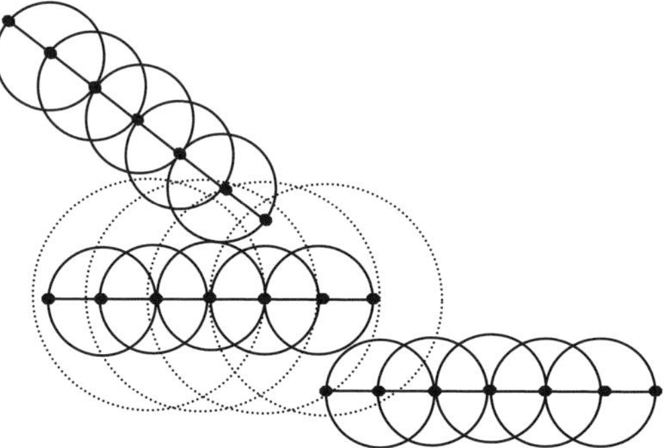

Figure 11.6
Graphical illustration of the pup contact algorithm. The length of a pup is calculated, and each solid circle has a diameter of 1/3 the length of the pup. If we consider the five inner points, we can draw a circle around each with radius r, which is the distance for contact. Thus, for any two pups, if the inner point is within $2r + \varepsilon$, the pups are in contact.

To regain detail and realism in the data, we have developed algorithms to reconstruct some of the lost realism and detail. Of course, these algorithms are models with simplifying assumptions and limitations. For example, we developed an algorithm to take these coordinates and determine which pups are contacting each other (figure 11.6). In this model, line segments represent pups, and points are spaced within line segments to assess contact (see figures 11.5 and 11.6). If we let r be the width of a pup, then for two pups, there are points of contact if an inner point of one pup is within $2r + \varepsilon$ of an inner point of another pup, where ε is an error term we use to tune the algorithm. To tune the algorithm, we systematically vary ε, and compare its classification of contact by visual inspection. We also have developed algorithms for motion, speed, orientation, wall contact, activity, and, given the identification of types of events (e.g., contact or position), conditional probabilities of behaviors.

This approach naturally gives rise to data model variants. There at least two ways in which variants arise. First, from the basic model depicted in figure 11.6, different types of data models can be generated, such as patterns of contact, compactness of aggregates, and contact with other objects and walls. Second, models that more accurately represent the body contours of pups can be developed, if necessary, from the basic model. However, the crucial point is that data as models are not assumption-free, and how good the data are, depends on the assumptions we make, as with any other model.

The Need for New Statistical Models

The need for greater integrativeness in studying behavior highlights the practical problems inherent in behavioral research. Unfortunately, the history of methodology within the behavioral sciences is one of near total stagnation.

After Fisher (1935) published his seminal *The Design of Experiments*, the statistics of agronomy took root in both psychology and behavioral research (reviewed in Gigerenzer, 1980). Why null-hypothesis statistical testing grew in popularity so rapidly, despite the intense and unremitting criticism of its utility (e.g., Cohen, 1994), is a mystery, but may be due simply to ease of use. The result of a null-hypothesis test, the p-value, is an index of the probability of the data, assuming the null hypothesis is true. That the p-value has no relationship to the truth of the null or a particular alternative, nor the probability that a given experimental result will be replicated, is well known but seldom believed (Oakes, 1986).

What we need are better statistical models for analyzing data and relating data directly to theoretical models. This need has been acutely felt within psychology (Cohen, 1994), and has prompted calls for reform of the way graduate students are educated in statistics (Schmidt, 1996). It has also fostered the development of methods such as meta-analysis, bootstrapping, and Monte Carlo simulation, which while not directly testing theoretical hypotheses, are an improvement over simple null-hypothesis testing. For example, meta-analysis permits the quantitative evaluation of theories by examining multiple independent experiments, and has been adequately reviewed elsewhere (e.g., Schmidt, 1996; Kirk, 1996). Although it is beyond the scope of this chapter, we believe that bootstrapping and Monte Carlo techniques are possible ways out of the Fisherian trap that will allow better assessment of the relationship between data and theoretical models.

To illustrate possibilities, we have combined statistical analysis with the agent-based probabilistic models of pup behavior described above (Schank and Alberts,

2000a). Specifically, we have used a regression-like approach in which a data set is fitted to a model by finding parameter values that minimize the proportion of the data not explained by the model. In the case of linear regression, the analytical solution for the fit of the data to the model is straightforward: minimize the mean-squared error and thereby minimize the proportion of variance not explained by the model. For probabilistic agent-based models, there are no analytical solutions, and it is not clear that minimizing the mean-squared error is the best approach (Schank and Alberts, 2000a).

However, fit can be achieved using genetic algorithms to fit data to models (genetic algorithms are, of course, models of evolutionary processes; Schank, 2001a). Once a good fit is found, simulated experiments are run (with the best evolved model), data from the model are collected, and summary statistics are calculated. Thousands or millions of simulation experiments are run, allowing us to estimate the expected values for the predictions of the model with confidence intervals (Schank and Alberts, 2000a). This allows us to assess not only the fit of the model to the data, but also and more important, to estimate the probability of different aspects of the data, given different components of the model (Schank and Alberts, 2000a).

Conclusions

Biobehavioral research is unavoidably multimodel in nature. With the development and addition of new methods and theoretical modeling techniques, it will become more complicated. Nevertheless, we believe this is the only way an integrative and dynamic understanding of behavior and its mechanisms is possible. To deal with the complexity of multimodel research, we described nine dimensions we think are important to models. Models in any active area of research can be seen as phylogenetically related. We argued that there are several useful features of constructing phylogenies of models. First, they may be the best characterization of theories in a given area of research. They also allow apportionment of praise and blame through the evolution of models. We have illustrated that models exist at all levels of analysis in behavioral research. There are no epistemically privileged models, and models are not necessarily hierarchically ordered. Models at any level of analysis can lead to the revision or even rejection of models at other levels. Thus, paradoxically, theoretical models may lead to the reevaluation and ultimate rejection of data.

Acknowledgments

This work was supported in part by NIH grant 5R01MH065555-02 to Schank and NIH grant T32 MH18273 to Koehnle.

References

Alberts JR, Motz BA, Schank JC (2004) Positive geotaxis in infant rats (*Rattus norvegicus*): A natural behavior and a historical Correction. J Comp Psychol 118: 123–132.

Altmann J (1974) Observational study of behavior: Sampling methods. Behaviour 48: 227–265.

Arkin A, Shen P, Ross J (1997) A test case of correlation metric construction of a reaction pathway from measurements. Science 277: 1275–1279.

Barnett SA (1975) The Rat: A Study in Behavior (rev. ed.). Chicago: University of Chicago Press.

Bateson P (1988) The active role of behavior in evolution. In: Evolutionary Processes and Metaphors (Ho M-W, Fox SW, eds), 191–207. New York: Wiley.

Beer RD (2003) The dynamics of active categorical perception in an evolved model agent. Adapt Behav 11: 209–244.

Belovsky GE (1978) Diet optimization in a generalist herbivore: The moose. Theoret Pop Biol 14: 105–134.

Bothe GWM, Bolivar VJ, Vedder MJ, Geistfeld JG (2004) Genetic and behavioral differences among five inbred mouse strains commonly used in the production of transgenic and knockout mice. Genes, Brain Behav 3: 149–157.

Cohen J (1994) The earth is round ($p < .05$). Amer Psychol 49: 997–1003.

Collier GH, Hirsh E, Hamlin PH (1972) The ecological determinants of reinforcement in the rat. Physiol Behav 9: 705–716.

Cooper JJ, Mason GJ (2001) The use of operant technology to measure behavioral priorities in captive animals. Behav Res Meth Instrum Comput 33: 427–434.

Coulson T, Rohani P, Pascual M (2004) Skeletons, noise and population growth: The end of an old debate? Trends Ecol Evol 19: 359–364.

Darwin C (1882) The Descent of Man and Selection in Relation to Sex (2nd ed). London: John Murray.

Dewey J (1896) The reflex arc concept in psychology. Psych Rev 3: 357–370.

Duncan AJ, Gordon IJ (1999) Habitat selection according to the ability of animals to eat, digest and detoxify foods. Proc Nutr Soc 58: 799–805.

Fisher RA (1935) The Design of Experiments. London: Oliver and Boyde.

Flatt T, Scheuring I (2004) Stabilizing factor interact in promoting host-parasite coexistence. J Theoret Biol 228: 241–249.

Giere RN (1999) Using models to represent reality. In: Model-Based Reasoning in Scientific Discovery (Magnani L, Nersessian NJ, Thagard P, eds), 41–57. New York: Kluwer/Plenum.

Gigerenzer G (1980) The Empire of Chance: How Probability Changed Science and Everyday Life. New York: Cambridge University Press.

Gigerenzer G (2000) Adaptive Thinking. New York: Oxford University Press.

Griesemer JR, Wimsatt WC (1989) Picturing Weismannism: A case study in conceptual evolution. In: What Philosophy of Biology Is: Essays Dedicated to David Hull (Ruse M, ed), 75–137. Dordrecht: Kluwer.

Hassel MP (1978) The Dynamics of Arthropod Predator-Prey Systems. Princeton, N.J.: Princeton University Press.

Hemple C (1966) Philosophy of Natural Science. Englewood Cliffs, N.J.: Prentice-Hall.

Hesse M (1966) Models and Analogies in Science. Notre Dame, Ind.: University of Notre Dame Press.

King AA, Hastings A (2003) Spatial mechanisms for coexistence of species sharing a common natural enemy. Theoret Pop Biol 64: 431–438.

Kirk RE (1996) Practical significance: A concept whose time has come. Educ Psychol Measure 56: 746–759.

Koehnle TJ, Gietzen DW (2005) Modulation of feeding behavior by amino acid-deficient diets: Present findings and future directions. In: Nutritional Neuroscience: Overview of an Emerging Field (Prasad C, ed), 145–160. Boca Raton, Fla.: CRC Press.

Koehnle TJ, Schank JC (2003) Power tools needed for the dynamical toolbox. Adapt Behav 11: 291.

Laudan L (1977) Progress and Its Problems. Berkeley: University of California Press.

Laudan L (1996) Beyond Positivism and Relativism: Theory, Method, and Evidence. Boulder, Colo.: Westview Press.

Levins R (1966) The strategy of model building in population biology. Amer Scient 54: 421–431.

Lewontin RC (1982) Organism and environment. In: Learning, Development, and Culture (Plotkin HC, ed), 151–170. New York: Wiley.

Lotka AJ (1956) Elements of Mathematical Biology. New York: Dover.

May CJ, Schank JC, Joshi S, Tran J, Taylor RJ, Scott I (2006) Rat pups and random robots generate similar self-organized and intentional behavior. Complexity 15: 53–66.

Maynard Smith J, Price GR (1973) The logic of animal conflict. Nature 246: 15–18.

Moran TH, Schulkin, J (2000) Curt Richter and regulatory physiology. Amer J Physiol Regul Integr Comp Physiol 279: R357–R363.

Murphy ME, King, JR (1989) Sparrows discriminate between diets differing in valine or lysine concentrations. Physiol Behav 45: 423–430.

Murphy ME, Pearcy SD (1993) Dietary amino acid complementation as a foraging strategy for wild birds. Physiol Behav 53: 689–698.

Nepomnyashchikh VA, Podgornyj KA (2003) Emergence of adaptive searching rules from the dynamics of a simple nonlinear system. Adapt Behav 12: 245–265.

Oakes M (1986) Statistical Inference: A Commentary for the Social and Behavioral Sciences. Chichester, UK: Wiley.

Oftedal OT (1991) The nutritional consequences of foraging in primates: The relationship of nutrient intakes to nutrient requirements. Philos Trans Roy Soc Lond B334: 161–169.

Orzack SH, Sober E (1993) A critical assessment of Levin's "The strategy of model building in the social sciences" (1966). Quart Rev Biol 68: 533–546.

Overton WF (1994) Interpretationism, pragmatism, realism, and other ideologies. Psychol Inquiry 5: 260–271.

Pulliam HR (1975) Diet optimization with nutrient constraints. Amer Naturalist 109: 765–768.

Schank JC (2001a) Beyond reductionism: Refocusing on the individual with individual-based modeling. Complexity 6: 33–40.

Schank JC (2001b) Dimensions of modelling: Generality and integrativeness. Behav Brain Sci 24: 1075.

Schank JC, Alberts JR (1997) Self-organized huddles of rat pups modeled by simple rule of individual behavior. J Theoret Biol 189: 11–25.

Schank JC, Alberts JR (2000a) The developmental emergence of coupled activity as cooperative aggregation in rat pups. Proc Roy Soc Lond B267: 2307–2315.

Schank JC, Alberts JR (2000b) A general approach for calculating the likelihood of dyadic interactions: Applications to sex preferences in rat pups and agonistic interactions in adults. Anim Learn Behav 28: 354–359.

Schank JC, May CJ, Tran JT, Joshi SS (2004) A biorobotic investigation of Norway rat pups (*Rattus novegicus*) in an arena. Adapt Behav 12: 161–173.

Schmidt FL (1996) Statistical significance testing and cumulative knowledge in psychology: Implications for training of researchers. Psychol Meth 1: 115–129.

Schoener TW (1971) Theory of feeding strategies. Ann Rev Ecol Sys 2: 369–404.

Stephens DW, Krebs JR (1986) Foraging Theory. Princeton, N.J.: Princeton University Press.

Stigler SM (1986) The History of Statistics: The Measurement of Uncertainty Before 1900. Cambridge, Mass.: Harvard University Press.

Stone JR (1996) The evolution of ideas: A phylogeny of shell models. Amer Naturalist 148: 904–929.

Suppes P (1962) Models of data. In: Logic, Methodology and Philosophy of Science (Nagel N, Suppes P, Tarski A, eds), 252–261. Stanford, Calif.: Stanford University Press.

Thompson DB, Tomback DF, Cunningham MA, Baker MC (1987) Seed selection by dark-eyed juncos (junco-hyemalis): Optimal foraging with nutrient constraints? Oecologia 74: 106–111.

Timberlake W (2002) Niche-related learning in laboratory paradigms: The case of maze behavior in Norway rats. Behav Brain Res 134: 355–374.

van Baalen M, Sabelis MW (1999) Nonequilibrium population dynamics of "ideal and free" prey and predators. Amer Naturalist 154: 69–88.

von Bertalanffy L (1968) General System Theory. New York: George Braziller.

Webb B (2001) Can robots make good models of biological behaviour? Behav Brain Sci 24: 1033–1050.

Whiten A, Byrne RW, Barton RA, Waterman PG, Henzi SP (1991) Dietary and foraging strategies of baboons. Philos Trans Roy Soc Lond B334: 187–195.

Wimsatt WC (1987) False models as means to truer theories. In: Neutral Models in Biology (Nitecki M, Hoffman A, eds), 23–55. New York: Oxford University Press.

Wimsatt WC (1997) Aggregativity: Reductive heuristics for finding emergence. Philos Sci 64, Supplement: Proceedings of the 1996 Biennial Meetings of the Philosophy of Science Association. Part II: Symposia Papers (Dec. 1997), S272–S384. Philos Sci 64 (Proceedings) S372–S384.

12 Information-Theoretic Modeling of Sensory Ecology: Channels of Organism-Specific Meaningful Information

Chrystopher L. Nehaniv, Daniel Polani, Lars Olsson, and Alexander S. Klyubin

Information theory developed by C. Shannon and his followers in the mathematical theory of communication surprisingly but successfully abstracted away from two questions: (1) the origin and maintenance of information channels, and (2) the meaning of information. However, in understanding the evolutionary sensory ecology of the many information channels used by particular organisms, we are confronted with exactly these issues. What are the evolutionary and developmental origins of particular channels in an embodied organism? How do they benefit the organism? Why this type of sensor and not another? Sensors are costly to build, maintain, operate, and carry. The costs and benefits of access to particular information have an impact on survival and reproductive sucess within a particular ecological context. We introduce a framework in which channels of information meaningful to an organism (sensors, actuators, and internal channels within it, and between it and other organisms, and/or between an organism and its environment) can each be treated using an extended Shannon information theory. Rigorous information metrics relate informational channels to organism-specific relevance and utility.

The study of such organism-specific measures of relevant information is aimed at the development of a predictive theory applicable to the evolution of sensory channels in both biological and artificial systems. We also briefly consider the issues of the temporal horizon and interaction in this framework. Computer simulation and robotic and biological models illustrating the approach yield insight into the selection of sensory channels in relation to their relevance to organismal goals such as reproductive success. We seek to understand the origin, maintenance, and evolutionary and adaptive changes in the sensory apparatus of living things. Sensors are not arbitrary in construction nor, evidently, are they informing an organism of most of the vast quantities and kinds information that could possibly be extracted from its environment.

What are the benefits and costs of being able to perceive particular *sources of information*? There are costs both of development of and of support of the metabolism required by the corresponding sensory organs, as well as due to processing of the information they gather. What determines which type of sensory channels are selected and which are not? And why are certain sensors elaborated in certain ways (e.g., to have high resolution or to be more sensitive in a particular stimulus range), while others are not?

Constructive Biology can be pursued as an approach to modeling and understanding biological systems. As a complement to describing biological organisms, it is desirable to develop predictive theory and to test our understanding by building realizations of the biological processes we seek to understand. These realizations may be considered *models*, but in some cases they may also be actual *instances* of the phenomena under investigation. For example, seismographs, electric eyes, radar, and other electronic sensors are doing real sensing, independent of whether or not we regard them as models of any particular kind of biological sensing. Some insightful examples of such constructive approaches include those of Webb (1995) to modeling cricket phonotaxis; the evolution of communication (technically defined as signaling that tends to increase reproductive success of both sender and recipient) demonstrated by MacLennan (1992) in a synthetic approach to ethology; while social recognition and release of interactive behaviors have been shown in minimal social robots (e.g., Dautenhahn, 1995).

Similarly, evolution that is implemented on populations of RNA molecules or "digital organisms," as well as possibly serving as a model of evolution of life on Earth, is itself an instance of evolution to the extent that it satisfies Darwin's requirements of heritability, variability, and differing reproductive success in a struggle for existence. Self-replication has been constructively demonstrated in automata models (von Neumann, 1966; Langton, 1984), as has evolution in populations of such self-reproducing "digital organisms" (Ray, 1991; Sayama, 1999).[1]

This viewpoint of Constructive Biology is one of a generalized biology, including biology on Earth but also all of life as it could be, even in currently undiscovered cases such as unexplored environments and distant worlds as studied in Astrobiology (Lunine, 2004), or in other, nonorganic media as studied in Artificial Life (Langton, 1989), including studying the possibilities of artificially created protocells, digital organisms in software, and instances of evolution other than the one instance we know from Earth biology. In this chapter, we overview current steps toward a mathematical, predictive, constructive theory for understanding the evolution of sensors based on information theory and some explorations by building systems in which sensory processing and organization literally evolve.

Bringing Information Theory to Biology?

It is not always clear how to bring rigorous notions of information and information-processing into biology. How information is involved in growth and development is, despite remarkable advances of developmental regulatory genetics and evolutionary developmental biology, still little understood. There are, however, at least two obvious inroads for information theory. One is the study of heritable genetic information, to the extent that this information is encoded in digital strings of nucleic acids, DNA and RNA, although the levels of processing and importance of epigenetic marking (and so on) are considerably more complex than originally could be envisioned by any string-level analysis. The second obvious area is biological sensing. Clearly, any sensor involves an information channel to the organism and a set of possible values that the sensor can assume, such as all possible images represented in an eye by states of activation of its component rods and cones.[2] This latter situation is near to that formalized in Shannon information theory, although we will not be able to apply this theory without some care (see below).

Information as the "Currency" for Embodied Computation

One of the unsolved problems of biology is to understand how nature achieves the enormously powerful performance of its information-processing mechanisms. Since the emergence of the first attempts to construct artificial models for information-processing and self-organization in cybernetics and artificial intelligence (Turing, 1936, 1952; McCulloch and Pitts, 1943), this question has been at the forefront of research in these fields. Different paradigms have been explored to tackle that question: symbolic approaches (see Russell and Norvig 2002); the power of parallel processing and, with it, neural computation (Rumelhart et al., 1987); the artificial modeling of evolutionary search (see Fogel, 1998). Common to each of the approaches is the choice of a particular modeling architecture. This provides a mechanism to develop substantial understanding of the selected architecture and ways to improve it; on the other hand, it also provides for a prolific variety of architectures and mechanisms, the most successful of which can be brought to solve many given scenarios to a similar degree.

At this stage, however, problems still remain. Optimizing a particular architecture for a certain task arena may provide an understanding of extracting high performance from the specific architecture selected. It does not, however, provide fundamental insights into the principles that guide the discovery of suitable solutions by self-organized processes. Furthermore, a significant amount of "hand-tuning" and preselecting the processing models still remains, which makes it difficult to understand how a "blind" evolutionary and self-organization process could have achieved that in nature. In fact, the particular selection of models corresponds to picking a "coordinate system"[3] in physics. It is possible to do appropriate calculations in a world ruled by a particular coordinate system, but significant principled insights, as well as new paradigms and perspectives, can best emerge in a coordinate-free view of systems.

In the following, we will argue that such a view can indeed be constructed for organisms (and other agents) and that it can give novel insights into the fundamentals that guide the evolutionary emergence and development of their sensory processing. The central concepts required here are the notions of *entropy* and *information.* Introduced by Shannon (1948) and with intimate connection to thermodynamics, mathematics, engineering, and coding theory, information theory provides a universal quantity to measure the "amount" of data transmitted in a channel between two nodes, a "source" and "target." Information processed by an organism can be treated with this formalism (see below) with the help of some additional notions.

Precursors: Meaningful Information

We will need several ingredients beyond information theory before we can apply it fruitfully to the understanding of sensor evolution. These are hinted at by the insights of many scientists who worked in different fields:

• A deep insight of C. S. Peirce, the father of semiotics is that signs *mediate* meaning, which makes sense only in the context of systems of signs, and that something (an *interpretant*) must link a sign or interpreted signal (a *signifier*) to what is *signified* in an embedded and embodied process (*semiosis*). (See Peirce, 1965.) This situated and embodied nature of agent semiotics highlights the meaninglessness of signals, signs, and sign systems in isolation, without agents and thus without uses. For organisms, the genesis, perception, and expression of signs is shaped by evolution in an ecological context.

• Wittgenstein emphasized that "meaning" of words and other signs lies in their *use* by agents engaged in context in *language games* (including artificial and everyday language). (See Wittgenstein, 1968.)

• Von Uexküll pointed out the crucial role of the *Umwelt*, the sensory world of an animal (von Uexküll, 1909).

• Shannon formalized information as the (average) reduction in uncertainty when a signal is received, and developed the theory of information channels, transmission rates, and capacities, as well as coding in the presence of noise. Meaning of information was, however, completely excluded from the discussion. (See Shannon, 1949, and below.)

• Animal communication studies have investigated coding, game-theoretic analyses of signaling in particular species, and connections to optimization theory, and have developed signal detection theory for organisms (see, e.g., Bradbury and Vehrencamp, 1998).

• To understand an animal, one must understand its world: embodiment, sensors and actuators, ecological context, predators/prey, the lighting conditions of its environment, and those creatures that can sense it or that it can sense (von Uexküll, 1909; Endler, 1992). As von Uexküll wrote in the early twentieth century:

> Our anthropocentric way of looking at things must retreat further and further, and the standpoint of the animal must be the only decisive one.
>
> When this occurs, everything that we hold to be self-evident disappears: all of nature, earth, heaven, stars, indeed, all the objects that surround us. Only those effects remain as factors of the world that, corresponding to [its] construction plan, exert an effect on the animal. Their number and the way they fit together is determined by the construction plan of the animal. Once this correspondence between construction plan and the external factors has been carefully researched, then a new world takes shape around every animal, completely different from our world: its *Umwelt*. (von Uexküll, 1909)

- Evolution of signalers and receivers may be influenced by related processes of *sensory drive* (Endler, 1992), *sensory exploitation* and *receiver bias* (Ryan, 1998), and by *receiver psychology* (Guilford and Dawkins, 1993).

Shannon Information Theory and Some Unanswered Questions

> *Signal to noise ratio [in biology] can only be defined with reference to receiver's perception.*
> —(Innes Cuthill, pers. comm., 2003)

The "mathematical theory of communication" (Shannon, 1948) completely abstracts away (1) the origin and maintenance of information channels and (2) the meaning of information.

The paradigm considers a given arbitrary configuration of *sender* or *source* (information source, and transmitter which encodes and sends the message in a discrete, continuous, or mixed signal); the channel (signal medium, upon which noise acts); a source of noise; and *recipient* or *target* (receiver which decodes the message and passes it to the destination; see figure 12.1). Such an information channel is often called a "communication channel" by information theorists, which can lead to confusion. The channel might not satisfy an ethologist's definition of communication (see, e.g., MacLennan, 1992; Bradbury and Vehrencamp, 1998): neither source nor target need benefit from the transmission (even in a statistical sense of benefiting on average).

Indeed, neither the source nor the target need be an organism (generally required by any ethological definition), but can be any entity (even an inanimate entity in the environment may be a source: the nighttime moon as a source of electromagnetic waves; the magnetic field of the Earth; falling water initiating pressure waves interpreted as an acoustic signal; the sun emitting light with polarization visible, for instance, to hymenopterans. Inanimate targets may be a set of concentrations of enzymes within a living cell changed by a channel of protein biosynthesis, or a configuration of objects in the environment in stigmergy).

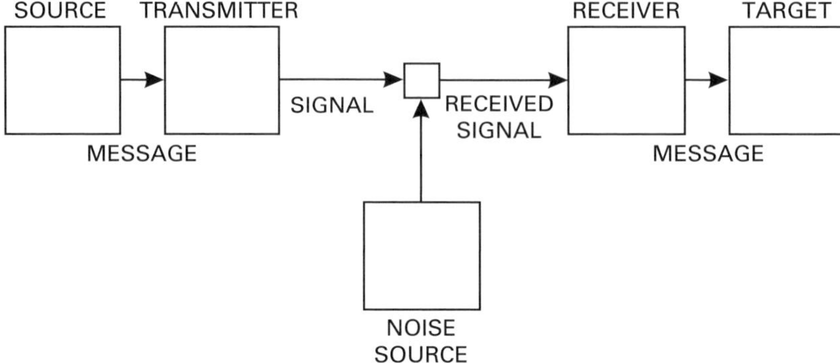

Figure 12.1
Schematic diagram of a general, abstract information channel, the classical setting of information theory. Issues of meaning and relevance, origins and maintenance of channels, and rationale for signaling intermittency or channel nonusage are left open in the framework, which completely abstracts away from them. (Modified from Shannon, 1948.)

The paradigm considers an arbitrary sender-channel-plus-noise-receiver configuration, and defines and relates entropies of sources and channel capacities to coding theorems that address minimization of errors. This theory views *information* as reduction in uncertainty (entropy) concerning what signals a given source will produce. Numerous applications exist in telecommunications, as well as in many areas of science and engineering. Formally, if an information source, modeled as a random variable X can emit values $\{x_1, \ldots, x_n\}$, and the probability that value x_i occurs is $p(x_i)$, then the *entropy* of X is $H(X) = -\sum_{i=1}^{n} p(x_i) \log p(x_i)$ and, as mentioned, measures the amount of uncertainty in the value of X in *bits*[4] (note: logarithms throughout this chapter are taken to base 2). This is also the amount of information obtained—the reduction of uncertainty—when the value of X has been determined.[5]

Both the power and the weakness of this seemingly innocuous quantity arise from the fact that Shannon information explicitly ignores any semantics of the data transmitted. This has made it difficult for a long time to endow the concept of information with a "meaning" or "relevance," something that would be of essence in intelligent information-processing. In fact, Gibson (1979:242) goes as far as to argue that in spite of the attractiveness of its universality, Shannon information is inadequate to address these issues for perception precisely because of this indifference to semantics.

Classically, information theory, quite intentionally, does not at all discuss several key questions[6] that it is natural for a biologist to ask when attempting to apply the mathematical theory of information to organisms:

1. Where does such a channel come from?
2. When and why should an informational channel be used?

3. How can such channels change over time?
4. Who or what are the source and recipient?
5. How regularly and how often should a channel emit any signal at all?
6. What are the costs and benefits of constructing and using the channel?

The fact that organisms are engaged in communication and signaling, sensing, and acting on their environments (including other organisms) entails the existence of myriad channels of information. These channels are not arbitrary in their origin, structure, source, and targets, but arise and change in the course of evolution. They provide channels by which a particular organism can perceive and manipulate its world; it seems reasonable, therefore, to assume these carry information that is meaningful for an organism—useful for its survival and reproduction (Nehaniv, 1999).

"Biological semantics"—or, terminologically better, *biosemiotics*—cannot be developed with absolute notions of meaning and relevance that are somehow independent of any organismal embodiment.[7] The huge differences between organisms' bodies and environments entail that these notions are well defined only when considered with respect to specific organisms or at least specific species.

Dropping Some Conceptual Baggage

For purposes of understanding biological sensing, signaling, and communication, classical information theory is applicable, but by itself is incomplete. To bring rigorous information theory into biology, it is necessary to violate some of the "received wisdom," dropping baggage based on Western philosophical notions of absolute Platonic truth and meaning that are rooted in the assumed centrality of human language and logic. Human language is highly evolved and complex, and formal logic is abstracted from it. It is necessary to explicitly reject some tacit assumptions commonly used in naive (and not-so-naive) theories of meaning and representation that have been based on the misleading step of taking human language and formal logic as central models rather than as the highly derived, nonprimitive, powerful, and specialized systems that they are (see Millikan, 2004a, 2004b; Harms, 2004 for illuminating discussions). In this regard, the following contrary paradigmatic views are adopted:

• "*No agent, no meaning!*" (Nehaniv and Dautenhahn, 1998). That is, it makes no sense to speak of meaningful information except with respect to a specific agent or organism. Meaningful information for one organism might well be meaningless for another.

• "Meaning" can begin to be approached only as related to agent-specific information (1) that may occur in interaction games between the agent and its environment or between agents, mediated by their own sensors and actuators (sensor/actuator channels, or internal state) and (2) that is *useful*, in a statistical cost-benefit sense, for satisfying homeostasis and other drives, needs, goals, or possibly intentions, especially evolutionary fitness (survival, fecundity, reproductive success). (Nehaniv et al., 1999, 2002)

• Truth-values and truth conditions of biological representations are not in general separate from their purposeful role in guiding the actions and responses of organisms (Millikan, 2004a).

Dynamics of Sensor Evolution: Origin, Maintenance, and Evolution

Of that vast morass of potentially measurable physical phenomena in an organism's environment, most are not perceived. For instance, human and other animal senses do not detect the myriad television and radio broadcasts, and other signals and noise, carried on potential channels along most of the electromagnetic spectrum. Our bodies and resources are finite. We have sensors for only an infinitesimal fraction of what could potentially be sensed with appropriate sensory apparatus.

These limitations force the restriction of senses to be finite in number, with limited capacity, and to tune in to only a small subset of all potential channels of information. Existing chemicals and structures in organisms, such as opsins and jawbones, may acquire rudimentary sensory functionality, such as vision and hearing, via exaptation. If an organism has access to some channel providing a reproductive advantage, then it is likely to be preserved in evolution. If evolutionary advantage is to be had by varying the range of sensitivity of a channel, or its resolution, then, if evolutionary variability can generate individuals who can do so[8], these are likely to persist and proliferate. Evolution roams a vast *design space* of body plans and their respective sensory apparatus with regions of differing viability and accessibility to evolution for given ecological contexts (see chapters 5 and 14 in this volume).

Diversity of Sensors in Nature

Sensors are important. They determine what information an organism can have regarding its environment, including other organisms. By actively manipulating its environment, an organism can change what it perceives, and such action, followed by new perception in the *perception-action loop*, can guide further action. Nature has produced a wide variety of sensory organs in different modalities that are well adapted for the specific animals and their respective environments.[9]

Sensor Types

• Chemical

• Olfactory

- Tactile/haptic
- Auditory
- Sonar
- Photoreceptive
- Electrical
- Magnetic

Sensors for these modalities have evolved via frequent exaptation, where an existing actuator/body part becomes involved in sensory function, followed by elaboration and tuning to useful channels, or sometimes via the loss of sensors that are costly to produce and no longer useful, as in the well-known loss of vision in dark-dwelling fish and salamanders.

Eyes

Eyes have evolved forty to sixty times in independent lines of descent, with different levels of differentiation and specialization, exhibiting convergence phenomena, appropriate to the carrier. Typical vertebrates have five receptor types (rods, UV [ultraviolet], SWS [short wavelength sensitivity], MWS [middle wavelength sensitivity], and LWS [long wavelength sensitivity]). Birds and reptiles have all five types. Most placental mammals have SWS, LWS, and rods. Many marine mammals have rods and LWS. In the evolution of nocturnal animals, many lost some chromatic vision, but primates reinvented trichromacy, many of them having dichromatic males and trichromatic females. Cave- and dark-dwelling animals frequently evolve reduction or loss of eyes. Many insects have UV, green, and blue photoreceptors. There are also numerous invertebrate types of photorecptors (see figure 12.2). For a review, see Kelber et al. 2003).

Interactions

Channels of Meaningful Information

What kind of informational channels can carry organism-specific meaningful information? The types of channels of information meaningful to an organism may include sensing, actuating, communication/signaling—offloading, active perception, stigmergy—and actively creating the conditions for relevant information acquisition by other organisms. If we write O for an organism, O' for another organism, and E for environment, then schematically we can indicate the source and target of different types of channels used by the organism: motor channels, $O \to E$; sensory channels, $E \to O$.

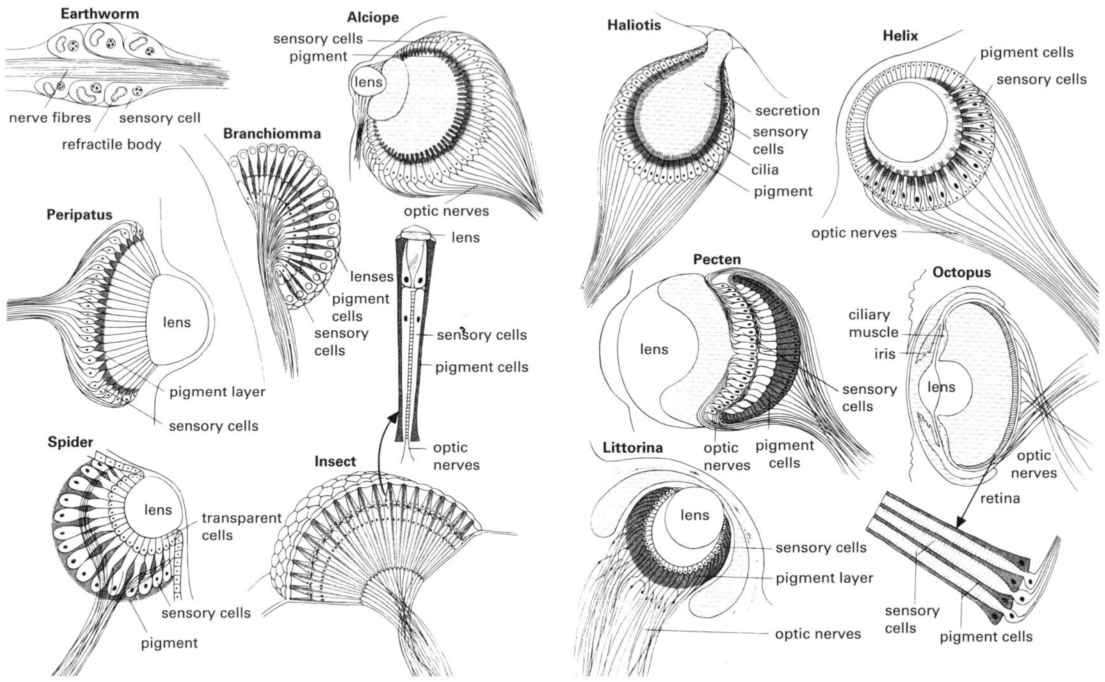

Figure 12.2
Diversity of eyes illustrated by a selection of invertebrate eyes. Eyes in evolution on Earth have arisen with forty to sixty independent lines of descent, with different levels of differentiation and specialization; display convergence phenomena; and are appropriate to their carrier. (From Wells, 1968; used by permission of the author.)

These have special cases, communicative or signaling channels: $O \to O'$, $O \to O'$. Finally, an organism may have access to channels whereby it "sends" information to itself $O \to O$, including memory and internal state (e.g. hormonal) changes, or via the external environment in self-stigmergy and information offloading. Stigmergy and other indirect communication not targeted at any particular O' can be factored into two or more channels that proceed via the environment (see the section on information flow, below).

An intuition on how to bring organism-specific meaningful information within the framework of information theory is to formalize it roughly as mutual information between a sensed variable (such as an image currently registered in the eye) and a variable used to achieve the goals of the agent (e.g., action to be taken that maximizes reproductive success, fitness, fecundity, or survival).

Useful sensors will have high mutual information (measured in bits) between their own values and another random variable whose value, if available, allows achievement of the goals of the organism (e.g., action to be taken that maximizes reproductive success, fitness, fecundity, survival). Meaningful information need not be carried in a sensory channel, but can exist in various types of channels: sensing, actuating, communication/signaling, interaction games, and remembering. Meaningful information can also be stored in or pass through the external environment, as in offloading, active perception, and stigmergy. It is to be expected that evolved organisms will actively create the conditions for facilitating their own relevant information acquisition.

Interaction Games: Signaling

Tremendously rich animal-animal interactions involving channels of signaling and communication that employ strange modalities and sophisticated strategies occur every day. For instance, direct neural modulation of skin patterning, coloration, and texture occurs in many cephalopods. Although probably color-blind, some can reflect and match ambient colors, match background patterning and motion, and perhaps see and display patterns of polarized light to each other on their bodies (a channel invisible to teleost fishes). They adopt communicative coloration indicating or hiding their intentions from potential rivals, mates, predators, and prey (figure 12.3), often with multiple (conspecific and/or interspecific) targets at once (e.g., Moynihan, 1985; Hanlon and Messenger, 1996). Social signals in many other animals require exquisite sensory and motoric apparatus, as well as sophisticated neural processing (Hauser, 1997; Bekoff, 1977; Smith, 1995).

Generalizing Wittgenstein's considerations of language use in context, the term *interaction games* refers to the particular patterns of usage of information channels by organisms of the same or different species in particular classes of interactions (Nehaniv, 1999).

This area of signaling in interaction games is related to network information theory where many senders and receivers inhabit an environment with interference and noise (see, e.g., Cover and Thomas, 1991), although—unlike this theory—the evolution of organisms is not necessarily concerned with maximization of achievable transmission rates. Moreover, in many cases, as with potential prey or predators, avoiding transmission or obfuscating the relevance of signals to potential recipients may be highly desirable (cf. Moynihan, 1985). Nevertheless, network information theory, extended to networks of channels changing over organismal life time and evolutionary time, may eventually be developed to yield deeper insights into social and biological interaction games.

Costs and Benefits of Sensory Processing

There are, for any organism, costs and benefits of building, maintaining, and operating sensors (or access to other channels of information, such as those involved in memory and in motor activity). Organisms build, maintain, carry, and operate

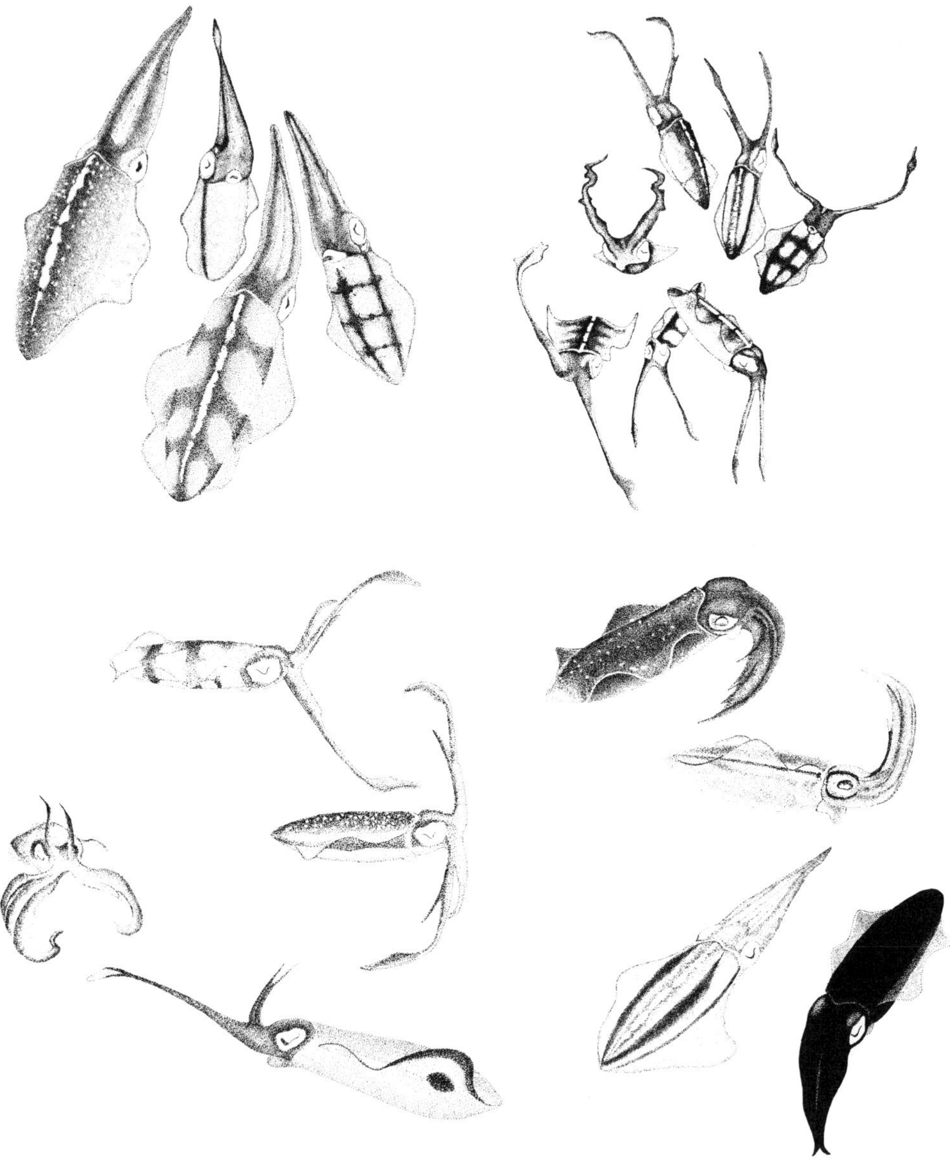

Figure 12.3
Complex sensing and signaling interactions of the Caribbean reef squid *Sepioteuthis sepioidea* may involve deception, manipulation, communication, signaling mate attraction, threatening a rival, camouflage, and confusing prey and predators. (Redrawn from Moynihan and Rodaniche, 1982.)

sensors at significant cost. This involves many trade-offs in the acuity of sensors (e.g., frequency vs. temporal resolution in vision).

Laughlin et al. (1998) have studied the information channel of the phototransduction in the *Calliphora* (blowfly) photoreceptor and its cost in terms of the energy currency molecule ATP (adenosine triphosphate). This provides approximately 1,000 bits/second at a cost of 10^6 ATP molecules/second to operate a vigorous sodium/potassium pump in the large monopolar secondary photoreceptor (LMC). Overall, the fly photoreceptor consumes 10^{10} ATP/second, while coding about 1,000 bits at a unit cost of 10^7 ATP/bit (comprising 10 percent of the basal metabolic rate in the eye, not to mention the cost of building, maintaining, and carrying the eye). This represents an exertion of 480 μmoles ATP per gram minute—similar to human muscle working at an unsustainable rate (Laughlin et al., 1998).

Mutual information of light intensity signal and different types of stimuli (naturalistic, Gaussian noise, or random) has been measured experimentally in these flies (Juusola and de Polavieja, 2003). It is worthy of note that in some flies, males have three times as many facets as females, using 50 percent more ATP/photoreceptor, and pay 4.5 times as much as females, apparently to acquire more relevant information (Hornstein et al., 2000). Investment of this energy suggests fitness is advanced despite the additional costs to males.

Conversely, in the absence of useful information coming through sensory channels, investment of energy, processing power, or any other resources into these channels is not repaid, and so, over evolutionary time, they tend atrophy or may be lost completely. This is seen from examples of eyeless organisms (e.g., cave fish that live in total darkness; see Tian and Price, 2005) and also in the sensory evolution of digital organisms (Ray, 1998).

Clearly there are trade-offs involved between balancing the cost of sensors with the acquisition of useful information, relevant information, and resulting benefits. There are roles for sensor resolution, sensitivity, and information rate here.

Information Theory for Understanding Sensors

A central issue for being able to use information theory for studying sensor evolution is the capability to distinguish between relevant and irrelevant parts of information. Shannon's original theory does not cater for that; in fact, it is one of the remarkable properties of Shannon's approach to do away with any concept of semantics and meaning, which opens it up for rigorous mathematical analysis (Shannon, 1948). However, if we deal with agents that have to pursue selected strategies (whether for survival or to achieve some more specific goal), this approach is no longer sufficient without extension. It is necessary to separate *relevant* parts of (Shannon) information from the information "bulk."

Sensors as an Information Bottleneck

The data coming to an organism via its sensors—or a given sensor—can be represented by the values of a variable that depends on the state X of the ambient universe. As discussed above, most of the information in X cannot be extracted via sensors, nor would it be desirable to have all that data, most of which is wholely irrelevant to the organism.

The key for the organism is to extract information from X via values in some new sensory variable X^* that are useful for the organism. That is, the organism needs sensors that give a (probabilistic) mapping from the state of world X to the readings of its sensors X^* that help maximize some kind of organism-specific utility (such as its fitness, fecundity, survival, reproductive success, or other "goals"). The variable X^* can take only a relatively small, finite number of values compared with the incircumscribable world state X. The sensors (or each sensor) are a *bottleneck* for compressing information from the world; however, it should not attempt to compress all that information, but only information *relevant* for the organism.

Tishby et al. (1999) show how selected aspects of information can be "squeezed" out through a filter, a "bottleneck," directed purely by information-theoretic criteria. The approach chosen there requires a *relevance indicator variable* which corresponds to an externally provided ideal variable Y depending on the world state X, whose value, if known, would provide information on everything relevant at the current state of X. This information, selected through the information bottleneck principle, is called *relevant information* by the authors.

Relevant information is of particular importance when studying agents that capture sensory information from their environment. The richness of typical environments compared with the agent's information-processing capability makes it necessary to limit the information that can be processed. To apply the information bottleneck notions, it is then necessary to instantiate the relevance indicator variable for this purpose. In Nehaniv (1999), relevance is considered with respect to *actions*—in particular, we are concerned with information that is useful in helping the organism determine how to *manipulate* circumstances in order to help it achieve its goals.

This approach qualifies information by distinguishing relevant and irrelevant parts, and thereby resolves Gibson's aforementioned dilemma. While in Polani et al. (2001) there is still the need to specify a suitable utility function explicitly, this approach nevertheless provides a way to translate Gibson's dilemma (i.e., applying information theory to perception and its semantics) into the language of information theory and, moreover, it paves the way to conceptualize distinct types and classes of Shannon information and alleviate the "bulk view" of classical information theory.

Thus, we can view (Shannon) information not just as a bulk quantity measured over a system, but as a commodity to be processed, if necessary in components, not

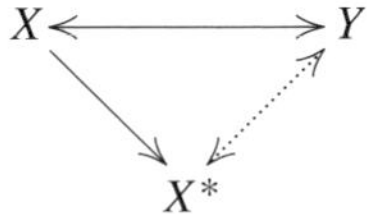

Figure 12.4
The information bottleneck method. Given X and Y, find a random variable X^* that results from a probabilistic mapping $p(x^* \mid x)$ from X such that X^* captures as much information about Y as possible. Note, however, that X^* has access to Y only via X. The dotted double arrow indicates that the connection is only indirect. For details see text.

entirely unlike flow of matter, where different components may be processed by different parts of a metabolism. Of course, Shannon information is not additive in general—in the same way that mass or energy is—since multiple copies and sources might provide the same or overlapping information, so we might expect new kinds of effects to be discovered under this approach.

We have argued that information plays an important role in determining what an organism should and could do. This is not just a "soft" informal statement but, in fact, can be cast into precise mathematical language by suitable adaptation of Shannon's original notion of information. To be able to do so, there are a number of issues to clarify, notions to establish, and formalisms to develop. That is the purpose of the present section.

The Information Bottleneck Method Tishby et al. (1999) propose a formalism, the information bottleneck method, to tackle this problem. This is done by endowing the random variable (from which relevant parts are to be elicited) with a "marker" via a jointly distributed relevance indicator variable. Formally, let X be a random variable that is jointly distributed with Y, its relevance indicator variable. The information bottleneck method (figure 12.4) is interested in only that part of the information in X that is relevant to identifying the value of Y.

This is achieved by constructing a suitable probabilistic mapping (implemented as a conditional probability) $p(x \mid x^*)$ from X to an "extractor" random variable X^*. $p(x^* \mid x)$ is selected such as to maximize the mutual information $I(X^*; Y)$ that the extractor variable obtains about Y—in our case, the amount of relevant information captured by the sensors—while $I(X; X^*)$—in our case, the amount of information extracted from the environment by sensors—is kept fixed.[10] This corresponds to extracting from X as much as possible of what is relevant (i.e., information about Y) while ignoring as much as possible about the rest. Note that X^* has access to Y only via X.

Relevance Through Action The information bottleneck method provides a perspective for the treatment of partial informational components. In the original approach, the relevance is indicated by the "labeling" variable Y, which has to be specified externally (e.g., by human intervention). However, in general such a labeling is not available.

True to Goethe's (or Faust's) principle of "in the beginning was the act," Nehaniv et al. (1999) emphasize that meaning and relevance emerge from actions available to agents. This can be cast rigorously in the language of the information bottleneck by instantiating the relevance indicator variable as an (optimal) action selection strategy (Polani et al., 2001). In this approach, the existence of a utility function is assumed that assigns a utility to an action selected in a given agent state and that thus determines the optimal action. Such utility functions can often be given for agents, and thus provide a generic way of quantifying the amount of environmental information relevant to an agent. The advantage of the approach is that it gives a straightforward, operational quantification of relevant information in the spirit of Shannon, something that was believed to be a considerable challenge, if not impossible, for a long time (Gibson, 1979).

Evolution of organisms with complex useful sensors apparently entails that they are "solving," via the embodied computation of the evolutionary processes, an information bottleneck problem.

Information Ecology of Sensors

The relationship between the ideal relevant variable Y and the state of the world X is modeled as a probabilistic mapping $p(y \mid x)$ describing how the value Y should depend on X. Such a mapping can be found by minimizing $I(X; Y) - \alpha E[u(x, y)]$, using negative expected utility in place of distortion in the iterative Blahut-Arimoto algorithm from information theory, where $E[u(x, y)]$ is the expected *utility* of mapping to $y \in Y$ when the world state is $x \in X$, and where $\alpha > 0$ is a parameter controlling the trade-off between the amount of compression in Y and its resolution in distinguishing among different utilities.[11] Concretely, one can instantiate the relevance indicator variable using the actions to be selected according their optimality with respect to some utility (Polani et al., 2001).

Since X is a random variable for the state of the world, knowledge of this mapping is not directly available to the organism. Instead, the organism needs to access information about X via signals registered at its sensors X^*. These should compress the information in X as much as possible, subject to providing as much information about Y as possible. Again, this can be modeled as the search for another probabilistic mapping $p(x^* \mid x)$ from states of the world to sensor states, such that mutual information with Y is maximized.

This is the same as minimizing $I(X^*; X) - \beta I(X^*; Y)$ over all probabilistic mappings from X to X^*, for the given nonnegative parameter β. There is a trade-off between compression—the informational "efficiency" of sensors—and the amount of useful information $I(X^*; Y)$ provided by the sensors, whose relative importance can

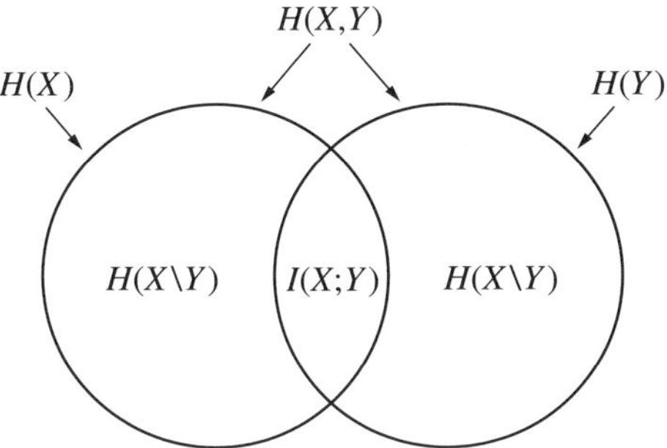

Figure 12.5
Diagrammatic view of information in two information sources, X and Y. $H(X)$ and $H(Y)$ are the entropy of X and Y, respectively, visualized as areas. $H(X \mid Y)$ is the entropy (uncertainty) about the value of X if that of Y is known and is visualized as the left oval excluding the area of overlap with the right oval. Conversely, $H(Y \mid X)$ is the uncertainty about Y, given that X is known. The overlap area of mutual informational $I(X; Y)$ relates to how much information is held in common when values of the variables are read. Information distance is the joint entropy $H(X, Y)$ of the pair (visualized as the union) minus the mutual information (overlap).

be controlled by the parameter β. Now one can apply the information bottleneck methodology as developed by Tishby et al. (1999) to address this problem, using a natural extension of the iterative Blahut-Arimoto algorithm.

In the case of sensors, the evolutionary ecology of organisms should maximize relevant information about Y coming through the sensor channel X^*, while at same the time balancing this against the costs of the sensors. This can be described by finding a mapping from world states to sensor states that minimizes $I(X^*; X) - \beta I(X^*; Y) + \gamma C(X^*)$, where γ is nonnegative, and C gives the cost of building, maintaining, and running the sensory apparatus X^*. In general, the cost C is a function of the kind of sensors constructed with the given probabilistic map $p(x^* \mid x) : X \to X^*$. The current information bottleneck method treats the the case of cost-free sensors ($\gamma = 0$). It is currently an open problem to model appropriate cost functions and to solve this problem algorithmically for positive γ.

Metrics: The Distance between Information Sources

How similar or different is the information provided by two information sources? This question has important consequences for understanding whether novel sensory channels or redundant ones are selected—the latter providing tolerance to error and noise. Similarity of sensory sources is also related to their placement on the body (e.g., due to spatial continuity of stimuli at nearby sensors), as well as to the potential to integrate sensory input from different modalities (see below).

For a measurement to be a metric, the following should hold for all information sources X, Y, and Z:

- $d(X, Y) = d(Y, X)$. (Symmetry)
- $d(X, Y) = 0$ if and only if $X \equiv Y$. (Equivalence)[12]
- $d(X, Z) \leq d(X, Y) + d(Y, Z)$. (Triangle Inequality)

Some Metrics Various metrics can be used to estimate the (dis)similarity of sensors from the time series $X(t)$ and $Y(t)$ of readings given by sensors X and Y, where t is time (Pierce and Kuipers, 1997):[13] In a continuous world, adjacent sensors of similar type have similar values over time t. One metric d_1 is based on the average closeness of values of the sensors:

$$d_1(X, Y) = \frac{1}{T} \sum_{t=1}^{T} |X(t) - Y(t)|,$$

where T is total time. Another measure, d_2, which is nearly a metric, is based on the intuition that the frequency distribution of two close sensors should be similar:

$$d_2(X, Y) = \frac{1}{2} \sum_{k=1}^{n} |dist_k(X) - dist_k(Y)|,$$

where $dist_k(X)$ is the percent of observations within the kth subinterval of possible values of the sensor and n is the number of intervals.

The Information Metric Information theoretic metrics also exist. The distance between two information sources (e.g., two sensors) is defined as

$$d(X, Y) = H(X \mid Y) + H(Y \mid X), \tag{12.1}$$

where $H(Y \mid X)$ is the conditional entropy for Y, given X.[14] The information distance is visualized in figure 12.5 as the nonoverlapping portion of area of the two ovals representing the entropies of X and Y, respectively.

This measure is discussed in detail and shown to satisfy the metric axioms on the space of information sources by Crutchfield (1990). It has several advantages: it is domain independent, and it can find informational structure not found by d_1 and d_2, including nonlinear correlations.

Models: Information Methods in Sensor Evolution

In robotics, sensors are still usually seen as something "given" and fixed. What we would like to achieve is (1) robots that can adapt and evolve their sensors to solve a certain task as efficiently as possible, and (2) ultimately building machines that can "discover" new sensory channels (by hardware or wetware evolution).

This is a pursuit in constructive biology, as a model of sensor evolution in nature, but it is not without obvious applications to science and engineering. Where should we start to start? Assume that a robot or organism receives a number of streams of sensory data with no knowledge of its structure. In order to compare different sensory channels, we need a method to do so. This is provided by the information metric. Given the possibility to compare sensory channels, it should be possible to build a model of the sensory input. We call this the problem of sensory reconstruction. Its solution is described below, and yields a homunculus-like picture of sensoritopic organization, including motor variables.

The Sensorimotor Structure of Robots

The conceptual power of information-theoretic methods is often considered to be marred by the difficulties they offer to practically relevant implementations (e.g., on robotic systems). Numerical and statistical problems are often considered inevitable with purely information-theoretical approaches to real-world implementations. We wish to argue that this is not the case if done properly. In fact, it provides additional insight into possible principles guiding the organization of sensorimotoric capabilities in living beings, and thus allows us to narrow the gap between theoretically deep insights from studies such as those described here and actual biological observation.

Pierce and Kuipers (1997) have developed an approach to create sensory maps from uninterpreted sensors. Among other steps, they measure the distances of the temporal signatures of all pairs of sensors. They then project the sensors into a lower-dimensional space so as to make the pairwise distances between individual sensors in the projection match as closely as possible the distances of their signatures. In their original approach, Pierce and Kuipers use Euclidean and similar metrics, which work well in many cases but presuppose a orderedness and continuity in the sensory input. No such assumption is needed if one uses information-theoretic distance that is sensitive to informational similarity of any kind.

We therefore use the information distance advocated by Crutchfield (1990) for information sources in place of the other measures (Olsson et al., 2004b). Important advantages of using the information metric are that (1) it allows us to apply the mathematical framework of information theory, and (2) it removes any assumptions about ordering or continuity of sensory values, placing sensorimotor variables together exactly when they are informationally correlated, no matter how far their temporal signatures are in Euclidean space.

Sensory Channel Selection, Integration, Placement

We study here the use of information distance in the integration of uninterpreted sensory channels. The methods use the information metric to guide and evolve sensor channel selection and placement on the body, produce integration of uninterpreted

Figure 12.6
Model: The Sony AIBO robot.

sensor streams into cortexlike somatosensory maps, and achieve descrambling of visual sensory input and the learning of sensorimotor maps. A useful measure is $coverage = \alpha H(X, Y) + \omega I(X; Y)$, which we have used as a selection criterion to reward novelty (with weight α) or redundancy (with weight ω) among selected channels. If X is a sensory variable, then another Y and all others can be evaluated for their novelty and redundancy with respect to X. In addition to sensory ecology of organisms including the development and unfolding of sensory maps, applications to sensor networks and space probes may be possible.

The platform for this set of studies was the Sony AIBO robot (figure 12.6). We collected data from a Sony AIBO chasing a ball in a university lab or engaging in other behaviors in various environments.

The sensory map reconstruction method, first described and illustrated in simulation by Pierce and Kuipers (1997), is applied to raw, uninterpreted data from sensor streams of the robot. The method has two major parts: (1) Given a number of sensors, find their relative positions and dimensionality. (2) The result is a metric projection of the sensors into a space of appropriate dimensionality. The projection attempts to minimize the distortion of distance between every pair of sensors.

Sensorimotor variable data is gathered by the robot as follows. Let the agent move more or less at random for t time steps. For each time step the value of each sensor is saved. After the t time steps, perform the following steps: (1) Compute the distance between each sensor pair using the distance metric d. (2) Compute the dimensionality and relative positions of the sensors using this metric. The appropriate dimensionality for a group of sensors is found using a "scree-diagram," which shows the amount of variance in the data that is accounted for by each dimension. Then positions of the sensors are assigned in a space of this dimension that reflect the distance metric d. This is a constraint-solving problem that can be solved with a number of different methods for which we have used metric scaling and relaxation.

Nearness of sensory variables and motor variables in the data suggests that such a reconstruction can guide the learning of sensorimotor control (Olsson et al., 2005b, and current work).

A Simple Simulation Model The method first is illustrated here by a two-dimensional grid world (not toroidal), 200×200 pixels. The agent is a two-dimensional square, 10×10 pixels, moving over this world. Each pixel is a sensor, and hence the agent has 100 sensors. For each time step the agent can move at the most one step in the x-direction and one step in the y-direction; that is, displacements Δx and Δy can assume values in $D = \{-1, 0, 1\}$. Each timestep there is a 15 percent probability that Δx or Δy changes to another value in D. The sensory reconstruction method based on this experience allows recovery of the relative position of sensors to each other (figure 12.7).

Metric Projections of AIBO Sensors

Allowing the AIBO robot to wander in a university laboratory environment and recording time series data from 151 sensorimotor variables, including a 10×10 array of visual sensors, and applying the sensory reconstruction method yields a homunculus-like representation of the sensorimotor variables allowing recovery of much information concerning their relative positions.

Combining the metric projection with the information metric on the sensorimotor equipment of the Sony AIBO robot yields the map shown in figure 12.8 (Olsson et al., 2004b). Here, we use the complete sensory spectrum of the AIBO, including a 10×10 portion of the visual field; the actuator duties are treated exactly like sensors. Without having any prior input regarding the structure of the robot being represented, the method reveals both the bilateral symmetry of the AIBO robot and the natural two-dimensional structure of the visual field portion.

Sensoritopic Maps Such reconstructed sensorimotoric maps are highly reminiscent of somatosensory and sensoritopic maps ("homunculi") in the brains of humans and other animals. This suggests that the informational organization such as we detected using informational geometry—induced by the information metric on sensorimotor variables—may be reflected in the cortex in its spatial organization, with informationally similar sensorimotor variables located nearby. Such proximity need not respect sensory modality as it occurs, e.g., in the optic tectum of rattlesnakes, where bimodal neurons receive input from the retina and infrared-sensing pit organ cross-modally integrating sensory information (Newman and Hartline, 1981).

Environments with Oriented Contours Many experiments have been performed with animals in which their visual experience is somehow restricted to contours of a

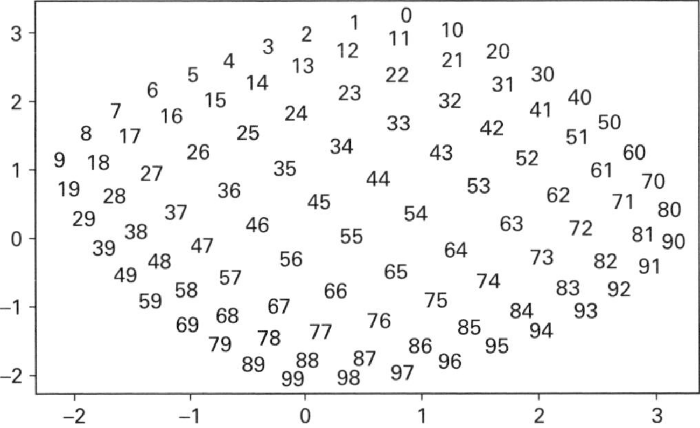

Figure 12.7
Simulation model agent senses a 10×10 pixel region (enlarged in lower left corner) in the upper image and moves randomly but continuously over the image. Sensoritopic reconstruction of sensor relationships according to the method outlined in the text that is based on the information metric shows the correct relative layout of the pixel sensors (lower image).

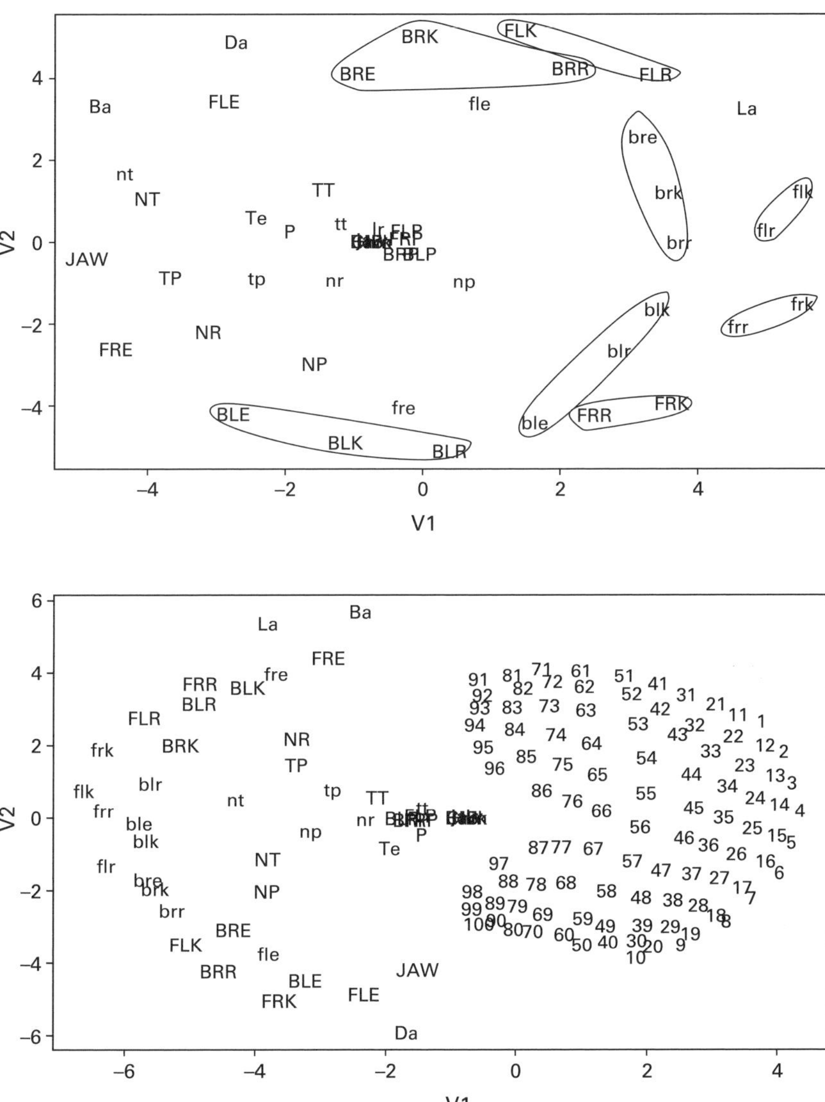

Figure 12.8
Sensoritopic maps of the AIBO robot derived from the information metric. (From Olsson et al., 2004b.) These are maps of the informational relationships between sensors. The bottom figure shows all sensors including visual sensors, and the top figure shows all sensors excluding visual sensors. Circled groups of sensorimotor variables in the top map correspond to natural groupings in the environment (e.g., front right leg). Bilateral symmetry is evident in these recovered sensoritopic representations of the informational relationships of sensory and proprioceptive motoric variables.

(a)

Figure 12.9
Sensoritopic maps in changing environments. (a) Aibo robot in environment with contours in only one orientation. (b) Sensoritopic map constructed from experience of this "impoverished" environment; sensors in the same visual column have very similar "experience" and tend to group. (c) Unfolded sensoritopic map after transfer to an "naturalistic" university lab environment with contours in multiple orientations and subsequent development. (Olsson et al., 2004a.)

certain orientation to study the effects of this on cortical development. See, for example, Wiesel (1982).

Experiencing only an environment with vertically oriented contours (figure 12.9, left), the robot's visual experience is similar in sensors that are in the same column of its visual field. The sensory reconstruction groups these together, as shown in the resulting sensoritopic map (figure 12.9, center). If the robot is then allowed to experience a more natural environment with contours in many directions, sensory experience of sensors in the same column becomes less similar, and the resulting sensoritopic map gradually *unfolds* so that the visual sensors assume a rectangular distribution (as in figure 12.9, right).[15]

Experience of the richer environment permits separation of the clumped-together sensors, reflecting the new informational relations between sensors in experience of the new environment.

The results strikingly show the impact of the structure of environmental experience on the informational relationships between sensors. Developmental processes based on information relations can be expected to be strongly influenced by environmental experience.

Descrambling of Vision Sensors

If the topological arrangements of visual sensors are randomly scrambled so that a pixel in vision is no longer near previously nearby pixels, can an organism recover? Having scrambled the sensory input in this way, it may seem hopeless to recover a

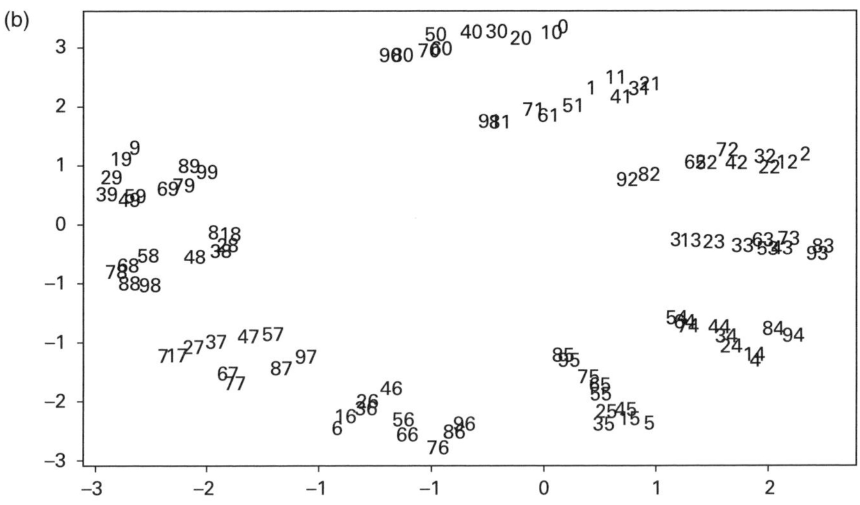

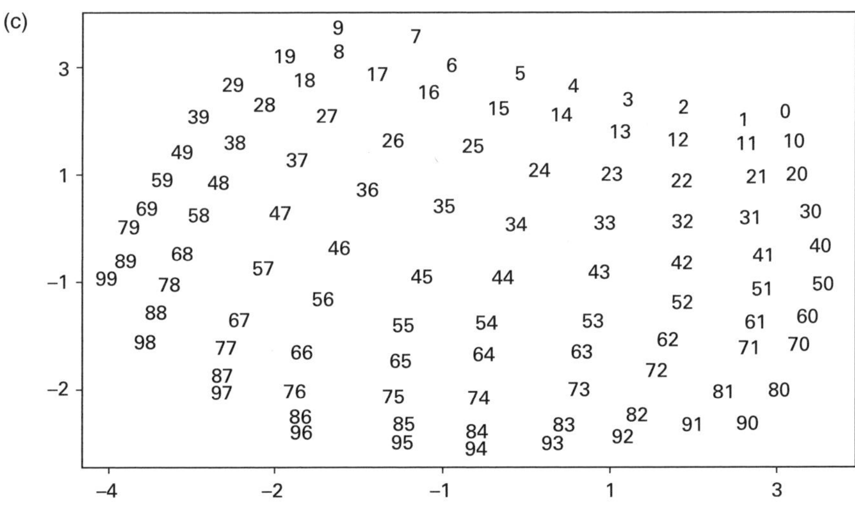

Figure 12.9
(continued)

coherent continuous visual field. By detecting informational similarity, it is possible to do this fairly quickly. How to find the applied map that has scrambled vision? In Olsson et al. (2004b) the uninterpreted sensory streams for an AIBO robot have been compared, using the information distance (Crutchfield, 1990) that makes no assumptions about the spaces of the sensory inputs.

The problem can be solved in the same way as in the construction of the sensoritopic map of the robot, but restricting consideration to visual sensors: Compute a two-dimensional metric projection of the vision sensors. This metric projection can be used to approximate a discrete map and then recover the original image (up to orientation) (see figure 12.10).[16]

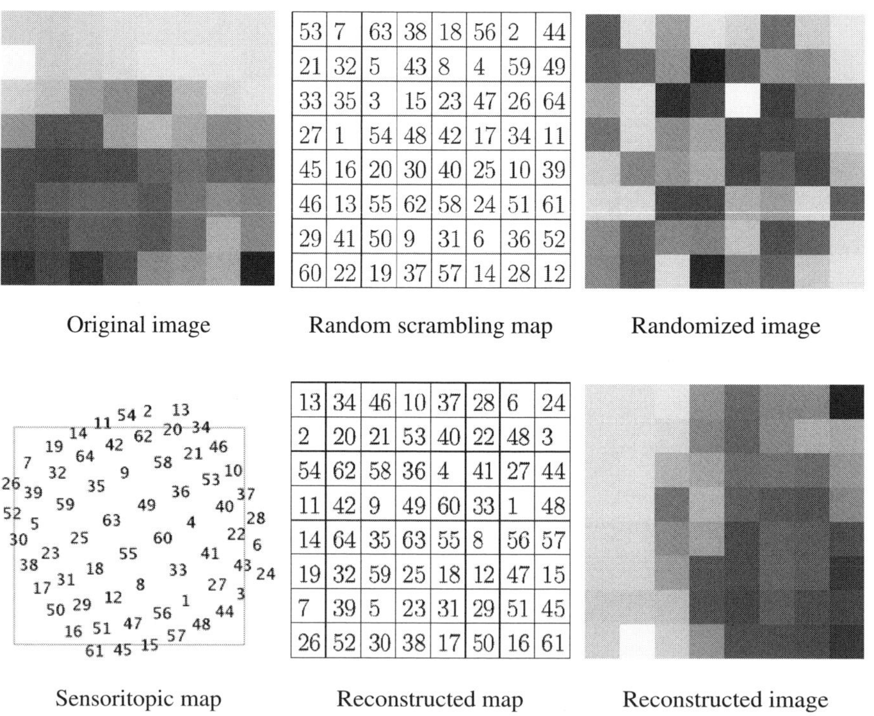

Figure 12.10
Discrete reconstruction of scrambled visual field. The top row shows the normal image in a 10 × 10 pixel array of visual sensors (right) and a random scrambling map (center) that is applied to yield a chaotic image (left) in which nearby pixels no longer correspond to similar stimuli given by the piecewise continuity of usual imagery. The bottom row shows the corresponding situation after the sensory reconstruction method is applied to the visual sensors, giving a map of sensors according to information similarity—in this case the identity mapping (left and center) and the resulting reconstructed image (right). In general, reconstruction depends on a sufficient experience in a rich enough piecewise continuous sensory environment, and can recover relative sensor positioning—in this case, only up to one of the eight symmetries of the square visual field. (Olsson et al., 2004b.)

Creating a two-dimensional map that attempts to match the sensory distances reveals a good part of the relationship of the sensors and actuators. These can be interpreted by a human observer and distinguished into visual, tactile, and other sensors. However, the ultimate goal is the achievement of an autonomous organization of sensorimotor structure. The approach is general enough to allow for the accommodation of additional sensory modalities which are related, but not identical to each other. Olsson et al. (2005c) show how optical inputs from the same scenery but with different color channels (corresponding to overlapping sensory modalities) can be naturally fused to attain a sensory mapping that projects them into a similar area. A biological analogy for this phenomenon is found in pit vipers, where the infrared pit sensors project their information into the optical tectum as to match corresponding stimuli.

The universality of the concept of information distance can be seen in the experiments from Olsson et al. (2005a), where, using the AIBO robot in a simple scenario, it was possible to derive the versatile concept of optical flow from information flow-like quantities.

The power of using information theory as a basis for the organization of the sensorimotoric variables (extero- and proprioceptors) allows treatment of a number of real robot scenarios in a unified way:

1. (UNSCRAMBLING) We can recover the original topological structure of garbled visual inputs (Olsson et al., 2004b)

2. (UNFOLDING OF SENSORITOPIC MAPS) Information distance sensoritopic mapping allows us to obtain a biologically plausible robotic model for developmental failures in impoverished environments (Olsson et al., 2004a)

3. (FLOW IN A SENSORY FIELD; SENSORIMOTOR LAWS) In some simplified scenarios, we were able to derive quantities related to the central concept of optical flow from purely informational quantities (Olsson et al., 2005a), and then sensorimotor laws describing how certain motor actions induce changes in optical flow (Olsson et al., 2005b)

4. (INTEGRATION OF SENSORY MODALITIES) Using the information distance method together with adaptive binning techniques, it is possible to emulate the convergence of different sensory modalities in a common brain region (Olsson et al., 2005c); this corresponds to phenomena observed in biologically relevant cases (Newman and Hartline, 1981).

As one strives to understand the organizational principles underlying the sensorimotoric development in biological organisms, the above examples show that a variety of phenomena can be explained from the common ground of information theory. The relevance of the results stems from the facts that a wide spectrum of biologically relevant motifs can be achieved using the limited set of information-theoretic principles and that crisp phenomena are obtained even under the limited sensor and actuator quality of a real-world robot.

Information Flow and Stigmergy

Here we consider information channels that target the environment or the organism's relationship to it (e.g., its position and orientation). These will be motoric effector channels rather than sensory channels, but use of these effector channels facilitates the employment of sensors and thus is intimately connected with them in the perception-action loop.

Flows via the Environment

Epistemic actions (as opposed to conventional, pragmatic ones) uncovering hidden information using the perception-action loop are discussed and demonstrated by Kirsh and Maglio (1994). There, advanced Tetris game players are shown to quickly rotate a falling block while it still is not completely visible. This active modification of the environment allows the players to discover the shape of the block before it becomes completely visible on the screen.

Offloading of information into the environment can be illustrated by the fact that we often write notes or reminders which we later look at to reacquire some information. A good account of such information flow (although not analyzed in the Shannon sense) between several people is given in the analysis of how members of an airliner crew communicate indirectly, using cockpit controls as a medium (Hutchins, 1995). For example, long before landing, one of the pilots calculates proper flap settings for various speeds and then, based on the results, sets special markers on the airspeed indicator. Later, during the landing phase, the markers allow the crew to quickly and reliably find out what flap settings to use for the speed at that time. In a wider context, indirect communication via the environment is of high importance in distributed systems (Hollan et al., 2000; Bonabeau et al., 1999).

Stigmergy is usually considered as the indirect communication between agents via the environment without targeting a specific recipient. Stigmergy has been used to explain nest-building, sorting, and foraging in social insects (Theraulaz and Bonabeau, 1995; Bonabeau et al., 1997, 1999).

Tracking Information Flow

Treating information as a universal commodity in perception and action is the central philosophy behind the work in Klyubin et al. (2004a, 2004b). Here, the perception-action loop of given agents is modeled using a Bayesian network unrolled

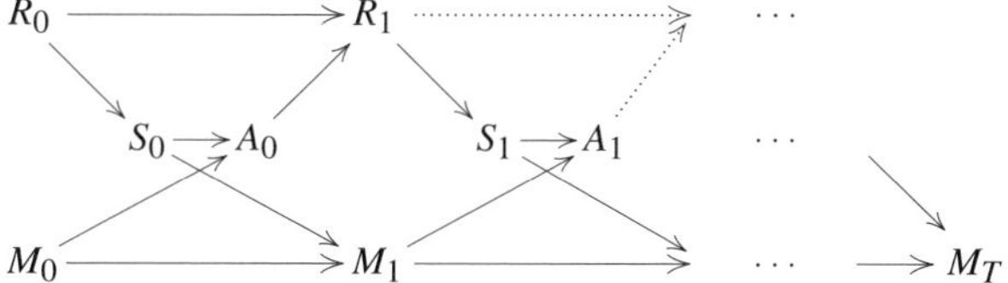

Figure 12.11
Bayesian representation of an agent acting in an environment, used to compute the information flows as described in the text. R_t denotes the variable describing the environment at time t, S_t and A_t describe sensor and actuator states, and M_t is the agent's memory state at time t (with t ranging from 1 to T).

in time. Shannon information enters the agent via sensors, and the actuators imprint information coming from the agent onto the environment (figure 12.11). The internal state of the organism or other agent can be perturbed by information coming from its sensors, resulting, for instance, in "storage" or "memory" of the information. This perturbed internal state is modeled as an internal conditional probability, and can be simplified in the given scenario into a finite state machine.

In Klyubin et al. (2004a) this model is used to maximize the information flow from the environment at the beginning of the run to the agent memory state later on, via an evolutionary algorithm. In the concrete scenario, agents in the evolving population are finite automata that, depending on their current internal state and on sensory input, transition to a new state and move in the world; the information flow is given in this case just by the mutual information between the initial state of the agent and the internal memory state of the agent.[17] Maximizing this information flow gives rise to the—at first—surprising result that the final memory representation of the initial state becomes quite well structured and recovers geometrically and structurally relevant aspects of the world. This is particularly intriguing because, once given the agent's embodiment (the structure of the coupling of sensors and actuators to the environment), the effect arises naturally from the information flow maximization. It is entirely unsupervised, and a different embodiment will give rise to a different representation.

This model can be seen as a powerful generalization of Linsker's infomax principle (Linsker, 1988). There, neural networks have been optimized to maximize the information transmission through the network, causing biologically plausible features to emerge. However, a strong restriction of that approach is that Linsker's model already assumed localized input fields for the neurons, and thus the concept of space was already part of the model in a semi-explicit way. Our information-flow maximization approach, however, does not assume anything beyond the agent embodiment, which anyway is given a priori and is natural to the problem at hand. What is even more interesting is that the structures obtained through the approach can be decomposed into independent components which yield natural generalizations of spatial concepts in an unsupervised and self-organized manner.

It should be noted that nowhere in this model has an assumption been made as to how information should be represented. In other words, the whole concept is "coordinate-free" and independent of a particular choice of representation. Nevertheless, specific representations may be favored in a given setting, corresponding to "informational eigenstates" of the system.

These results suggest that it would be a good idea to use this perspective of information as a structurable concept and a commodity for information-processing in general.

Organization of Information Flows in Perception-Action Loops in Informational Ecology

The considerations on information theory and sensors lead naturally to more general methods of how agent behavior can be studied. In fact, they highlight two issues: (1) that via a suitable treatment, Shannon information can be "subpackaged," split and selected; (2) they provide a straightforward approach to formulating the amount of information relevant to an agent. These developments open a novel perspective for the study of agents: instead of considering some specific architectural properties of the agents or metabolic aspects of ecology alone, we can endeavor to study the *informational ecology* of agents. This means that we no longer need to concentrate on specifics of particular physical constraints and implementations, as well as particular processing architectures. This gives us a "coordinate-free"[18] probe into the informational properties of the perception-action loop of agents; there are good reasons to believe this provides a deeper and more stable understanding about what an agent can and should do in a given environment. Also, it turns out that it depends only on the embodiment of the agent (i.e., on the way perception and action of the agent are coupled to the environment). The importance of embodiment for agent studies has been strongly emphasized informally for some time (Varela et al., 1991; Quick et al., 1999). The information flow approach discussed next gives a more precise and formalized handle on the relevance of embodiment to understanding the perception-action loop.

The Information Flow Formalism for Agents

Klyubin et al. (2004a) study the information flows through the perception-action loop of agents and the organization induced in them by optimization. We give here a brief outline of the method. To formalize the approach, we cast the perception-action loop of the agent in its environment into the language of Bayesian graphs, as in figure 12.11. Bayesian graphs are a way of specifying structured probability distributions (Pearl, 2000). Directed edges in the graph, as in $X \rightarrow Z \leftarrow Y$ indicate that the probability of Z is given by $p(z \mid x, y)$ (i.e., Z is independent of any ancestors of X and Y in the graph). In other words, any collection of directed edges ending in a given node Z can be interpreted as a mechanism by which the parents of Z affect the distribution of Z.

Specifically, in figure 12.11, we unrolled the perception-action loop of an agent: from the environment R_t at time t, information is captured via the agent sensors S_t. The sensors, in turn, affect the actions A_t (which are also affected by the current agent memory state M_t). The actions A_t then imprint information onto the environment R_{t+1} at the next time step (the next memory state M_{t+1} is also determined by the sensor readings S_t and the current memory state M_t), and the cycle begins anew.

This is the basic Bayesian model for an agent's perception-action loop and can easily be extended for other scenarios (e.g., for a study of stigmergy, see Klyubin et al., 2004b). The central purpose of the Bayesian formulation is that it provides the necessary foundations to study (Shannon) information flows in the system. For simplicity, Klyubin et al. (2004a) discuss only the simplest case of information flow, mutual information $I(R_0; M_t)$; this is the amount of information that the agent memory captures at time t about the initial state R_0 of the world (in that particular scenario represented by the position of the agent in the world).[19]

Shannon information has been recognized as a valuable resource for biological systems. It has been found that they tend to optimize its transmission in the absence of other constraints or pressures (Barlow, 1989). Its optimization, in turn, provides powerful organizational principles such as Linsker's infomax principle (Linsker, 1988, 1989; Haykin, 1999). Various biologically relevant observations support this general thesis (Becker, 1996; Laughlin et al., 1998; Baddeley et al., 2000).

Klyubin et al. (2004a) therefore extended Linsker's infomax principle to the unrolled perception-action loop and maximized the value of information $I(R_0; M_T)$ about the initial state of the world R_0 captured by the agent's memory at the end of the run M_T. For this purpose, the agent's behavior $p(a_t \mid s_t, m_t)$ and its memory storage strategy $p(m_{t+1} \mid s_t, m_t)$ were evolved, using the mutual information $I(R_0; M_T)$ between the initial state and the later memory state as a fitness criterion.

More concretely, we considered an agent moving in a simple square grid world, the agent being able to sense a local gradient directed toward the center of the world (only four discrete directions, north, east, south, and west, could be sensed), and to perform a single move from one cell to a neighboring cell in one of these four directions. (See Klyubin et al., 2004a for details.)

This evolutionary process led the optimized agent strategies to separate out different structured components of the world state information into the final memory state. Final memory states would encode different groupings of the initial world state in a geometrically relevant way (figure 12.12). In addition, it turns out that the memory representations that end up characterizing the initial world state can be further

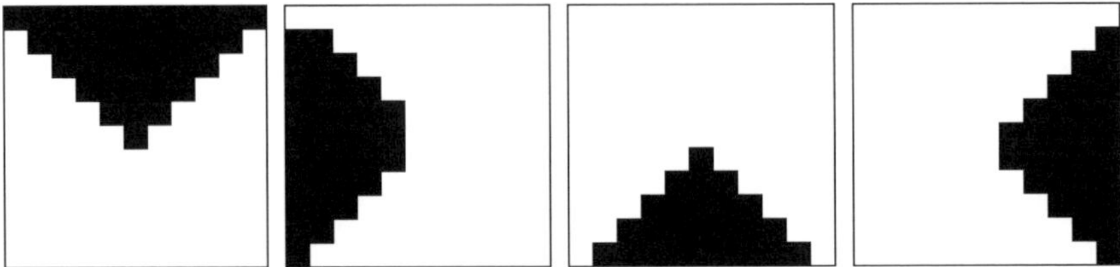

Figure 12.12
The probabilistic map for a four-state memory agent with optimized information flow. It shows $p(m_T \mid r_0)$, i.e., for each initial state of the grid world it visualizes to which of the four final states of the agent memory it will be mapped at the end of the agent run. Each square represents one of the four memory states, and the black part of each square, those initial agent positions in the world which cause that memory state to be assumed at the end of the run. For instance, let us number the memory states in the figure from 1 to 4 (left to right). Then final memory state 1 is assumed with probability 1 if the agent starts with its initial position r_0 in the northern triangle of the grid world, state 2 is assumed with probability 1 if it starts in the western triangle of the grid world, and so on.

grouped into "concepts" relevant to the world structure, using purely information-theoretic criteria without having to resort to any external or human-made semantics.[20]

This evolved correspondence between these internal agent states and the state of affairs and appropriate actions in this world is an example of what Millikan calls a *pushmi-pullyu representation*, in which *what is the case* and *what is to be done* are both present. Such representations are considered to be much more fundamental, biologically more primitive, and more common in nature than the highly derived ones seen in human language or the formal semantics of traditional logic (Millikan, 2004a; Harms, 2004).

The result is particularly interesting in light of the original results from Linsker (1988), because of two points. First, unlike Linsker's work, we do not have to assume any a priori spatial structure in the agent. All the structure arises from the agent's interaction with the environment (its embodiment) and is given naturally. Second, it does not stop at a single-level organization, but it provides mechanisms for further organization in a natural way.

In a study of stigmergy, evolving populations of similar agents, simply by maximizing information flow from an initial state to a later state of their memory, despite their memory being wiped clear, evolved the capacity to act (via manipulating the position of a box in the environment, or by running in a particular direction, depending on their initial state) to offload and reacquire information through channels going via the environment (Klyubin et al., 2004b).[21]

In addition, the information-theoretic view is particularly powerful, because it allows expansion of the scenarios and treatment of a number of phenomena in a principled, unified, interpretable, and mathematically sound way. It provides a consistent semantics for the mechanism of information-processing; it is independent of particular agent architectures—it is, in a way, a "coordinate-free" view into given information-processing mechanisms. At the same time, its quantitative character offers a natural path toward (self-)organization via optimization of suitable quantities.

There are strong indications that these results will extend to more general scenarios. In particular, the universality of Shannon information allows the derivation of a wide variety of quantities that could easily be acting in natural systems as quantities that could characterize the development and evolution of information-processing systems.

A first step toward such quantities is taken in Klyubin et al. (2005a, 2005b), where we define a notion of *empowerment* as the informational capacity of the channel between sensors and actuators of an agent. This inconspicuous definition has a number of powerful consequences. It allows identification of interesting places in the environment, and it gives rise to the concept of different relevant world contexts in a way that is entirely intrinsic to the agent and does not require any supervision. Furthermore, it provides a natural gradient for the development of sensorimotoric equipment relevant to the agent's niche. All that does not require any external intervention beyond the definition of the agent's embodiment, and even that has the freedom to evolve with respect to sensor and actuator equipment.

The use of information to understand the interaction of agents with their environment had been already put forward by Ashby (1956), but has only been taken up systematically by Touchette and Lloyd more recently (2000, 2004). In conjunction with the methods developed by Tishby et al. (1999), Olsson et al. (2004b), and Klyubin et al. (2004b, 2005b), we believe that the treatment of Shannon information as a quantity that can be partitioned and separated, and the separate parts manipulated, provides a powerful and coordinate-free approach toward information-processing, and thus toward a more systematic understanding of how adaptation can emerge in organisms and other adaptive systems.

Remembering and the Temporal Horizon

Temporal Scope of Relevant Information

The *temporal horizon* in which an organism operates depends on whether it uses information related to past and future, and the scope of the information in time (cf. Heidegger, 1972; Nehaniv et al., 1999, 2002). For example, homeostatic drives, hormones, and affective state may comprise aspects of the internal state of the organism that modulate perception and action. Their effects are not restricted to the instantaneous, but persist over a broader temporal scope. Memory, remembering, and planning all involve information whose persistence and relevance are not limited to the present but influence channels over a broader time scale (smaller, e.g., in the case of short-term memory, as opposed to long-term memory in humans).

States (variables) with temporal scope carry relevant information in values having high mutual information with relevant variables extended over time (past or future states and/or events). Different organisms may or may not use channels of meaningful information of broader temporal scope. Types of temporal horizons include reactive, learning/affective, autobiographic, and biographic; internal channels of relevant information may support empathic resonance, social learning, episodic memory and communication of episodes (generalized "narrative"), and biographical reconstruction. (See Nehaniv et al., 2002.)

Currently, a broader temporal horizon is not explicitly modeled in existing information formalisms for understanding the evolution of sensors and other channels of information useful for an organism in manipulating its world (but see Nehaniv, 1999). For instance, in the information bottleneck, the world state X can be construed as including any internal state (including memory) of the organism, but by itself this provides no means to approach problems such as how such memories should be organized and what information about the past they should retain. It is an open problem to extend the information bottleneck and information ecologies methods to models with explicit organismal memory modeled, say as another random variable changing over time in the organism's interaction with the environment. The sensory value and action an organism takes could then be modeled as depending not only on X and X^*, respectively, but also explicitly on the state of the organism's memory.

Internal states, such as memory, provide *representations* (in the sense of Millikan) that an organism can use purposefully in manipulating the world (Nehaniv, 1999). External states, via the offloading of information into the environment that is later accessed by the organism (or a conspecific nestmate, as in stigmergy), provide another instance of such representation. Regarding the internal state, say of memory, or the external state as a random variable, we can consider this variable as a potential source or target of a channel of information relevant to the particular organism "reading" and/or "writing" from it.

Discussion

We have argued that information theory is a natural candidate to study the operation of agents and agent-environment interaction. The reason we study information theory in this context is that it applies universally for any system that transmits and processes information, and, more important, it provides limits on its central resource quantity, Shannon information. These limits have to be respected by any system, including biological systems. More than that, it turns out that biological systems do, in fact, optimize their utilization of available Shannon information as a resource.

With such a principle at hand, one can a priori restrict the universe of solutions available for a given information-processing system, whether artificial or biological. The philosophy of the past few sections was to probe the universe of available solutions for relevant scenarios. Since information theory is universal and appears to be relevant to biological systems, it suggests that the organization phenomena observed by us could provide insight into analogous phenomena in biology. In the following, we wish to discuss the above findings on informational sensorimotor structure of robots and organization in perception-action loops.

Extension of Infomax to Perception-Action Loops

We have seen that the optimization of the information flow for an agent moving in its world gave rise to features representing topologically/geometrically relevant structures of the original world. Unlike Linsker's original infomax application, no a priori restrictions (such as spatially restricted receptive fields) had to be implemented in the information-processing mechanism to achieve this. The constraints that made the pattern detectors emerge was the embodiment of the agent (i.e., its coupling to the environment). The embodiment of an agent, though, is naturally given, and thus the approach has a significant advantage with respect to the original infomax principle.

A key for the success of this model was the fact that we incorporated the whole perception-action loop into the information processing model; the passive part of information-processing (measuring the information captured by sensors) is a common theme in scientific literature. However, we wish to emphasize two aspects that are by far not as common fare. The first is the splitting of the (Shannon) information bulk into "subpackages" and their selected treatment as propagated most prominently by the information bottleneck principle (Tishby et al., 1999). The use of "portions of information," and the ability to mark them as "relevant" where useful (see above), increase the applicability of Shannon information considerably. The second is the treatment of choice of action as a random variable that can imprint information on the environment. In hindsight, this is a relatively natural notion. However, as far as we are aware of, this treatment is surprisingly rare in the literature (Ashby, 1956; Touchette and Lloyd, 2000, 2004; Polani et al., 2001). The inclusion of actions in the framework allows us to close the perception-action loop using active elements, and thus strongly improve over the purely passive original infomax model.

One thing should be emphasized, though: as straightforward as the treatment of actions is in the model from figure 12.11, they still remain special to some extent as compared with the more passive parts such as environment, sensors, and memory. In fact, one can consider models that range from actions being predetermined by everything that happened before in the system (as in figure 12.11) to agent models with "free will" actions[22] that inject new randomness (i.e., fresh, independent information) into the system. It turns out that this latter perspective gives rise to the powerful notion of *empowerment*; the notion of empowerment was introduced in Klyubin et al. (2005a, 2005b) and provides a universal utility function for agents.

Relevance of Information Theory to Real-World Scenarios

A further question about the relevance of the information theoretic approach is whether information theoretic quantities can form the basis for the development of structured information-processing in noisy, ill-defined, and computationally demanding real-world scenarios. The robotic experiments show that this is quite possible. In fact, they demonstrate that information theoretic principles could easily underlie the organization of biological systems, covering several important parts of the information-processing needs of a biological organism.

Principles, Not Mechanisms

In the preceding sections, we have argued that information is a powerful *principle* from which various organizational needs of agents (biological or artificial) can be derived. However, the information theoretic approach does not say anything about the *mechanisms* involved. While mechanisms are often at the core of biological research, they are entirely missing from our model.

We would like to emphasize that this is in fact a strength of the approach rather than a weakness. Biological systems are not just exceedingly complex, but provide mechanisms to solve the same or similar problems in a large variety of forms. It involves significant effort to separate the different mechanisms of biological systems and the purposes they serve—even worse, sometimes one mechanism serves multiple purposes. In other words, focusing on mechanisms can sometimes obscure what is actually happening in the system.

Information theory gives us an insight into what is being "computed" in an organism's perception-action loop, independent of the particular mechanism used. This is why we refer to the information theoretic method as "coordinate-free." The analogy is that a particular mechanism or computational architecture is like the choice of a particular "coordinate system." Other architectures ("coordinate systems") may be used to perform the same informational transformations ("describe the same object"). However, the particular architecture ("coordinate system") is not the essence of things, but rather the shaping and exploitation of information that is performed. Information theory identifies the latter and ignores the former.[23]

The large selection of mechanisms that can perform the same or similar functions means that they are often incidental to contingency in the history of evolution (whether artificial or natural). We thus believe that in addition to the complex transformations of matter that organisms are involved in, understanding the information-processing by sensors, in internal or external memory representations, and via actuators, in essence is what we should strive for, and that information theory is the right tool to do this.

Furthermore, we believe that by studying information theoretical models of agents, one will be able to identify a fine-grained network of local utilities that may provide powerful selection pressures toward increasing organization in a fundamentally unsupervised way. An example is the information theoretic notion of *empowerment* introduced in Klyubin et al. (2005a, 2005b) and mentioned above. If this hypothesis can be corroborated, it will provide us with a new understanding of the drives that allow complexity and intelligence to emerge in living systems.

Notes

1. There is doubt whether these entities should be regarded as actual living organisms, since although they are self-replicating members of evolving populations, they do not engage in self-maintenance and may lack other essential properties of life. If these properties could be added, then proponents of the "strong view" in Constructive Biology would say we must then regard them as not just models of life, but as living.

2. These can be regarded as the values assumed by a random variable.

3. We are indebted to Peter Dauscher for suggesting this analogy.

4. Analogous definitions and theory are possible for continuous sensor values, but for simplicity we consider only the discrete case here.

5. Other fundamental quantities are *conditional entropy*, (i.e., the average amount of uncertainty remaining in a random variable X, given that the value of another (jointly distributed) random variable Y is known:

$$H(X \mid Y) = -\sum_{i=1}^{n}\sum_{j=1}^{m} p(x_i, y_j) \log p(x_i \mid y_j),$$

where Y can take values among $\{y_1, \ldots, y_m\}$, $p(x_i, y_j)$ is the probability that x_i and y_j occur as the values of X and Y, respectively, and $p(x_i \mid y_j)$ is the probability that X has value x_i given that the value of Y is y_j; and *mutual information* $I(X; Y)$, which measures how much X and Y have in common: $I(X; Y) = H(X) - H(X \mid Y) = H(Y) - H(Y \mid X)$.

6. The first author is indebted to Professor Otto Kegel for emphasizing in his lectures on the subject at the Albert-Ludwigs University in Freiburg, Germany, in the 1980s that there are many questions about information channels that are not asked or answered by information theory.

7. The capacity of formal logics and denotational semantics to ignore *context* and embodiment is a hard-won property of these abstract logics developed by human philosophers and mathematicians that makes them applicable across many settings, but it seems likely that this context-freeness property probably never occurs in natural systems.

8. This important point concerns *evolvability* of the sensory channel. See Nilsson and Pelger (1994), Hansen (2003), and *BioSystems* (2003) for further discussion.

9. The phenomena of sensing are of course not limited to animals. Similarly, responses and functioning of plants, fungi, single-cell free-living organisms, and individual cells in multicellular bodies all crucially rely on sensing and responding to the ambient environment.

10. An iterative construction based on the Blahut-Arimoto algorithm in the area of information theory, known as rate distortion theory (see Cover and Thomas 1991; Tishby et al., 1999), is used to obtain the probabilistic mapping from world states to extractor values. Other combinations of constraints and optimization can be implemented (Tishby et al., 1999; Friedman et al., 2001).

11. This observation is due to the second author, D. Polani (unpublished, 2002).

12. $X \equiv Y$ means that X and Y are recoding *equivalent*—that is, one can map the values of X into Y to determine them completely, and vice versa. Intuitively speaking, sources that are *recoding* equivalent provide exactly the same information.

13. Sensory values are scaled to map their range into the unit interval $[0, 1]$.

14. All quantities here can in practice be computed from the joint distribution of X and Y as estimated from time series data of their readings.

15. See online video.

16. For demo of descrambling vision sensors, see online video.

17. In general, it is a more complicated expression.

18. We discuss the term "coordinate-free" in the final section of this chapter.

19. Information flow is not, in general, the same as mutual information, but tracks information that flows from one particular part of a system to another. In the example discussed here, however, it becomes the same as mutual information, because the initial world state variable R_0 has no parents and, via random initialization, can be seen as "injecting" entirely new information into the system. We do not have the space here to discuss the concept of information flow for more general cases.

20. This is work in progress, but the central result has already been established.

21. See online videos.

22. While we should be reluctant to use such a drastically loaded term as "free will," we believe it evokes a helpful picture in the reader's mind, as long as one remembers to take it with a grain of salt.

23. The term "coordinate-free" as used here refers to completely leaving open the particular mechanisms that implement information-processing for a given a set of relevant informational variables. On the other hand, the choice via evolution (or development) of the particular informational variables that the particular sensory channels of an organism monitor (i.e., the choice of informational variables and channels themselves) does in a sense comprise a choice of "coordinates" for the organism's activity, and is not free, but highly contingent on evolutionary constraints. However, information theory is completely neutral as to the mechanisms by which these informational variables are processed.

References

Ashby WR (1956) An Introduction to Cybernetics. London: Chapman & Hall.

Baddeley R, Hancock P, Földiák P (eds) (2000) Information Theory and the Brain. Cambridge: Cambridge University Press.

Barlow HB (1989) Unsupervised learning. Neur Comput 1: 295–311.

Becker S (1996) Mutual information maximization: Models of cortical self-organization. Network 7: 7–31.

Bekoff M (1977) Social communication in canids: Evidence for the evolution of a stereotyped mammalian display. Science 197: 1097–1099.

Bonabeau E, Dorigo M, Theraulaz G (1999) Swarm Intelligence: From Natural to Artificial Systems. New York: Oxford University Press.

Bonabeau E, Theraulaz G, Deneubourg J-L, Aron S, Camazine S (1997) Self-organization in social insects. Trends Ecol Evol 12(5): 188–193.

Bradbury JW, Vehrencamp SL (1998) Principles of Animal Communication. Sunderland, Mass.: Sinauer.

Cover TM, Thomas JA (1991) Elements of Information Theory. New York: Wiley.

Crutchfield JP (1990) Information and its metric. In: Nonlinear Structures in Physical Systems: Pattern Formation, Chaos and Waves (Lam L, Morris HC, eds), 119–130. New York: Springer Verlag.

Dautenhahn K (1995) Getting to know each other—artificial social intelligence for autonomous robots. Robot Auton Syst 16: 333–356.

Endler JA (1992) Signals, signal conditions, and the direction of evolution. Amer Naturalist 139 Suppl: S125–S153.

Fogel DB (ed) (1998) Evolutionary Computation: The Fossil Record. New York: IEEE Press.

Friedman N, Mosenzon O, Slonim N, Tishby N (2001) Multivariate information bottleneck. In: Uncertainty in Artificial Intelligence, 152–161. San Francisco: Morgan Kaufmann.

Gibson JJ (1979) The Ecological Approach to Visual Perception. Boston: Houghton Mifflin.

Guilford T, Dawkins MS (1993) Receiver psychology and the design of animal signals. Trends Neurosci 16(11): 430–436.

Hanlon RT, Messenger JB (1996) Cephalopod Behaviour. Cambridge: Cambridge University Press.

Hansen TF (2003) Is modularity necessary for evolvability? Remarks on the relationship between pleiotropy and evolvability. BioSystems 69(2–3): 83–94. (Special issue on evolvability).

Harms WF (2004) Primitive content, translation, and the emergence of meaning in animal communication. In: Evolution of Communication Systems: A Comparative Approach (Oller DK, Griebel U, eds), 31–48. Cambridge, Mass.: MIT Press.

Hauser MD (1997) The Evolution of Communication. Cambridge, Mass.: MIT Press.

Haykin S (1999) Neural Networks: A Comprehensive Foundation (2nd ed). Englewood Cliffs, N.J.: Prentice-Hall.

Heidegger M (1972) On Time and Being (Stambaugh J, trans). New York: Harper Torchlight.

Hollan J, Hutchins E, Kirsh D (2000) Distributed cognition: Toward a new foundation for human-computer interaction research. ACM Trans Comput-Human Interact 7(2): 174–196. (Special issue on human-computer interaction in the new millennium, part 2).

Hornstein EP, O'Carroll DC, Anderson JC, Laughlin SB (2000) Sexual dimorphism matches photoreceptor performance to behavioural requirements. Proc Biol Sci 267(1457): 2111–2117.

Hutchins E (1995) How a cockpit remembers its speed. Cog Sci 19(3): 265–288.

Juusola M, de Polavieja GG (2003) The rate of information transfer of naturalistic stimulation by graded potentials. J Gen Physiol 122(2): 191–206.

Kelber A, Vorobyev M, Osorio D (2003) Animal colour vision: Behavioural tests and physiological concepts. Biol Rev 78: 81–118.

Kirsh D, Maglio P (1994) On distinguishing epistemic from pragmatic action. Cog Sci 18(4): 513–549.

Klyubin AS, Polani D, Nehaniv CL (2004a) Organization of the information flow in the perception-action loop of evolved agents. In: Proceedings of 2004 NASA/DoD Conference on Evolvable Hardware, 177–180. Los Alamitos, Calif.: IEEE Computer Society.

Klyubin AS, Polani D, Nehaniv CL (2004b) Tracking information flow through the environment: Simple cases of stigmergy. In: Proceedings of Artificial Life IX, 563–568. Cambridge, Mass.: MIT Press.

Klyubin AS, Polani D, Nehaniv CL (2005a) All else being equal be empowered. In: Proceedings of the Eighth European Conference on Artificial Life, 744–753. Berlin: Springer-Verlag.

Klyubin AS, Polani D, Nehaniv CL (2005b) Empowerment: A universal agent-centric measure of control. In: IEEE 2005 Congress of Evolutionary Computation, 128–135. Piscataway, N.J.: IEEE Computer Society Press.

Langton CG (1984) Self-reproduction and cellular automata. Physica D 10: 135–144.

Langton CG (1989) Artificial life. In: Artificial Life: Proceedings of an Interdisciplinary Workshop on the Synthesis and Simulation of Living Systems, Los Alamos, New Mexico, September 1987 (Langton CG, ed), 1–47. Redwood City, CA: Addison-Wesley.

Laughlin SB, de Ruyter van Steveninck RR, Anderson JC (1998) The metabolic cost of neural information. Nat Neurosc 1(1): 36–41.

Linsker R (1988) Self-organization in a perceptual network. Computer 21(3): 105–117.

Linsker R (1989) How to generate ordered maps by maximizing the mutual information between input and output signals. Neur Comput 1(3): 402–411.

Lunine J (2004) Astrobiology: A Multi-Disciplinary Approach. San Francisco: Benjamin Cummings.

MacLennan B (1992) Synthetic ethology: An approach to the study of communication. In: Artificial Life II: The Second Workshop on the Synthesis and Simulation of Living Systems (Langton CG, Taylor C, Farmer JD, Rasmussen S, eds), 631–658. Redwood City, CA: Addison-Wesley.

McCulloch WS, Pitts W (1943) A logical calculus of ideas immanent in nervous activity. Bull Math Biophys 5: 115–133.

Millikan RG (2004a) On reading signs: Some differences between us and the others. In: Evolution of Communication Systems: A Comparative Approach (Oller DK, Griebel U, eds), 15–29. Cambridge, Mass.: MIT Press.

Millikan RG (2004b) Varieties of Meaning: The 2002 Jean Nicod Lectures. Cambridge, Mass.: MIT Press.

Moynihan M (1985) Communication and Noncommunication by Cephalopods. Bloomington: Indiana University Press.

Moynihan M, Rodaniche AF (1982) The behavior and natural history of the Caribbean reef squid *Sepioteuthis sepioidea* with considerations of social, signal and defensive patterns for difficult and dangerous environments. Adv Ethol 25: 1–151.

Nehaniv CL (1999) Meaning for observers and agents. In: Proceedings of the IEEE International Symposium on Intelligent Control/Intelligent Systems and Semiotics, 435–440. Washington, DC: IEEE Computer Society Press.

Nehaniv CL (ed) (2003) BioSystems 69(2–3). (Special issue on evolvability.)

Nehaniv CL, Dautenhahn K (1998) Embodiment and memories: Algebras of time and history for autobiographic agents. In: Cybernetics and Systems '98: Proceedings of the Fourteenth European Meeting on Cybernetics and Systems Research, vol. 2 (Trappl R, ed), 651–656. Vienna: Austrian Society for Cybernetic Studies.

Nehaniv CL, Dautenhahn K, Loomes MJ (1999) Constructive biology and approaches to temporal grounding in post-reactive robotics. In: Sensor Fusion and Decentralized Control in Robotics Systems II (McKee GT, Schenker P, eds), 156–167. Bellingham, WA: SPIE.

Nehaniv CL, Polani D, Dautenhahn K, te Boekhorst R, Cañamero L (2002) Meaningful information, sensor evolution, and the temporal horizon of embodied organism. Art Life 8: 345–349.

Newman EA, Hartline PH (1981) Integration of visual and infrared information in bimodal neurons of the rattlesnake optic tectum. Science 213(4509): 789–791.

Nilsson DE, Pelger S (1994) A pessimistic estimate of the time required for an eye to evolve. Proc Roy Soc Lond B256: 53–58.

Olsson L, Nehaniv CL, Polani D (2004a) The effects on visual information in a robot in environments with oriented contours. In: Proceedings of the Fourth International Workshop on Epigenetic Robotics 2004 (Berthouze L, Kozima H, Prince CG, Sandini G, Stojanov G, Metta G, Balkenius C, eds), 83–88. Lund: Lund University Cognitive Studies.

Olsson L, Nehaniv CL, Polani D (2004b) Sensory channel grouping and structure from uninterpreted sensor data. In: Proceedings of the 2004 NASA/DoD Conference on Evolvable Hardware, 153–160. Los Alamitos Calif.: IEEE Computer Society.

Olsson L, Nehaniv CL, Polani D (2005a) Discovering motion flow by temporal-informational correlations in sensors. In: Proceedings of the Fifth International Workshop on Epigenetic Robotics: Modeling Cognitive Development in Robotic Systems (Berthouze L, Kozima H, Prince CG, Sandini G, Stojanov G, Metta G, Balkenius C, eds), 117–120. Lund: Lund University Cognitive Studies.

Olsson L, Nehaniv CL, Polani D (2005b) From unknown sensors and actuators to visually guided movement. In: Proceedings of the Fourth International Conference on Development and Learning, 1–6. Los Alamitos, Calif.: IEEE Computer Society Press.

Olsson L, Nehaniv CL, Polani D (2005c) Sensor adaptation and development in robots by entropy maximization of sensory data. In: Proceedings of the Sixth IEEE International Symposium on Computational Intelligence in Robotics and Automation, 587–592. Helsinki: IEEE Computer Society Press.

Pearl J (2000) Causality: Models, Reasoning and Inference. Cambridge: Cambridge University Press.

Peirce CS (1965) Collected Papers, vol 2: Elements of Logic. Cambridge, Mass.: Harvard University Press.

Pierce D, Kuipers B (1997) Map learning with uninterpreted sensors and effectors. Art Intell J 92: 169–229.

Polani D, Kim JT, Martinetz T (2001) An information-theoretic approach for the quantification of relevance. In: Advances in Artificial Life: Proceedings of the Sixth European Conference on Artificial Life (Kelemen J, Sosík P, eds), 704–713. New York: Springer.

Quick T, Dautenhahn K, Nehaniv C, Roberts G (1999) On bots and bacteria: Ontology independent embodiment. In: Advances in Artificial Life: Proceedings of the Fifth European Conference on Artificial Life (Floreano D, Nicoud J-D, Mondada F, eds), 339–343. Berlin: Springer.

Ray TS (1992) An approach to the synthesis of life. In: Artificial Life II (Langton CG, Taylor C, Farmer JD, Rasmussen S, eds), 371–408. Redwood City, CA: Addison-Wesley.

Ray TS (1998) Selecting naturally for differentiation: Preliminary evolutionary results. Complexity 3(5): 25–33.

Rumelhart DE, McClelland JL, the PDP Research Group (1986) Parallel Distributed Processing: Explorations in the Microstructure of Cognition, 2 vols. Cambridge, Mass.: MIT Press.

Russell S, Norvig P (2002) Artificial Intelligence: A Modern Approach (2nd ed). Upper Saddle River, N.J.: Prentice-Hall.

Ryan MJ (1998) Sexual selection, receiver biases, and the evolution of sex differences. Science 281(5385): 1999–2003.

Sayama H (1999) A new structurally dissolvable self-reproducing loop evolving in a simple cellular automata space. Art Life 5(4): 343–365.

Shannon CE (1948) A mathematical theory of communication. Bell Syst Tech J 27: 379–423, 623–656.

Shannon CE (1949) The mathematical theory of communication. In: The Mathematical Theory of Communication (Shannon CE, Weaver W, eds), 29–125. Urbana: University of Illinois Press.

Smith WJ (1995) Communication and expectations: A social precess and the cognitive operation it depends upon and influences. In: Readings in Animal Cognition (Bekoff M, Jamieson D, eds), 243–256. Cambridge, Mass.: MIT Press.

Tian NMM-L, Price DJ (2005) Why cavefish are blind. BioEssays 27(3): 235–238.

Theraulaz G, Bonabeau E (1995) Modelling the collective building of complex architectures in social insects with lattice swarms. J Theor Biol 177(4): 381–400.

Tishby N, Pereira FC, Bialek W (1999) The information bottleneck method. In: Proceedings of the 37th Annual Allerton Conference on Communication, Control, and Computing, 368–377. Monticello: University of Illinois.

Touchette H, Lloyd S (2000) Information-theoretic limits of control. Phys Rev Lett 84: 1156.

Touchette H, Lloyd S (2004) Information-theoretic approach to the study of control systems. Physica A331: 140–172.

Turing AM (1936) On computable numbers, with an application to the Entscheidungsproblem. Proc Lond Math Soc., Ser. 2 42: 230–265. Correction ibid. 43(1937): 544–546.

Turing AM (1952) The chemical basis of morphogenesis. Philos Trans Roy Soc Lond B237: 37–72.

Varela FJ, Thompson E, Rosch E (1991) The Embodied Mind. Cambridge, Mass.: MIT Press.

von Neumann J (1966) Theory of Self-Reproducing Automata (Burks, AW, ed and completer). Urbana: University of Illinois Press.

von Uexküll J (1909) Umwelt und Innenwelt der Tiere. Berlin: Springer. Excerpts in translation, Environment [Umwelt] and inner world of animals. In: Foundations of Comparative Ethology (Burghardt GM, ed; Mellor CJ, Gove D, trans), 222–245. New York: Van Nostrand Reinhold, 1985.

Webb B (1995) Using robots to model animals: A cricket test. Robot Auton Syst 16: 117–134.

Wells M (1968) Lower Animals. New York: World University Library.

Wiesel TN (1982) Postnatal development of the visual cortex and the influence of environment. Nature 299(5884): 583–591.

Wittgenstein L (1968) Philosophical Investigations (3rd ed., repr) (Anscombe, GEM, trans). Oxford: Basil Blackwell.

13 Human-Robot Interaction as a Model for Autism Therapy: An Experimental Study with Children with Autism

Iain Werry and Kerstin Dautenhahn

Human-robot interaction is a quickly growing area of research. While most projects deal with typically developing subjects, the particular study that we report on in this chapter investigates interactions of autistic children with a mobile robot. We briefly motivate this approach, and then present results from an experimental study. This comparative study provides evidence that the robot is able to engage the children in interaction better than a conventional toy does. These results are important because they highlight the potential of using robots in therapy and education of children with autism. Moreover, the work emphasizes the emergent, proactive, and embodied nature of interaction as it is applied to human-robot interaction. These studies, carried out as part of human-robot interaction research in assistive technology, might serve as a model for future medical intervention using robots in autism therapy.

Perhaps one of the greatest goals of technology is that of allowing inclusion for those people who are disadvantaged or disabled in some way. Technology has the potential to allow this group of the population to participate in ways that suit their needs. Developments also hold the potential to explore new ways of rehabilitation, and provide new methods for people to compensate for individual disabilities. In particular, those disabilities which can be classified as learning disabilities, involved with the way in which the individual is able to process information and develop, pose many challenges. Often, they also present some of the greatest benefits and contributions, since developments and advances often affect the ability of the individual to develop further, and they open further opportunities. Several conferences and workshops are dedicated to advancements in assistive technology, such as AAATE (Association for the Advancement of Assistive Technology in Europe, AAATE 2006). Other conferences, such as RESNA (Rehabilitation Engineering and Assistive Technology Society of North America, RESNA, 2006) and CWUAAT (Cambridge Workshop on Universal Access and Assistive Technology), also promote the advancement of assistive technology.

Our particular research interest is the use and development of robots that could potentially play a role in therapy and education of children with autism. Before we describe the project in more detail, we briefly discuss autism.

Autism

Autism is a developmental disorder which has been prevalent for a long time, but has only recently been identified and classified as a separate disorder rather than being associated with other disorders. It often manifests openly in development from an early age, but diagnosis can be troubled by the effects of its being similar to other disabilities, such as deafness. The *Diagnostic and Statistical Manual* is used by doctors and psychologists in order to remain consistent in their diagnoses. The fourth edition of this manual states that a diagnosis of autism should consist of a number of symptoms which are drawn for a variety of categories, including the following:

- An impairment in social interaction such as nonverbal behavior, eye contact, body gestures, an inability to develop peer relationships, and a lack of sharing of enjoyment and interests with others
- An impairment of communication, such as a delay in the development of verbal language and communication, an inability to engage in verbal communication even when the development of verbal skills is normal, and repetitive or idiosyncratic language behavior
- The display of stereotypical behavior, such as restricted patterns of interest, a reliance on routines and structure, repetitive motor behaviors such as hand flapping, and a persistent preoccupation with parts of objects
- A delay or abnormal functioning, before 3 years of age, in social interaction, language as used in social communication, and symbolic or imaginative play. (*DSM*, 2006)

Autism is classified as a social disorder because its primary effects often lead the individual to find social interaction and situations difficult to comprehend and to become involved in. While the individual effects of the disability vary considerably, the National Autistic Society classifies the effects of the disorder into three broad areas, which it terms the "triad of impairments":

- Impairments in social interaction
- Impairments in communication
- Impairments in imagination and generalization, and stereotypical behavior. (Wing, 1996)

The overall result of these is that the individual appears to find most kinds of social interaction difficult, and in particular, learning can often be extremely difficult. Generalizing experiences is frequently a great problem for people with autism. Situations that have been encountered in the past, in a slightly different form, may seem completely new and leave the individual totally unprepared. In particular, these effects can lead to children with autism displaying a lesser ability to imitate behavior displayed by other people. For example, Stone et al. (1997) discuss a group of autistic children who showed impaired imitative ability compared with typically developing children. Williams et al. (2001) examine the neurobiology of both the autistic person and imitation, and propose that autism may be connected with a disfunction in mirror neurons. Attwood et al. (1988) found that autistic children failed to understand and use gestures indicating basic emotional states, but that they were able to use basic instrumental gestures, such as pointing. In addition, a tendency toward stereotypical behavior and limitations in proactive (social) behavior are other common characteristics of people with autism.

Our primary research interest is to use robotics technology to help children with autism. Due to the nature of their impairments, we focus on encouraging basic social interaction skills, such as turn-taking, imitation, and joint attention. This work is pursued within the Aurora Project.

The Aurora Project

The Aurora Project was begun in 1998, with an aim of developing a robotic platform which would be able to function as an additional tool for therapists who aim to educate and aid children with autism, to help them to integrate into society in the future (Aurora, 2006). The Aurora Project is based on the philosophy of exploration through play (for example, Papert, 1980). Previous work has shown that computers, as well as virtual environments, can play a valuable role in teaching people with autism (e.g., Colby and Smith, 1971; Dautenhahn and Werry, 2004, for a review). Based on these encouraging results, we investigate the possible role of robots which, unlike computer software, afford full-body and unconstrained interaction, including touch and moving around the robot.

It has been proposed that robots be used for the study of child development (Michaud and Caron, 2002; Michaud et al., 2005) or rehabilitation (Plaisant et al., 2000), autism therapy (Weir and Emanuel, 1976; Dautenhahn, 1999; Werry and Dautenhahn, 1999; Michaud and Théberge-Turmel, 2002; Kozima, 2002; Kozima et al., 2005; Davis et al., 2005), and autism diagnosis (Scassellati, 2005).

The Aurora Project examines the use of a mobile robotic platform to encourage autistic children to explore potential social situations through interaction with the platform. It is proposed that this will provide a safe and stable environment in which the children are able to develop at their own pace and without the pressure of evaluation by a teacher. It is also postulated that the children will feel more comfortable interacting with a robotic platform, and that this platform will provide a stimulus

Figure 13.1
The robotic platform.

and set of social triggers which are stable across time, in ways that a human is unable to, while at the same time allowing the therapist to vary the stimulus at an appropriate speed.

The robotic platform is able to provide this social stimulus by limited expression through physical movement. This also allows the children to interact with the platform in any way that is natural to them, for example, by physically moving around the platform, or remaining stationary and interacting, using their arms and hands.

The robotic platform has been supplied by Applied AI Systems, and is 38 cm long by 30 cm wide, is 21 cm high, and weighs 6.5 kg. It is equipped with a single, rotating, four-segment pyroelectric heat sensor which can be seen in figure 13.1 at the front of the platform, and eight infrared sensors, four of which are at the front of the robot, one on each side of the platform, and two at the rear.

Initial goals of the project were to establish whether the robot has the potential to help children with autism. For the platform to promote social interaction, it must first be established that the platform is able to hold the attention and interest of the children, since if they are not interested in the platform, they will not be able to learn from and with it. In particular, we were interested in how children with autism play with a mobile, autonomous robot compared with a nonrobotic, passive toy.

Note that the robot in our studies does not serve as a model of human behavior, which often appears unpredictable and confusing to a person with autism. Instead, we consider the robot to be a more adequate participant in interactions with children with autism at the mid or lower end of the autistic spectrum, where children have great difficulty interacting with people. It is hoped that learning about simple interactive behaviors with a robot might be generalized to interactions with people and ultimately scale up in complexity, an issue that needs to be demonstrated in future work.

Experiments

The initial stage of the project was to ascertain the potential for the robotic platform to engage and interest the children. A number of experiments were conducted at a local school for children with autism (further details of these can be found in Werry et al., 2001a, 2001b; Werry, 2003; Dautenhahn and Werry, 2002). These experiments were performed at the school in order to minimize the disruption to the children's routine and environment. The trials were conducted in a small room which was assigned to the project. The children participating in the trials were selected by the teachers, and the trials fit in with their daily schedule. All of the children in the trials had been previously diagnosed with autism or Asperger syndrome, and were described as being midrange in their ability (only the chronological ages of the children were available; information about mental and verbal ages was not accessible). The room in which the trials were held was approximately 3 m long and 2 m wide, and contained only fittings such as light switches, a few chairs for the experimenters, and doors to the outside corridor and a next room; these doors did not have windows. Present with the child in the room were a teacher who was familiar with the child, and two experimenters with a video camera.

First Trials

For first trials, the robotic platform was programmed with basic behaviors such as avoidance and following behaviors, and a speech generation unit capable of uttering simple phrases and sentences. The behaviors could be selected by the experimenter, and consisted of two noninteractive and one interactive behavior programs:

- Forward and backward. The robotic platform moved in a steady forward and backward motion, and while doing so, it did not react to stimuli in the environment.
- In the dance behavior, the robot performed a sequence of preset motions. This behavior did not allow reactions to the outside environment.
- In follow and avoid, a relatively complex interactive behavior program, the robot was allowed to sense its environment through its heat sensor and infrared sensors. This behavior instructed the platform to approach the child by using the heat sensor, getting as close as it could while at the same time avoiding obstacles through the infrared sensors. A combination of these two factors led the platform to avoid features in the room and to approach the child without touching him or her.

However, during the trials it was discovered that the children quickly became bored with the two more basic and noninteractive behavior programs, and were more interested in the interactive "follow and avoid" behavior. This behavior was used for the trials reported in this chapter. The platform was equipped with a voice unit, allowing it to make vocal statements in response to a behavior-based trigger.

The trials were conducted using a passive toy truck as a control. It was chosen because it did not posses lights or sounds, and was of about the same size as the robotic platform. Trials were designed to last a total of ten minutes: four minutes of the child interacting with the robot, two minutes of the child present with both the toy truck and the robot (which was switched off), and further four minutes of the child playing solely with the truck. However, the trials were cut short on occasion, when it was clear that the child was no longer interested in the scenario or was distressed. In addition, the trials were randomized so that the child began with the robot in half of the trials and with the toy truck for the remainder of the trials. While the robot was present, the toy truck was hidden, in the alternative scenario, the toy truck was present and the robotic platform was hidden. Though the teacher was present during the trials, her primary roles were to observe and to reassure the children. The teacher did interact with the children occasionally, but this was kept to a minimum in most cases.

Seven children, all of whom were male, participated in the trials. All of them were of comparable ability and development, generally midrange on the autistic spectrum, and their ages ranged from six years and six months to nine years and one month at the time of the trial.

Second Round of Trials

A second round of testing was performed in the same school and with the same setup that was used in the initial round. In these trials, the robot's behaviors were more complex, in that robot progressed through a sequence of behaviors which grew in complexity as the children advanced in their interaction games. However, the children did not seem to recognize the difference in the complexity of the interactions, which is likely to be a result of the various stages not being defined explicitly enough. The robotic platform began the interaction with simple triggers such as moving in front of the sensor to provoke a simple movement reaction. After successful triggers, more complex reactions were displayed by the robot, such as "follow and avoid." The architecture set out with the robot allowed an advancement in complexity if the child triggered the robot's response within a time frame of two minutes (cf. figure 13.2). If this trigger was not given by the child, the robot progressed to a different behavior, utilizing a simpler trigger.

Other factors in the trials, such as the use of a passive toy, the segmentation of the trial, and the room and environment, were the same as for the initial trials.

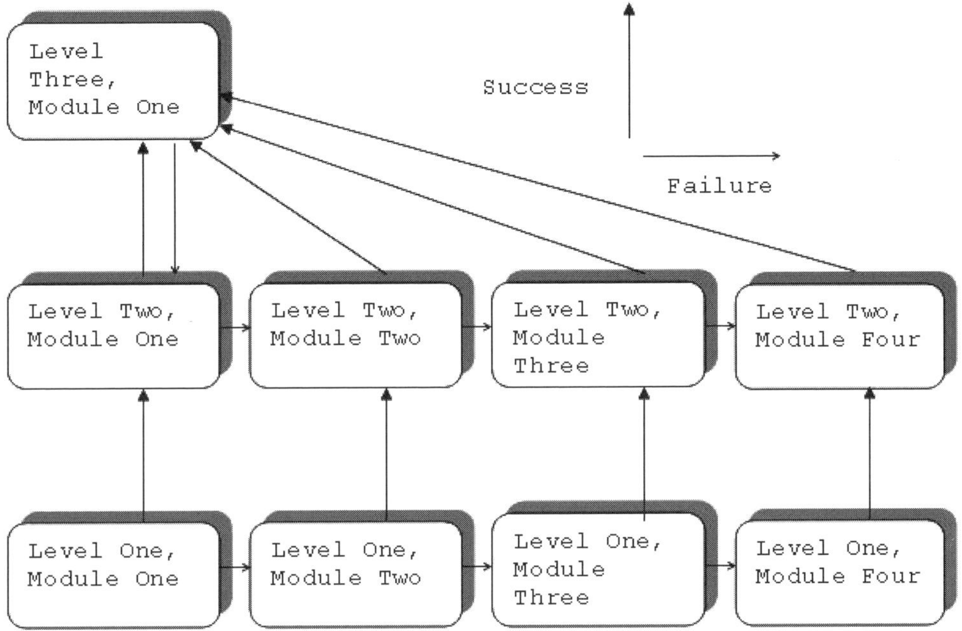

Figure 13.2
The robot architecture for the second single-child trial sessions.

In this study, twelve children participated. They included children who had seen the robot in the previous trials and children who had not encountered the robot before. The group also had a wider age range than the first group, from fours years up to nine years, and consisted of ten males and two females.

Evaluation Methodology: Quantitative Behavioral Analysis

In order to evaluate the trials, each session was videoed continuously, with the main focus being on the child. These records were converted into a digital format in order to preserve the exact time stamp on the frames and to archive the data for later detailed evaluation. The children were able, and encouraged, to interact with the platform in any way they were comfortable, including moving around the room. A variety of different types of interactive behaviors were observed during the trials that needed later careful evaluation.

Each trial session is viewed as a sequence of linked behaviors displayed by the child. In this way, the evaluation method records and scores a specific subset of behaviors which are examined and determined prior to the trials. This methodology is inspired by Tardiff et al. (1995) and methods of behavior analysis in ethology, modified for our own purposes, that are widely used in the Aurora Project (see Dautenhahn and Werry, 2002; Robins et al., 2004a). The specific behaviors are selected in order to provide information about the child's interaction, and are scored by the child's actions only, although other behaviors, such as a limit set produced by the robot platform, are also recorded in order to provide a context for the child's responses.

Each session is broken into one-second intervals. Within each interval, it is recorded if a specific microbehavior (also termed behavior atom, a basic behavior building block or component such as "reach out") is exhibited by the child in this interval. In addition, the focus of the behavior is recorded where this is clear, so as to make a distinction between, for example, touching the robot and touching a wall. Also recorded is what triggered the action (for example, whether the action, such as "look at the robot," was triggered by the robot itself or by the teacher. In some behaviors, the focus of the behavior is relatively clear (for example, "touch" or "move toward"). However, in others, such as vocalization and speech, it may not be so easy to determine whom or what the behavior was aimed at, if it was indeed aimed at a specific object or person at all. Since each one-second interval is evaluated individually, duration is recorded as a by-product of the methodology. In addition, behaviors are generally scored for their duration. However, in the case of the vocal behaviors, the onset is used for evaluation purposes. This is because the duration of the vocal behavior is based more on the actual phrase uttered than on the intention of the child.

In order to account for the variety of different interactive behaviors that emerged during the trials (see figures 13.3–13.6), the following microbehaviors were initially considered:

- *Eye gaze* for example, whether the child was looking at the robot, the environment, or an experimenter.
- *Eye contact* whether the child met the eye gaze of another agent. Where the robotic platform was involved, it was scored when the child looked into the platform's heat sensor, which seemed to be identified by many of the children as its "head."
- *Operate* this behavior is possible only with the robot, since it involves robot movement as a result of the child's behavior, and therefore as at some level an understanding of the way in which the robot is utilized. For example, a child may operate the robotic platform by causing it to move in a desired direction by obstructing any other route.
- *Handling* physically pushing and manipulating the object (for example, pushing it across the floor or pushing the heat sensor to point in a certain direction).
- *Touch* physical contact with an object, but not in a desire to move it.

• *Approach* physically moving toward the object in order to close the distance between the object and the child. Moving toward the object in an effort to move away from another agent is not scored as this behavior, nor is a movement by the robot toward the child.

• *Move away* physical movement to increase the distance between the child and the robot.

• *Vocalization* producing a sound which is not recognizable as speech. This includes, but is not limited to, laughter, sighs, and yells.

• *Speech* An attempt, successful or otherwise, by the child to communicate using recognizable speech. This can be directed toward the robot, a teacher, or an experimenter, or can simply be rhetorical.

• *Stereotypical verbal behavior* Autism is often associated with certain verbal behavior, such as echolalia. Where vocalization or speech utterance is recorded as a stereotypical verbalization, both behaviors are recorded. In addition, it is also often difficult to identify stereotypical verbal behavior, since echolalia may be occur over an extended time period and therefore the child may be repeating phrases learned from a film many days ago. Thus this microbehavior is not used for evaluation purposes.

• *Repetition* this microbehavior is used to record instances of behaviors which are repeated a number of times (for example, the spinning of a wheel on the toy truck or the same phrase uttered a number of times). Like the stereotypical verbal behavior, it is recorded in addition to the actual behavior displayed, such as manipulation, and is an action often associated with autism.

• *Attention* this is an indication of what the child is attending to at any given instance. For example, it is possible that a child is looking at the robotic platform, and touching it, but that his attention is directed to sounds from the environment, and that the behavior involving the robot is idle behavior. It can be a difficult behavior to evaluate, but is often linked to eye gaze.

• *Other* this is a catchall category which allows the evaluator to record any other behaviors of interest or to make specific notes on the child involved in the trial.

• *Blank* occasionally, a child sits in the environment and does not interact or display behaviors. This is recorded as a separate behavior in order to differentiate it from behaviors which are displayed but are not otherwise recorded.

In addition, the vocal production of the robot was noted, specifically, when each phrase was produced and whether the phrase was classified as a question (i.e., a statement which would otherwise solicit a response, such as "Where are you?") or as a statement (such as "There you are").

The above list indicates the variety of behaviors that can occur in this quite simple interaction scenario involving a mobile robot and a child. For the purposes of the evaluation, three behaviors are grouped together to create a single parameter which is termed "contact time." This group consists of operating, handling, and touching, and is generated because it is impossible to operate the toy truck, and therefore it is impossible to compare interaction with the robot and with the toy truck.

Results and Evaluation

Various human-robot interactions were observed during the trials (see figures 13.3, 13.4, and 13.5). For example, some of the children were able to understand that the robot attempted to avoid obstacles, and were able to use this knowledge to guide

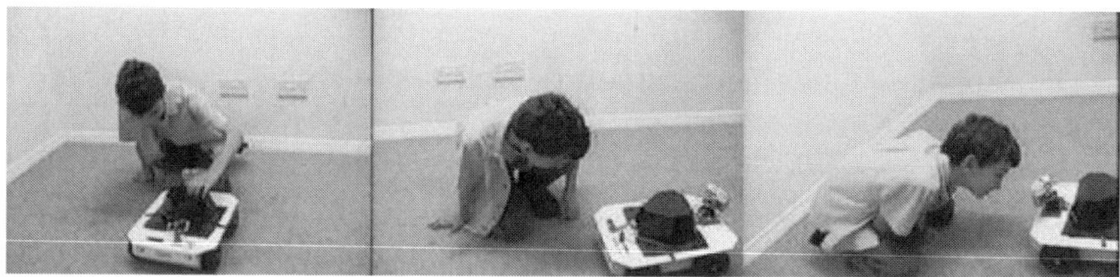

Figure 13.3
George interacting with the robot platform. (Names of all children in this chapter have been changed.)

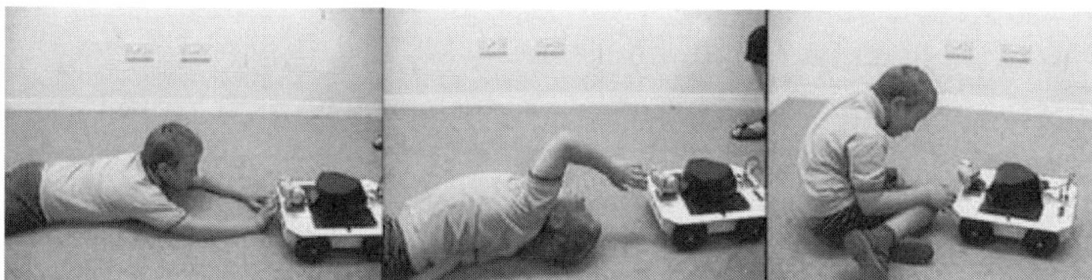

Figure 13.4
David playing with the robot.

Figure 13.5
Terry playing with the robot platform.

the robot into movements that they wanted, by positioning themselves so that the robot would move in a specific way. In addition, some of the children appeared to interact with the robot as an individual agent, attempting to converse with it and to get its attention by tapping it or making specific movements and noises.

For the two trial sessions which were conducted, three parameters were selected as the foundation for the evaluation: "contact time," "eye gaze," and "attention." It was felt that these parameters gave the best indication of the child's attitude toward and interaction with the robotic platform compared with the toy truck. Other parameters were not used for a variety of reasons. The microbehaviors "vocalization" and "speech" were not used because a number of the children were nonvocal. Microbehaviors such as "approach" and "move away" depended on the distance to the platform, and were therefore subject to extra factors such as the movement of the robot.

Over the two trials, a total of nineteen children participated, seven in the first set and twelve in the second set. The evaluation compared those parameters which had as their focus either the robotic platform or the toy truck. Parameters which were focused on other objects, such as the environment or an experimenter, were discounted from the comparison.

Behavior Evaluation

The data which resulted from the trials were compiled and analyzed. A summary can be seen in table 13.1, which states the average length of the behavior and states how many children had a higher average behavior length for the robotic platform and how many had a higher average behavior length for the toy truck, for each of the three parameters which were examined. Table 13.2 examines the same parameters, but displays how many children from each of the two sessions devoted a higher percentage of the total trial time to the robot or the toy, for each of the behaviors examined. In addition, due to the evolution of the behavior parameters and evaluation process, the "attention" parameter was evaluated for only five of the seven children

Table 13.1
The number of children showing higher average behavior lengths for the robot and the toy

		Number of children displaying a higher average behavior length for the robot	Number of children displaying a higher average behavior length for the toy
Contact time	First trial	3	4
	Second trial	0	8
Eye gaze	First trial	5	2
	Second trial	8	0
Attention	First trial	4	1
	Second trial	8	0

Table 13.2
The number of children showing higher percentage times for the robot and for the toy

		Number of children displaying a higher percentage time for the robot	Number of children displaying a higher percentage time for the toy
Contact time	First trial	2	5
	Second trial	1	7
Eye gaze	First trial	3	4
	Second trial	8	0
Attention	First trial	3	2
	Second trial	7	1

in the first trial. Eight of the twelve children in the second trial were evaluated for all parameters.

In both tables, contact time is higher for the toy truck, which contradicts the other statistical results that show that the robotic platform scores higher in attention and eye gaze. There are several possible explanations. First, this may happen because the children are more familiar with the toy—they have toys of the same type to interact with in their everyday environment. Alternatively, this may be due to the fact that the robot exhibits self-propelled, autonomous movements, which the toy does not, and therefore the children are not able to watch the toy unless they touch it. In addition, children were occasionally prevented from damaging the robotic platform by teachers and experimenters, necessarily restricting their interaction, which may affect their willingness to interact openly with it.

Overall, it was found that most of the children displayed a relatively low contact time with the robot as a percentage of the entire trial, with only two children from the initial trial and one in the second trial displaying a higher percentage contact time with the robot. Comparing this with the data in table 13.1, which shows that many of the children exhibited a longer average behavior length in terms of contact time with the toy truck, we can see that the general trend is for the children to physically interact more intensively with the toy than with the robot.

When examining the eye gaze behavior, table 13.2 shows that only four of the children spent a higher percentage of the trial session focusing eye gaze on the toy, despite the previous result that most of the children spent longer periods of time physically interacting with the toy. In addition, five of the children spent 70 percent or more of the trial focusing eye gaze on the robotic platform. Note that in the second trial session, all children displayed a higher average behavior length and a longer percentage of the total trial length gazing at the robotic platform. Similarly, for the "attention" behavior, we can see that more children displayed a longer average time attending to the robotic platform in both trial sessions, and that more of the children

Table 13.3
Significance levels of each microbehavior over both sets of trials for percentage of the trial time spent engaged in the behavior

Contact Time		Eye Gaze		Attention	
P < 0.05, significant (N = 15, T = 24)		P < 0.05, significant (N = 15, T = 24)		P < 0.001, highly significant (N = 13, T = 7)	
Mean percentage time for the robot: 29.98%	Mean percentage time for the toy: 58.30%	Total percentage time for the robot: 61.07%	Total percentage time for the toy: 52.96%	Total percentage time for the robot: 62.86%	Total percentage time for the toy: 44.96%

participating in the second trial session attended to the robotic platform for more of the session length than they did to the toy truck.

Statistical Evaluation

More in-depth analysis was performed, utilizing the Wilcoxon matched pair signed rank test (Siegel, 1956). A nonparametric test was selected because of the relatively small sample size. The data were tested for significance, and the results are displayed in table 13.3.

This table supports the results that overall, the children spent significantly more time focusing eye gaze on the robot, and attending to the robotic platform, but also had significantly more contact time interacting with the toy truck.

The evaluation was also tested for interrater reliability, using the Kolmogorov-Smirnov test; in all cases results were consistent. Overall, two secondary evaluators were used to check the robustness of the evaluation methodology, checking the recorded data from five of the children. These secondary scorers were volunteer Ph.D. students within the project who had not been present at the trial sessions and were not associated in any way with the outcome. Their evaluation consisted of scoring the trials in the same way that the initial evaluator had done, watching the individual video recordings of the sessions, and writing down the various behaviors and when they occurred. The two sets of raw data were then compared for consistency.

In addition to the direct evaluation achieved by the trials, the teachers who were involved filled in brief questionnaires about the children participating. While this information was not significantly examined, it was apparent that the three children who were judged by the teachers to have the most severe level of autism in the group, rated by the teachers as 7 or 8 out of 8, achieved individually significant results for their interaction with the robot over the toy. While further and more in-depth study is required, this may indicate that the robot is generally able to have a greater effect on children who have been diagnosed with more severe autism. It is interesting to note that these three children were also rated as having some of the lowest social ability in the group.

Discussion

The results of the trials reported in this chapter show that, in general, the children spent more of their trial session watching and attending to the autonomous robot than to the passive toy truck, but that contact time was generally higher for the truck. It is hypothesized that this is due to the requirement of direct, physical manipulation of the toy in order to engage with it, something which is not required with the robotic platform that moves autonomously. In addition, the toy is familiar to the children, and so is the method of interacting with it, meaning that the children are likely to interact without an initial period of discovery of the object.

These trials were initiated with the goal of examining whether robotic technology has the potential to contribute to the therapy of children with autism. For the technology to have such an effect, it first needs to be able to engage the children. Results presented in this chapter provide evidence for the robot's ability to engage the children better than a comparable nonrobotic toy.

Many different directions could be pursued in future trials. The discussed trials were initial studies in the area, aimed to establish a general groundwork, and have led to a number of more focused subsequent studies (e.g., Robins et al., 2004a, 2004b; Salter et al., 2004). Further work with a larger sample size could determine in which ways a robotic platform is able to contribute to the field, and in what cases the additional expense and development of a robot would warrant its use rather than established methods. In particular, the use of the voice production unit needs to be examined in more detail, in order to judge its potential to engage the children. Further development of the adaptive nature of the robot would increase its potential to become a more flexible device, able to interact with the children at their level of ability.

Also, the children who participated in the trial may benefit from further examination of their background and characteristics, since it is possible that some children would react better to a technological device such as the robot than others would. More longitudinal studies would examine the effect that a robotic platform can have on the therapy of the children over an extended period of time (see Robins et al., 2004a), and also would examine whether the children's attitude toward and interaction with the platform would alter with further exposure to it.

Observation of the trial session indicates the robot's ability to allow the children to explore important concepts of the interaction space, specifically involving relative distance in communication, and interaction and turn-taking.

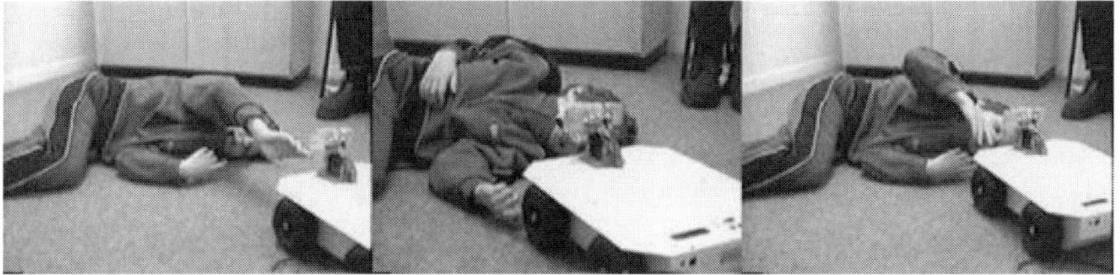

Figure 13.6
Brad playing with the robotic platform.

The study reported in this chapter, and subsequent studies carried out in the Aurora Project, highlight two fundamental properties of human-human interaction as applied to human-robot interaction: First, the *emergent nature of interaction*. A robot equipped with a simple behavior program interacting with autistic children can lead to a variety of fairly complex behaviors that cannot be reduced to the robot's behavior program. For example, turn-taking frequently emerged as a "global pattern" in our trials; many children quickly discovered how to "operate" the robot (e.g., making it approach or avoid by touching or reaching out toward the infrared sensors (triggering avoidance) or heat sensors (triggering approach) (see figure 13.6). Chasing and cornering the robot were other emerging patterns that could be observed. This sometimes resulted in such interactive games being sustained for five or ten minutes or longer, involving the common theme of approach and avoidance but manifested in highly varied behaviors. Note that due to the nature of autism, one must avoid reinforcing stereotypical tendencies in children with autism. Variations comprised different distances, body postures, timing, and so on.

Second, our studies highlight the *proactive nature of interaction*, as applied to human-robot interaction. Although some interactions may be ritualized, as shown by many examples across the animal kingdom (including humans), the majority of naturally occurring interactions are not prespecified; they do not follow a fixed scheme and they require, in dyadic interactions, proactiveness by both interactants. Encouraging proactiveness of behavior is therapeutically very important and part of many exercises in schools for children with autism. We are thus optimistic that, in combination with the *embodied nature of human-robot interactions*, robots potentially can make a valuable contribution to *understanding fundamental principles of embodied interaction* in general, and can contribute to the therapy and education of children with autism in particular.

For recent work continuing to use robots in autism therapy presented in this chapter, see work by Robins et al. (e.g., Robins et al., 2004a, 2004b). See also further recent publications on the Aurora Project Web site (Aurora, 2006). The work presented in this chapter, as well as other work on the Aurora Project, has been carried out as part of a research project in human-robot interaction in assistive technology. Future studies need to demonstrate the clinical benefit of this work. However, ultimately such studies, as exemplified in this chapter, may serve as a model for medical intervention using robots as therapeutic toys for children with autism and possibly with other developmental disorders.

Acknowledgments

The Aurora Project team owes a debt of thanks to the students and staff at Radlett Lodge School, a specialist school of the National Autistic Society, and in particular to Patricia Beevers. The robotic platform used in this work was provided by Applied AI Systems (donated by Takashi Gomi), and the work was supported by a grant from the EPSRC. The authors would also like to thank Dr. William Harwin, Department of Cybernetics, University of Reading, for his contribution to Iain Werry's Ph.D. work.

References

AAATE (Association for the Advancement of Assistive Technology in Europe). http://www.aaate.net. Last accessed December 6, 2006.

Attwood A, Frith U, Hermelin B (1988) The understanding and use of interpersonal gestures by autistic and Down's syndrome children. J Autism Develop Disorders 18(2): 241–257.

Aurora. http://www.aurora-project.com. Last accessed December 6, 2006.

Colby KM, Smith DC (1971) Computers in the treatment of nonspeaking autistic children. Curr Psych Ther 11: 1–17.

Dautenhahn K (1999) Robots as social actors: Aurora and the case of autism. In: Proceedings of the Third International Cognitive Technology Conference, 359–374 (Cox K, Gorayska B, Marsh J, eds). M.I.N.D. Lab, Michigan State University.

Dautenhahn K, Werry I (2002) A quantitative technique for analysing robot-human interactions. In: Proceedings of IEEE/RSJ International Conference on Intelligent Robots and Systems, vol 2, 1132–1138. IEEE Press.

Dautenhahn K, Werry I (2004) Towards interactive robots in autism therapy. Pragmat Cognit 12(1): 1–35.

Davis M, Robins B, Dautenhahn K, Nehaniv CL, Powell S (2005) A comparison of interactive and robotic systems in therapy and education for children with autism. In: Assistive Technology from Virtuality to Reality (Pruski A, Knobs H, eds), 353–357. Amsterdam: IOS Press.

DSM. Autism Biomedical Information Network—DSM-IV Criteria, Pervasive Developmental Disorders. http://www.autism-biomed.org/dsm-iv.htm. Last accessed December 6, 2006.

Kozima K (2002) Infanoid: A babybot that explores the social environment. In: Socially Intelligent Agents: Creating Relationships with Computers and Robots (Dautenhahn K, Bond AH, Cañamero L, Edmonds B, eds), 157–164. Amsterdam: Kluwer Academic.

Kozima H, Nakagawa C, Yasuda, Y (2005) Interactive robot for communication-care: A case-study in autism therapy. In: Proceedings of the IEEE Workshop on Robot remove and Human Interactive Communication, 341–346.

Michaud F, Laplante JF, Larouche H, Duquette A, Caron S, Masson P (2005) Autonomous spherical mobile robot to study child development. IEEE Trans Syst, Man, Cybernet 35(4): 471–480.

Michaud F, Théberge-Turmel C (2002) Mobile robotic toys and autism. In: Socially Intelligent Agents: Creating Relationships with Computers and Robots (Dautenhahn K, Bond AH, Cañamero L, Edmonds B, eds), 125–132. Amsterdam: Kluwer Academic.

Papert S (1980) Mindstorms: Children, Computers and Powerful Ideas. New York: Basic Books.

Plaisant C, Druin A, Lathan C, Dakhane K, Edwards K, Vice JM, Montemayor J (2000) A storytelling robot for pediatric rehabilitation. In: Proceedings of the Fourth International ACM Conference on Assistive Technologies 50–55. New York: ACM Press.

RESNA (Rehabilitation Engineering & Assistive Technology Society of North America). http://www.resna.org. Last accessed December 6, 2006.

Robins B, Dautenhahn K, te Boekhorst R, Billard A (2004a) Effects of repeated exposure to a humanoid robot on children with autism. In: Designing a More Inclusive World (Keates S, Clarkson J, Langdon P, Robinson P, eds), 225–236. London: Springer-Verlag.

Robins B, Dautenhahn K, te Boekhorst R, Billard A (2004b) Robots as assistive technology: Does appearance matter? In: Proceedings of the Thirteenth IEEE International Workshop on Robot and Human Interactive Communication, 277–282. IEEE Press.

Salter T, te Boekhorst R, Dautenhahn K (2004) Detecting and analysing children's play styles with autonomous mobile robots: A case study comparing observational data with sensor readings. In: Proceedings of the Eighth Conference on Intelligent Autonomous Systems (Groen F, Amato N, Bonarini A, Yoshida E, Kröse B, eds), 61–70. Amsterdam: IOS Press.

Scassellati B (2005) Quantitative metrics of social response for autism diagnosis. In: Proceedings of IEEE Workshop on Robots and Human Interactive Communication, 585–590.

Siegel S (1956) Nonparametric Statistics for the Behavioral Sciences. New York: McGraw-Hill.

Stone WL, Ousley OY, Littleford CD (1997) Motor imitation in young children with autism: What's the object? J Abnorm Child Psych 25(6): 475–485.

Tardiff C, Plumet M-H, Beaudichon J, Waller D, Bouvard M, Leboyer M (1995) Micro-analysis of social interactions between autistic children and normal adults in semi-structured play situations. Intl J Behav Develop 18(4): 727–747.

Weir S, Emanuel R (1976) Using Logo to catalyse communication in an autistic child. DAI Research Report no. 15. University of Edinburgh.

Werry I (2003) Development and evaluation of a mobile robotic platform as a therapy device for children with autism. Doctoral thesis, University of Reading.

Werry I, Dautenhahn K (1999) Applying robot technology to the rehabilitation of autistic children. In: Proceedings of the Seventh International Symposium on Intelligent Robotic Systems, 265–272.

Werry I, Dautenhahn K, Harwin W (2001a) Investigating a robot as a therapy partner for children with autism. In: Proceedings of the Sixth European Conference for the Advancement of Assistive Technology, 374–378.

Werry I, Dautenhahn K, Harwin W (2001b) Evaluating the response of children with autism to a robot. In: Proceedings of RESNA 2001 (Simpson R, ed), 14–16. Arlington, VA: RESNA.

Williams JHG, Whiten A, Suddendorf T, Perrett DI (2001) Imitation, mirror neurons and autism. Neurosc Biobehav Rev 25: 287–295.

Wing L (1996) The Autistic Spectrum. London: Constable Press.

VI EVOLUTION

Since evolution is a historical process spanning eons, evolutionary mechanisms are difficult to observe directly—with the exception of microbial and viral evolution and of long-term studies of adaptive responses to changing environments, such as those of Darwin's finches on the Galapagos Islands. Understanding of evolutionary mechanisms thus has always depended on models. Darwin conceptualized natural selection in analogy to artificial selection, with the latter acting as a model of the former. Darwin's canonical formulation of evolution by means of natural selection also implied that a complete theory of evolution has to be based on (1) an account of the generation of variation ("the laws of variation"), (2) the transmission of hereditary material, and (3) the population level consequences of differences in the performance of individual variants (fitness). Over the course of the last 150 years each of these areas of research has developed its own modeling strategies in order to address these questions.

In genetics, long before there was any concrete knowledge about the molecular mechanisms of inheritance, the theory of Mendelian inheritance consisted largely of a set of assumptions and models of transmission derived from pedigree studies and statistical analyses. These were further refined through the study of several model organisms, *Drosophila* foremost among them, that provided the foundation of successful research programs in genetics and eventually also in evolutionary biology. Population genetics combined the statistical descriptions of population variables with models of genetic effects and causation, thus creating a model-based mathematical foundation for evolutionary theory. In these models the creation of phenotypic variation was not explicitly problematized; it was assumed that a constant mutation rate provides sufficient raw material for natural selection to act upon. At the same time models produced in developmental biology, which was concerned with the explanation of embryonic differentiation, also did not specifically address the question of phenotypic variation. This situation is only now beginning to change (see part IV and chapter 16).

The idea that natural selection leads to adaptation through the optimization of individual traits has been another core assumption in evolutionary modeling. A whole class of models combined physical measurements of the performance of organisms with some basic assumptions about their reproductive success (fitness) to predict the outcome of natural selection. However, it soon became obvious that there are no necessary simple correlations between small changes in the organism's phenotype and its performance or fitness. Sewall Wright developed the concept of a fitness landscape as a rugged terrain with multiple peaks (optima) and valleys. In this model, natural selection can lead only to local optimization. The model of fitness landscapes was soon combined with the concept of a multidimensional morphospace or design space—a characterization of all possible forms—in order to explore possible evolutionary scenarios and to investigate in what way actual evolutionary history has led to an optimization of individual characters. The chapters in this part introduce several modeling approaches in this tradition.

Karl Niklas (chapter 14) combines the approach of theoretical morphology (see also chapter 5 in this volume) with several concepts and methods from evolutionary theory, such as fitness landscapes and optimization strategies, in a model of early land plant evolution. After colonizing land, plants experienced a rapid phenotypic radiation that produced all major extant lineages (with the exception of flowering plants) and all organizational grades within a period of roughly 46 million years. Niklas explores the question of how morphological types from the fossil record correspond to possible morphologies that are functionally viable. To this end he identifies four major challenges that land plants need to solve, and quantifies the ability of each morphological variant to perform these tasks.

The resulting model combines a theoretical morphospace containing all possible variants with fitness landscapes, which are a measure of the performance of morphological variants, and a search algorithm that mimics an optimization procedure. This model serves as a heuristic tool that permits testing assumptions about early land plant evolution that are derived from the fossil record as well as from evolutionary theory. It also leads to a set of general properties of this process that correspond to the currently available data and suggest new avenues for both empirical and theoretical research.

James Marshall and Nigel Franks (chapter 15) explore the possibilities of expanding the design space modeling approach to include the evolution of complex behaviors (see also chapters 5 and 14 in this volume). They present several models of nest area assessment and collective decision-making in ants (*Temnothorax albipennis*). Collective behavior in social insects is an emergent property of the actions of many individual agents (ants), and therefore lends itself quite naturally to a computer-based modeling approach. Individual behaviors are comparatively easy to observe and quantify. They can be implemented either within an agent-based framework or, as discussed here, embedded within the design space framework—a morphospace of behavioral variants. This approach enables Marshall and Franks to explore the evolution not only of behavior but also of complex social systems more generally. One of the perennial questions in evolutionary biology is how complex systems can evolve or, in the case of social insects, how social behaviors can first emerge. Marshall and Franks's models suggest that complex behaviors, such as the decision to move into a new nest site, can be the result of repeated interactions of individual ants.

James Collins et al. (chapter 16) discuss multiple modeling strategies in the newly emerging field of evolutionary developmental biology (EvoDevo). EvoDevo is a rich interdisciplinary research program focused on explaining patterns of phenotypic evolution through the integration of developmental mechanisms with evolutionary dynamics and, increasingly, with environmental factors. Currently, EvoDevo is a research program that has generated many important empirical results and several hypotheses with great potential to contribute to some of the questions left unanswered by traditional evolutionary theory—most notably the problem of evolutionary innovations—but it still lacks a foundational theoretical structure. In part this is a consequence of the many different types of data and levels of biological organization that are part of genuine EvoDevo explanations. At this stage of EvoDevo, models therefore represent important heuristic, conceptual, explanatory, and integrative tools. Collins et al. discuss the constitutive questions of EvoDevo and the new model systems that emerge in response to these questions, ranging from models that map genetic variation and the evolution of gene regulatory pathways to organism-environment interactions and the morphogenetic principles of individual organ evolution. They argue that a detailed analysis and subsequent integration of these separate modeling strategies can eventually lead to the theoretical integration of EvoDevo.

14 Modeling Optimization and Early Land Plant Evolution

Karl J. Niklas

The colonization of the terrestrial landscape by ancient plants paved the way for animal life onto land and irrevocably altered Earth's terrestrial ecosystems (Chaloner and Sheerin, 1979; Gensel and Andrews, 1984; Willis and McElwain, 2002; Berner et al., 2003). This pivotal event also sparked one of the most dramatic bursts of morphological and anatomical evolution in the history of life (Raven, 1984, 1985; Taylor and Taylor, 1993; Stewart and Rothwell, 1993; Niklas, 1997a). The most ancient land plants were characterized by simple, small, and leafless body plans, yet within 46 million years they diversified to encompass all major extant lineages with the exception of flowering plants. Additionally, the full spectrum of present-day organizational grades evolved within this time interval (Niklas and Kerchner, 1984; Knoll et al., 1984). (See figure 14.1.)

The objective of this chapter is to evaluate the use of computer models and biophysical principles to mimic this important episode in plant evolution and to explore why it involved such a rapid phenotypic radiation. The models used here are similar to the metaphor for adaptive evolution proposed by Sewall Wright (1931, 1932), who conceived of adaptation as the result of a series of "walks" over fitness "landscapes" with adaptive hills and maladaptive valleys. This metaphor draws sharp attention to the relationship between the number, location, and height of fitness peaks, and to the genomic transformations required to increase the relative fitness among neighboring variants. When it is translated into morphological rather than genomic landscapes and transformations, Sewall Wright's metaphor has been profitably exploited for plants as well as animals (e.g., Raup, 1961, 1962; Raup and Michelson, 1965; Niklas and Kerchner, 1984; Niklas, 1997b,c; Thomas and Reif, 1993; McGhee, 1999).

Importantly, it is comparatively easy to model the morphologies of the most ancient land plants and to quantitatively evaluate their abilities to perform the basic tasks required for growth, survival, and reproduction. For example, the perennial life form of all vascular plants (i.e., tracheophytes) is sporophyte generation. Those of the most ancient tracheophytes were composed of leafless branched cylindrical axes (Chaloner and Sheerin, 1979; Stewart and Rothwell, 1993; Taylor and Taylor, 1993).

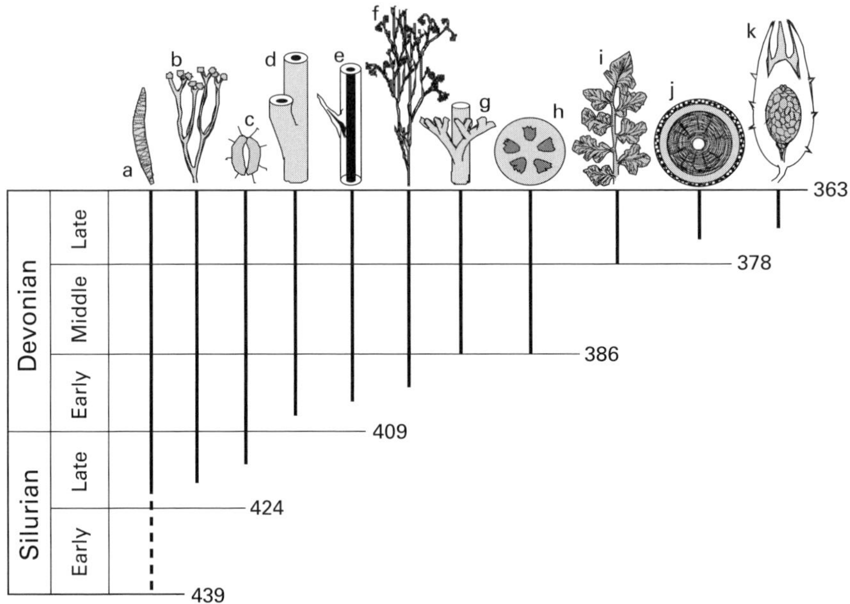

Figure 14.1
Time line for some major Silurian-Devonian phenotypic innovations. Vertical bars denote first appearance and range of figured features in the fossil record. (a) Tracheids. (b) Isobifurcate branching with terminal sporangia. (c) Stomata with guard cells. (d) Unequal branching of vascular axes. (e) Vascularized leaflike enations. (f) Overtopping with lateral branches. (g) Planated lateral branches. (h) Dissected primary vasculature. (i) Fernlike leaves (megaphylls). (j) Stems with periderm and wood. (k) Seeds.

Although the architectures of these sporophytes varied, in very general terms all mathematical variants can be modeled by varying axial length, diameter, the probability of branching, the angle of branching, and the rotation angle of axes with respect to ground level. The reproductive organs of the most ancient sporophytes were also simple in appearance. Although their number and mode of attachment to axes varied among taxa (see Gensel and Andrews, 1984), the sporangia of most ancient sporophytes freely shed their spores and relied exclusively or largely on wind for broadcasting reproductive cells. Therefore, reproductive efficacy (in the broadest sense of the phrase) can be modeled with the aid of only a few variables (i.e., sporangial elevation above ground, and spore number and size).

In terms of functional analyses, the basic requirements for plant growth and survival are well known and extremely amenable to quantitative evaluation (Gates, 1965; Nobel, 1983; Niklas, 1992, 1994, 1997b, 1997c; Taiz and Zeiger, 2002). With the exception of species that have secondarily adopted a parasitic lifestyle, all plants must intercept sunlight, cope with externally applied mechanical forces, and conduct and conserve water. Light interception can be qualified in terms of the surface area a plant projects toward the sun and how this surface area varies as a function of time of day, latitude, and season (Niklas and Kerchner, 1984). The exchange of atmospheric gases between the plant body and its external environment, as well as the ability to acquire and transport water, can be quantified in terms of the relationship between body surface area and volume in terms of comparatively simple equations from fluid mechanics (Nobel, 1983; Niklas, 1992). Likewise, the capacity to deal with bending and twisting mechanical forces can be quantified rigorously with the aid of engineering theory (Niklas, 1992). Finally, many aspects of plant reproduction, such as spore dispersal, can be modeled using simple aerodynamic principles (Okubo and Levin, 1989).

Indeed, the only real challenge to modeling early land plant evolution is to assess how all of these basic tasks can be performed simultaneously without imperiling the performance of any one task. The difficulty emerges because it is virtually impossible to maximize the performance of any one task without negatively affecting the performance of some other task. For example, the capacity to intercept sunlight and the ability to exchange carbon dioxide or oxygen between the plant body and its surrounding fluid increases as a function of surface area, whereas the ability to conserve water decreases (Gates, 1965; Nobel, 1983). Likewise, engineering principles show that cantilevered organs have the best orientation for light interception but the worst orientation in terms of inducing self-imposed mechanical stresses (Niklas, 1992). When they are viewed from a biophysical or engineering perspective, none of the basic biological tasks that plants perform can be maximized without decreasing the performance of another necessary task. It is reasonable, therefore, to suppose that the evolution of plant form-function relationships required some sort of optimization process (see Horn, 1979; Kauffman and Levin, 1987).

The Optimization Process: An Analytical Illustration

The role played by optimization processes in early land plant evolution is illustrated here in only a very general way. My purpose is to briefly explore the basic principles of optimization.

A basic feature of the optimization process is that it results from a trade-off between two or more conflicting but necessary requirements or processes that are attributable to a single system. Typically, the performance of one of these requirements increases as the performance of the other decreases with respect to some common parameter of interest (e.g., a metabolite, energy, or biomass). Under these circumstances, optimization can be conceptualized analytically by considering a hypothetical mathematical function which quantifies performance P that is the sum of two other performance functions, each of which is dependent on some sort of biomass investment M. We will assume that these investments have the form of power functions, which abound in the biological literature (e.g., annual growth increases as the 3/4 power of body mass). Denoting the scaling exponents of the two performance functions as m and n and the proportionality constants as k_0 and k_1, a simple example of P is given by the formula

$$P = \frac{k_0}{M^m} + k_1 M^n. \tag{14.1}$$

In this example, M^m could represent the scaling of a bending moment (created at the base of a tree trunk by wind-induced drag) with respect to trunk and leaf mass, whereas M^n could represent the ability to harvest light (which is some function of leaf area, and thus of leaf biomass). A trade-off between these two performance functions exists because any reduction in the bending moment (which increases k^0/M^m) requires a reduction in either stem or leaf mass (which reduces drag but also reduces the ability to support leaves or harvest light, respectively).

To find the optimal investment of biomass M^* (i.e., the extremum), we take the partial derivative of P with respect to M and set the partial differential equation (PDE) equal to zero:

$$\frac{\partial P}{\partial M} = -mk_0 M^{-m-1} + nk_1 M^{n-1} = 0. \tag{14.2}$$

Solving for M^* gives

$$M^* = \left(\frac{mk_0}{nk_1}\right)^{1/(m+n)}, \tag{14.3}$$

and combining equations (14.1) and (14.3) gives the optimal performance P^* exclusively in terms of the scaling exponents and proportionality constants:

$$\begin{aligned} P^* &= k_0\left(\frac{mk_0}{nk_1}\right)^{-m/(m+n)} + k_1\left(\frac{mk_0}{nk_1}\right)^{n/(m+n)} \\ &= \frac{k_0(m+n)}{n}\left(\frac{nk_1}{mk_0}\right)^{m/(m+n)}. \end{aligned} \tag{14.4}$$

This derivation draws attention to the fact that performance functions of the type illustrated here are analogous, albeit loosely, to fitness landscapes in which P^* and M^* specify the most cost-effective biomass investment. Figure 14.2A illustrates this feature for the hypothetical case $k_0 = 2$, $k_1 = 3$, $m = 1/2$, and $n = 3/4$, which obtains $M^* \sim 0.519$ and $P^* \sim 4.61$. The extent to which neighboring P-values differ numerically from P^* translates loosely into relative fitness. In this regard, there exists an inverse relationship between the magnitudes of the scaling exponents and the relative fitness of these other strategies, that is, fitness relative to the optimal condition

A

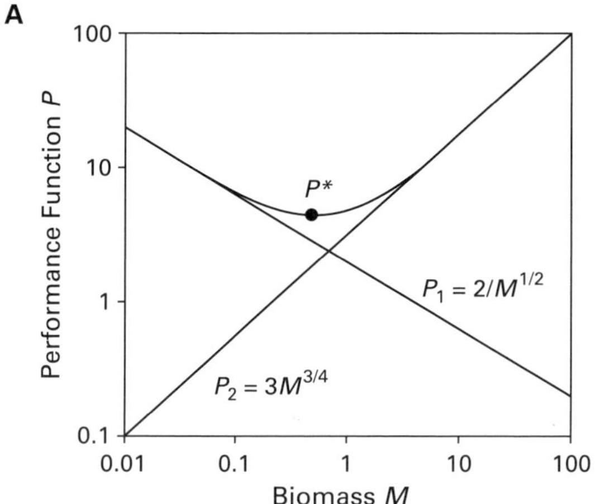

B

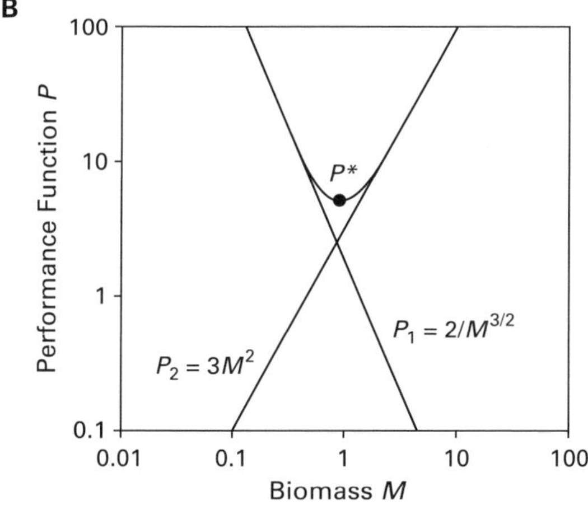

Figure 14.2
Two cases of optimization between two performance functions P_1 and P_2 (plotting as straight lines) measured in terms of biomass investment M. The optimal performance function P^* (plotting as a curved line) and optimal biomass investment M^* for each case are defined by the trade-off between P_1 and P_2. The "sensitivity" of the optimal performance function depends on the numerical values of the exponents of P_1 and P_2. In general, small exponents obtain less sensitive P^* than large exponents (case A and case B, respectively).

decreases as the magnitudes of exponents m and n increase (fig. 14.2B). To visualize the analogy between the performance functions and an adaptive landscape, invert the P^* functions in figure 14.2 to produce "adaptive hills." Likewise, inverting the P^* function in figure 14.2A results in an adaptive hill with gentle slopes and a broad top, whereas inverting the P^* function in figure 14.2B results in an adaptive hill with steep slopes and a narrow peak.

An interesting but unanswered question is whether evolution by natural selection gives rise to less "sensitive" performance functions. Certainly there are advantages to an organism if its performance functions are relatively insensitive, because this provides a greater latitude for energy or biomass investment. If this is true, then nature should abound in scaling functions with numerically small exponents.

Morphospaces of Simple Morphologies

An analytical approach illustrates in only the most general way how the optimization process and Sewall Wright's metaphor for adaptive evolution intersect conceptually. Here, I want to strengthen this impression by constructing a small morphospace, quantifying the ability of each morphological variant within it to perform some simple biological tasks, and evaluating some very simple "adaptive walks."

The variants in this morphospace have simple geometries (oblate and prolate spheroids, and unbranched and one-branched cylinders, with and without hemispherical ends). In addition to being easy to manipulate mathematically, these morphologies mimic, albeit very crudely, the basic body plans of thalloid gametophytes and the most simple of all known sporophytes. Accordingly, the morphospace explored here provides an opportunity to address why the gametophytes of many modern-day liverworts (and, presumably, the gametophytes of some ancient land plants) are dorsiventral and why this architecture is rarely used to construct the sporophyte body plan.

The morphospace is constructed by varying the size and shape of each geometric class of objects independently. Here, "size" is defined and specified by the volume occupied by each variant V, whereas "shape" is specified by the quotient of body length and diameter l/d. Note that size and shape are not synonymous. Every measure of size has units, whereas every measure of shape is a pure (dimensionless) number. Consequently, within any geometric class of objects (e.g., terete cylinders or oblate spheroids), shape can remain constant as size increases to obtain the well known 2/3 scaling "rule" governing the relationship of surface area to volume. In contrast, shape can be altered as size increases to obtain an S versus V scaling exponent that considerably exceeds 2/3.

Two biological tasks are used here to quantify the performance of each variant: the ability to intercept light and to conserve water. The ability to harvest sunlight can be quantified by integrating the area under the curve generated when a light interception index I is plotted against the incident angle of radiation θ. In the simplest case, we can assume daylight conditions at the equator such that, from dawn to noon, the angle range of incident sunlight is $0^\circ \leq \theta < 90^\circ$. Likewise, we can use the quotient of the projected and total surface areas of each variant as the light interception index; that is, $I = \sum_{i=0^\circ}^{90^\circ}(S_P/S)$, because it can be easily calculated for simple geometries and because S (in tandem with body size V) provides an extremely useful measurement for the ability of any variant to conserve water (in terms of reducing S with respect to V).

Not surprisingly, analyses of "walks" through this morphospace quickly reveal that the ability to harvest sunlight conflicts with the ability to conserve water for oblate spheroids, less so for prolate spheroids, and not at all for cylinders. Analyses further show that the ability to harvest sunlight is least sensitive to changes in the orientation, size, or shape of cylinders, and that this geometric class of objects provides the most effective geometry for elevating other structures above ground (e.g., sporangia).

These claims are illustrated first by examining the areas under the SA_P/S versus θ plots for four simple geometries as shape is varied (figure 14.3). Noting that a sphere results when $l/d = 1$, we see that the ability to harvest sunlight changes rapidly with respect to changes in θ as the shapes of cylinders and oblate and prolate spheroids change. Across all geometries and the majority of shapes, light interception is maximized when $\theta \sim 90^\circ$ (figure 14.3A–14.3D). Nevertheless, the geometry that absolutely maximizes the light-harvesting index is the oblate geometry, and the best shape for this geometry is an extremely flattened disklike spheroid (figure 14.3E). Therefore, if light interception were the only biological obligation of terrestrial plant life, the analyses presented here predict that the architecture maximizing this task is a prostrate, very large disklike one.

However, it cannot escape attention that oblate spheroids also maximize their surface area with respect to their volume as they continue to flatten and increase in size (figure 14.3F). Without some sort of external coating that is impermeable to water, this geometry has a high probability of dehydrating rapidly in comparison with the other simple geometries treated here. Although the gametophytes of many terrestrial nonvascular plants possess a dorsiventral (oblate-like) shape, these plants are susceptible to rapid dehydration and typically grow in hydric microenvironments. Indeed, the reproductive requirements of these gametophytes confine them to very wet habitats (for sperm dispersal and the fertilization of eggs).

In contrast, provided they maintain a constant girth as they elongate, cylinders are particularly insensitive to changes in their surface area with respect to volume. Cylinders are also insensitive to changes in the light-harvesting index as they increase in size (figure 14.3E). Additionally, they provide extremely good mechanical support, as opposed to flattened structures (such as oblate spheroids), because their geometric

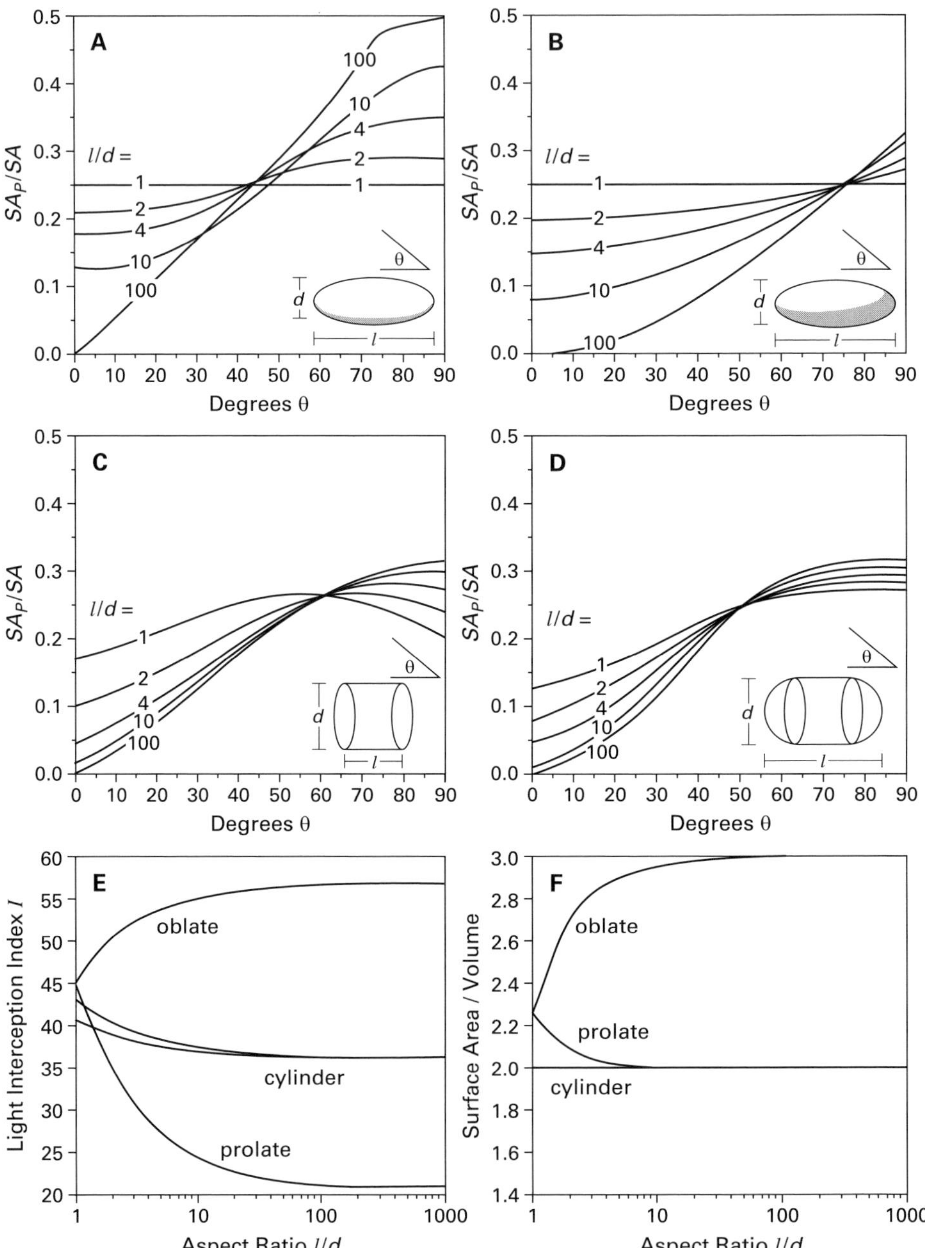

Figure 14.3
Light interception index I plotted as a function of the solar angle θ for oblate and prolate spheroids (A and B, respectively) and for terete cylinders without and with hemispherical ends (C and D, respectively), differing in shape as defined by the quotient of body length and diameter, l/d.

contribution to the ability to resist bending and twisting forces is insensitive to the direction of the application of forces with respect to their longitudinal axis (i.e., cylinders fall down and twist less easily than flattened objects).

The preceding shows that a cylindrical body plan confers many advantages to a terrestrial plant in terms of the trade-off between light interception and water conservation. It also provides a mechanically stable method to elevate aerial body parts. Nevertheless, there is a serious trade-off between the ability to harvest light and the ability to cope with the mechanical bending forces resulting from the weight of cylindrical axes when one or more cylindrical axes branch.

This trade-off is easily illustrated for a simple Y-shaped geometry. Here, the same protocol can be used to evaluate the light-harvesting ability of spheroids and cylinders, provided that a few more mathematical parameters are added to describe the shape and orientation of the Y with respect to incident sunlight: (1) separate solar incident angles for each of the branched axes and the one subtending them—that is, θ, θ', θ''; (2) the bifurcation angle between the two branched elements ϕ; and (3) the rotation angle with respect to the diurnal path of incident radiation γ (figure 14.4A). With these additional parameters, computer simulations readily show that a T-shaped orientation (i.e., $\phi = 180°$) and a rotation angle that projects the full profile of the T toward incident light (i.e., $\gamma = 90°$) maximize the capacity to intercept sunlight (see Niklas and Kerchner, 1984) (figure 14.4B).

However, simple engineering theory also shows that this configuration maximizes the bending moment M exerted at the base of the two branched elements (figure 14.4C). There is no single "optimal" bifurcation angle capable of reconciling this antagonism between the two functional obligations, because the magnitude of the bending moment depends on the diameters, lengths, and bulk tissue densities of the branched elements. However, calculations suggest that ϕ in the range of 90° provides a reasonable solution for this trade-off.

Morphospaces and Fitness Landscapes for More Complex Early Vascular Sporophytes

The Morphospace

The sporophytes of all known vascular plants (tracheophytes), both modern and ancient, are branched, typically may times. Therefore, the preceding morphospace has to be expanded considerably to include the morphologies of ancient and modern-day sporophytes. The procedure used to construct this multidimensional morphospace is far more complex than that used in the previous section, but it contains the same basic elements (Niklas and Kerchner, 1984; Niklas, 1994, 1997a, 1997b, 1997c). Briefly, each morphological variant must be assembled from cylindrical axes (branchlike

A

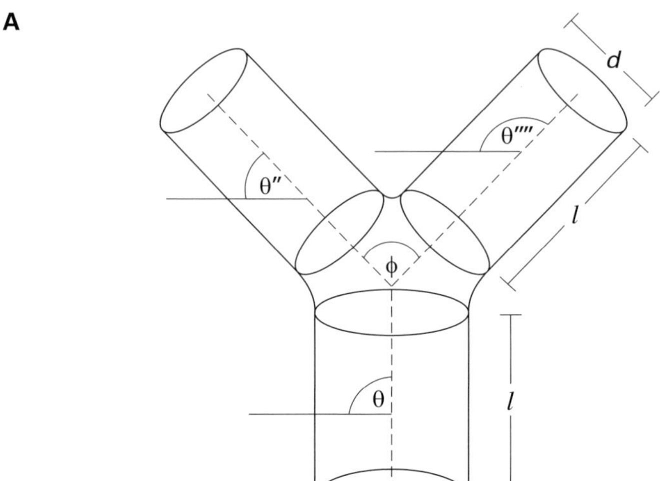

d
θ″″
θ″
l
ϕ
θ
l
γ

ϕ = 180°
30°
20°
10°
0°
SA_P/SA
Degrees θ
0
1
2
3
4
0
10
20
30
40
50
60
70
80
90

Light Interception Index I
50
40
Bifurcation Angle ϕ = 0°
30°
60°
90°
120°
180°
50
40
90°
45°
0°
Rotation Angle γ
0
1000
2000
3000
4000
Bending Moment M

B

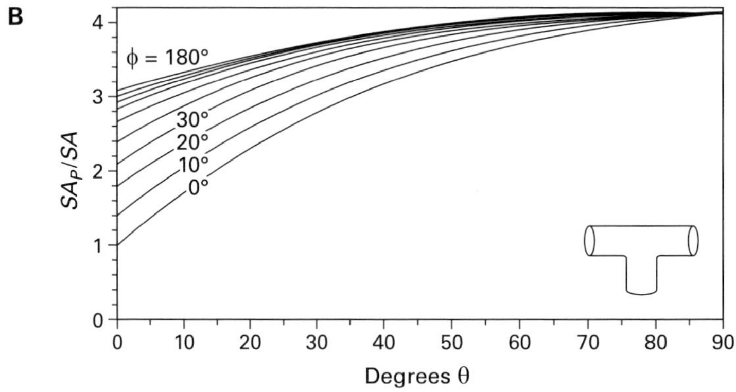

C

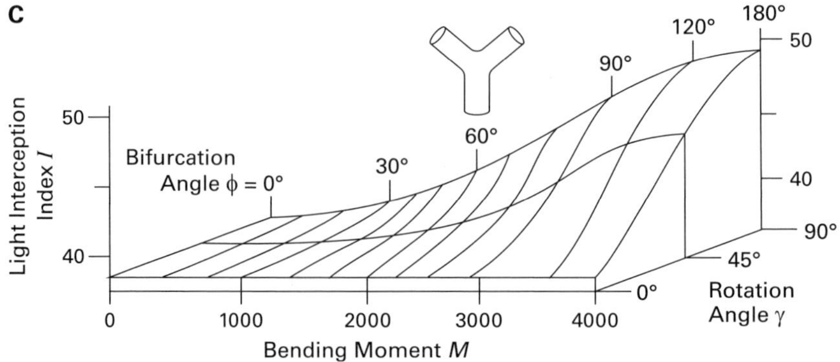

elements) with uniform girth and tissue density. As in the previous morphospace for Y-shaped objects, the orientation of each axis with respect to the vertical and horizontal planes of reference is specified by a bifurcation angle ϕ and a rotation angle γ (in this case, each ranges between 1° and 180°). The number of axes for a particular morphology is determined by assigning a probability value for apical bifurcation p, where $p = 0$ indicates no branching and $p = 1$ indicates branching (figure 14.5). Each morphology is restricted to ten levels of branching N; the maximum number of axial elements for each morphology is thus 2,047.

Mathematically, $p = 8p_{n-(k+1)}/(N+k)$, where $p_{n-(k+1)}$ is the probability of terminating branching at the next generated (higher) level of branching and $(N+k)$ designates the previously generated level of branching. The length of each axis L is stipulated to be directly proportional to the probability of branching (i.e., axial length is a linear function of p such that the basalmost axial element is always as long as or longer than any more distal [higher level] axial element).

The morphospace also has two subdomains, one containing all equally branched (isobifurcate) morphologies and another containing all unequally branched (anisobifurcate) morphologies. This requirement emerges from the fossil record, which reveals that both types of branching architectures evolved at roughly the same time. Mathematically, the isobifurcate subdomain contains all variants for which $\phi_1 = \phi_2$ and $p_1 = p_2$. The anisobifurcate subdomain contains all morphologies for which $\phi_1 \neq \phi_2$, $\gamma_1 \neq \gamma_2$, and $p_1 \neq p_2$ (fig. 14.5). Since L is a function of p, it follows that $L_1 \neq L_2$ for this subdomain.

The entire morphospace is constructed by independently varying each of the aforementioned variables in each of the two subdomains. The spatial ordering of morphological variants is predetermined by assigning ascending numerical values to each of the variables (e.g., the bifurcation angle varies in 1° increments; the probability of branching varies in 0.01 increments). In the isobifurcate subdomain, ϕ, γ, and p were plotted orthogonally (i.e., the location of any variant in the subdomain is specified by a unique set of Cartesian coordinates). The anisobifurcate subdomain

Figure 14.4
Trade-off between light interception and mechanical stability for a simple equally branched (isobifurcate) structure consisting of three cylindrical elements having equivalent lengths and diameters. (A) The angle of incident light θ differs among the three cylindrical elements as a function of the bifurcation and rotation angles (ϕ and γ, respectively). (B) Light interception (quantified as the area under the plot of projected area divided by total area, SA_p/SA, versus the angle of incident light at the base of the structure, θ) plotted as a function of θ for structures with different bifurcation angles (all other variables held constant). A bifurcation angle of 180° provides the best orientation for light interception (see insert drawing). (C) Light interception index I (in units of watt-hrs) versus bending moment M (in units of N-m) plotted as a function of the bifurcation and rotation angles (ϕ and γ, respectively). The orientation that maximizes I also maximizes M. Although no single optimal solution exists, structures with $\phi \sim 60°$ achieve a reasonable degree of efficiency (see insert drawing).

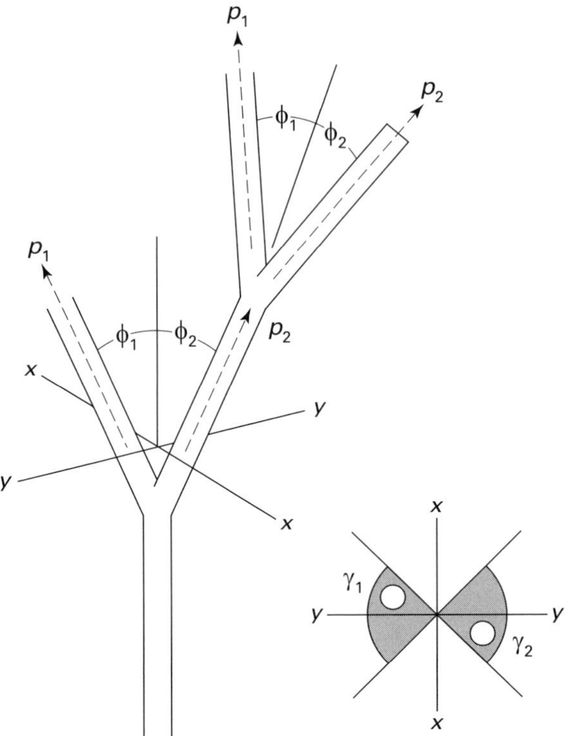

Figure 14.5
Mathematical variables used to construct the morphospace of ancient vascular plants. The orientation of each set of companion axes subtended by a single axis is specified by a branching angle ϕ and a rotation angle γ. The probability of an axis branching subsequent to its appearance is designated as p. Morphologies in the isobifurcate subdomain of the morphospace for early vascular plants have axes with equal branching angles ($\phi_1 = \phi_2$) and rotation angles ($\gamma_1 = \gamma_2$), and probabilities of branching ($p_1 = p_2$) such that these morphologies can be located in a simple Cartesian coordinate system (see lower right cube). Morphologies in the anisobifurcate subdomain have unequal angles and probabilities of branching (not shown).

is spatially far more complex: a six-dimensional space with ϕ_1, ϕ_2, γ_1, γ_2, p_1, and p_2 axes.

As in the simple morphospace, the relative fitness of each hypothetical variant must be evaluated using basic physics or engineering principles that describe the performance of each task designated to influence growth, survival, and reproductive ability. The ability of each variant to perform one or more of these tasks can then be divided by the maximum performance level in a particular landscape. To compute the relative fitness of morphologies performing two or more tasks simultaneously H, it is assumed that each task contributes equally and independently to overall fitness such that the fitness of a particular variant h is given by the formula $h = \wp^{1/N}$, where

$\wp$ is the product of the performance levels of the number of tasks performed N. For example, if the numerical performance of two tasks is designated by the quantities A and B, the relative fitness of a particular variant morphology is given by $H = (AB/\boldsymbol{AB})^{1/2}$, where $\boldsymbol{A}$ and $\boldsymbol{B}$ are the performance levels of the most efficient variant in the particular landscape.

The Fitness Landscape

As noted, all of the fitness landscapes are predicated on the performance of one or more of four tasks assumed to dictate survival and reproduction: light interception, maintaining mechanical stability, dispersing spores, and conserving water. These tasks can be considered in isolation or in sets of two or more tasks performed simultaneously. Thus, there are four one-task landscapes, six two-task landscapes, four three-task landscapes, and one four-task landscape. The methods for quantifying the performance of these tasks are provided in detail in Niklas and Kerchner (1984) and Niklas (1994, 1997b, 1997c). Each is reviewed briefly here.

As discussed in the context of very simple morphologies, light interception is quantified for each morphological variant in the entire domain by integrating the area under the curve generated by plotting the projected surface S_p divided by the total surface area S of the variant as a function of the solar angle θ for the values $1° \leq \theta \leq 180°$, using an algorithm that accounts for the self-shading of branchlike axes (see Niklas and Kerchner, 1984). Mechanical stability was calculated by computing the total bending moment on the single basal vertical axis of each morphological variant. Noting that a bending moment M is the product of a bending force (weight W) and the length of a lever arm l, the bending moment of each distal axis in a branching architecture m_i is a function of $W = \pi D^2 L/4\rho$, where D is diameter and ρ is the bulk tissue density (assumed to equal 1 kg/m^3 for all axes), and its specified bifurcation and rotation angles (which specify the orientation of an axis and thus its lever arm length). The total bending moment at the base of any variant, which is computed numerically, is a complex quantity to compute because it is the sum of all the bending moments of all axes, and because these axes are joined together and oriented differently as a function of the level of branching. Nevertheless, the maximum bending moment can be computed with the aid of a computer (see Niklas, 1997a,b,c).

The ability of each variant to conserve water is gauged to be a simple linear function of total plant surface area (i.e., it is assumed that all surface areas have equivalent rates of evapotranspiration). Spore dispersal is assumed to rely exclusively on wind. Therefore, a simple ballistic model can be used if it is assumed that all sporangia are at the tips of all distal axes and that all sporangia are equivalent in size (and thus in spore number per sporangium). This ballistic model is given by the formula $x = HU/T$, where x is maximum distance of spore dispersal, H is plant height, U is ambient wind speed, and T is the spore terminal settling velocity (see Okubo and

Levin, 1989). Plant height is a function of the number, length, and orientation of all axes comprising a particular morphology; the ambient wind speed depends on plant height (since wind speeds typically decrease exponentially toward ground level, and the terminal settling velocity is taken as 0.15 m/sec (the average velocity of *Lycopodium* spores). These assumptions and specifications obtain the scaling relationship $x \propto H^2$ across all morphologies, which indicates that the spore dispersal range is, on average, proportional to the square of plant height.

In terms of simulating "adaptive walks," the purpose is to identify the sequence of morphological variants with progressively higher relative fitness as defined by performing one or more designated tasks. Each walk begins at the same location in the morphospace. This location corresponds to the morphology of the most ancient vascular plants, such as *Cooksonia* and *Steganotheca*, which were isobifurcate with one or two levels of branching in which all terminal axes bore sporangia (see fig. 14.12A). From this location, a search algorithm evaluates the relative fitness of all neighboring variants using a specified fitness criterion (e.g., light interception, light interception and mechanical stability, or light interception, mechanical stability, and conservation of water). If one or more neighboring variants have an equivalent or higher relative fitness, the walk proceeds to their locations in the morphospace. This process continues until the relative fitness of the morphologies in the last iteration of a search is higher than that of all surrounding variants in the terminal steps. By definition, these morphologies occupy adaptive peaks on the landscape.

Fitness landscapes can be "stable" or "unstable." That is, an adaptive walk can proceed through the morphospace such that the criterion defining relative fitness remains constant to obtain a "stable" fitness landscape. Alternatively, the criterion used to define relative fitness can be changed arbitrarily at any time during a walk to mimic a change in selection pressure or an "unstable" fitness landscape.

Adaptive Peaks and Their Phenotypic Occupants

Stable Landscapes

Using the aforementioned protocols, computer simulations of early vascular plant evolution indicate that comparatively few hypothetical morphologies are capable of maximizing the performance of any one of the four single tasks. In contrast, the number of morphologies capable of optimizing the performance of two or more tasks simultaneously increases as the number of tasks increases. However, the overall (global) relative fitness of these multitask morphologies decreases significantly (table 14.1). These simulations indicate that adaptive evolution may be more rapid and easier when selection acts on the ability to perform multiple rather than single tasks.

Table 14.1
Relationships among the number of tasks used to qualify relative fitness, the number of morphologies identified as either maximizing or optimizing these tasks, and the relative fitness of these morphologies

Number of Tasks Defining Fitness	Number of Morphologies (Mean ± SE)	Relative Fitness (Mean ± SE)
1 ($n = 4$)	2.50 ± 1.0	35.3 ± 1.80
2 ($n = 6$)	3.33 ± 1.6	11.6 ± 0.68
3 ($n = 4$)	6.25 ± 0.5	7.5 ± 0.14
4 ($n = 1$)	20	2.4

n = number of fitness landscape permutations.

Turning to specifics, the few hypothetical morphologies capable of maximizing the performance of any one of the four biological tasks used to quantify relative fitness are, for the most part, comparatively simple in general appearance (figure 14.6). The simplest of these are those which maximize the capacity to conserve water as gauged by minimizing their total surface area (figure 14.6A). These hypothetical phenotypic variants are Y-shaped and similar in general appearance to the most ancient vascular plants, such as *Cooksonia*. Thus, water conservation may have been achieved early in the evolutionary history of land colonization. However, from a mathematical perspective, these Y-shaped morphologies are trivial, because they are those used to initiate each adaptive walk, whether on single- or multiple-task landscapes (see figure 14.4).

In contrast, the most complex single-task morphologies are those that possess lateral branching systems confined partly or entirely to the horizontal plane and elevated on a single vertical main axis (figure 14.6C). Some of these morphologies are located in the isobifurcate subdomain, whereas others are found in the anisobifurcate subdomain of the morphospace for ancient tracheophytes. Because the latter are reached only by an extensive series of hypothetical morphological transformations, the adaptive walks reaching these morphologies generally are highly branched.

The morphologies capable of optimizing the performance of two or more tasks are typically far more diverse in general appearance than those maximizing the performance of a single task (figures. 14.7–14.9). Some of these variants are simple Y-shaped *Cooksonia*-like morphologies (figure 14.7A–B), whereas others are strikingly reminiscent of the fossil remains of some zosterophyllophytes, such as *Zosterophyllum*, an ancient group of vascular plants (figure 14.8D). As noted, the number of morphologies identified as equally efficient at performing two or more tasks simultaneously increases as the number of tasks increases (table 14.1). For example, on average, 3.3 morphological variants are reached by adaptive walks on the six two-task landscapes (figure 14.7), whereas 6.5 variants are reached by walks on the four

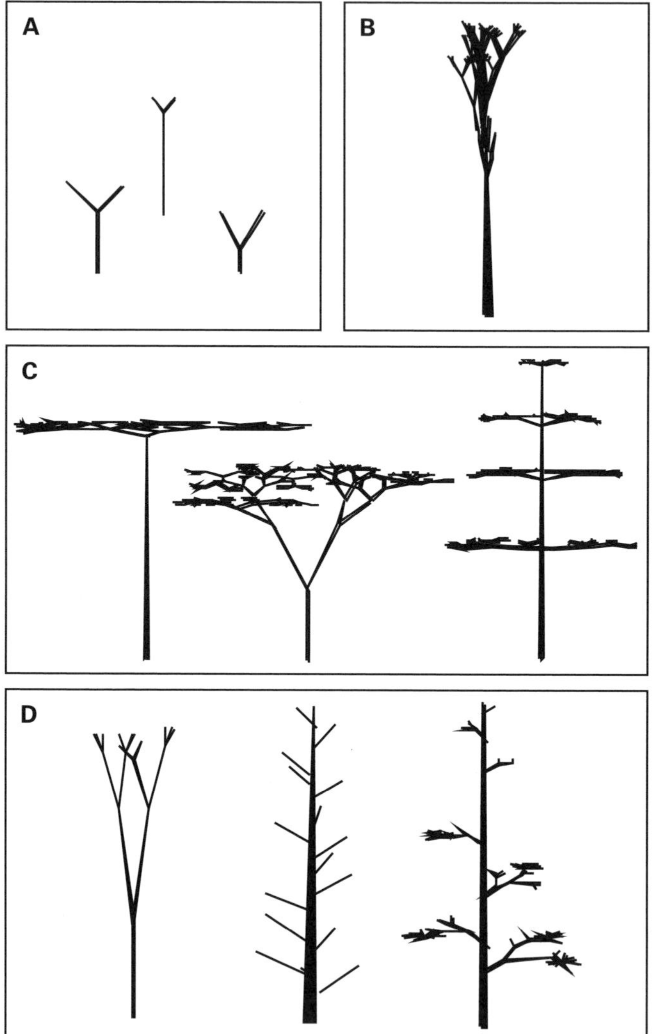

Figure 14.6
Phenotypes identified by adaptive walks on single-task landscapes capable of maximizing water conservation (A), spore dispersal (B), light interception (C), and mechanical stability (D).

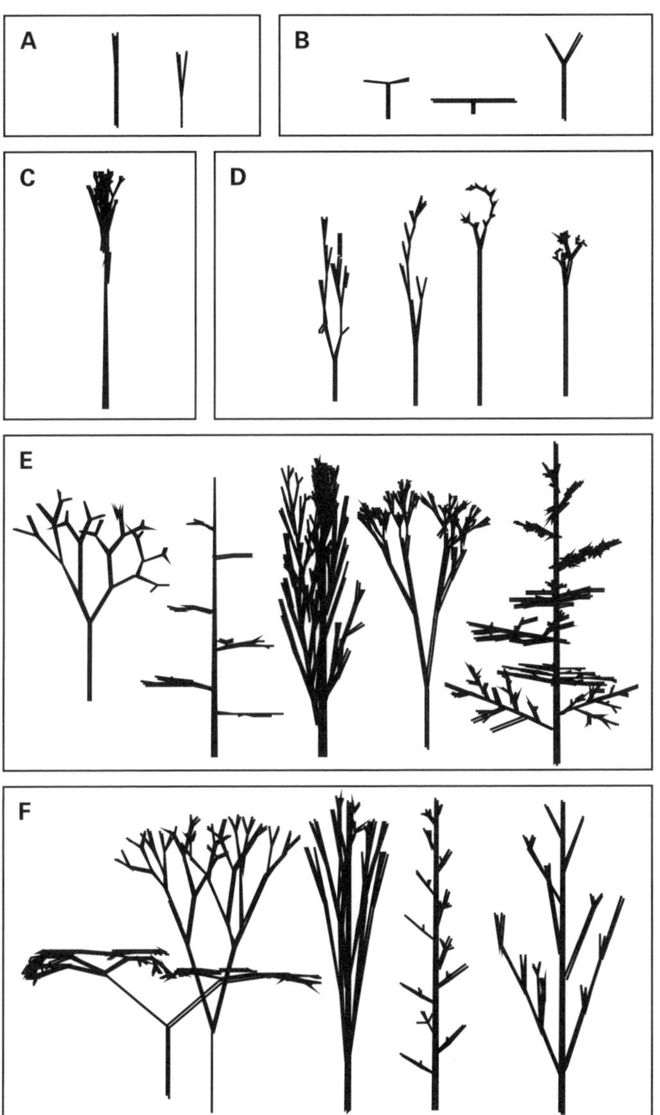

Figure 14.7
Phenotypes identified by adaptive walks on two-task landscapes capable of optimizing mechanical stability and water conservation (A), light interception and water conservation (B), mechanical stability and spore dispersal (C), spore dispersal and water conservation (D), light interception and mechanical stability (E), and light interception and spore dispersal (F).

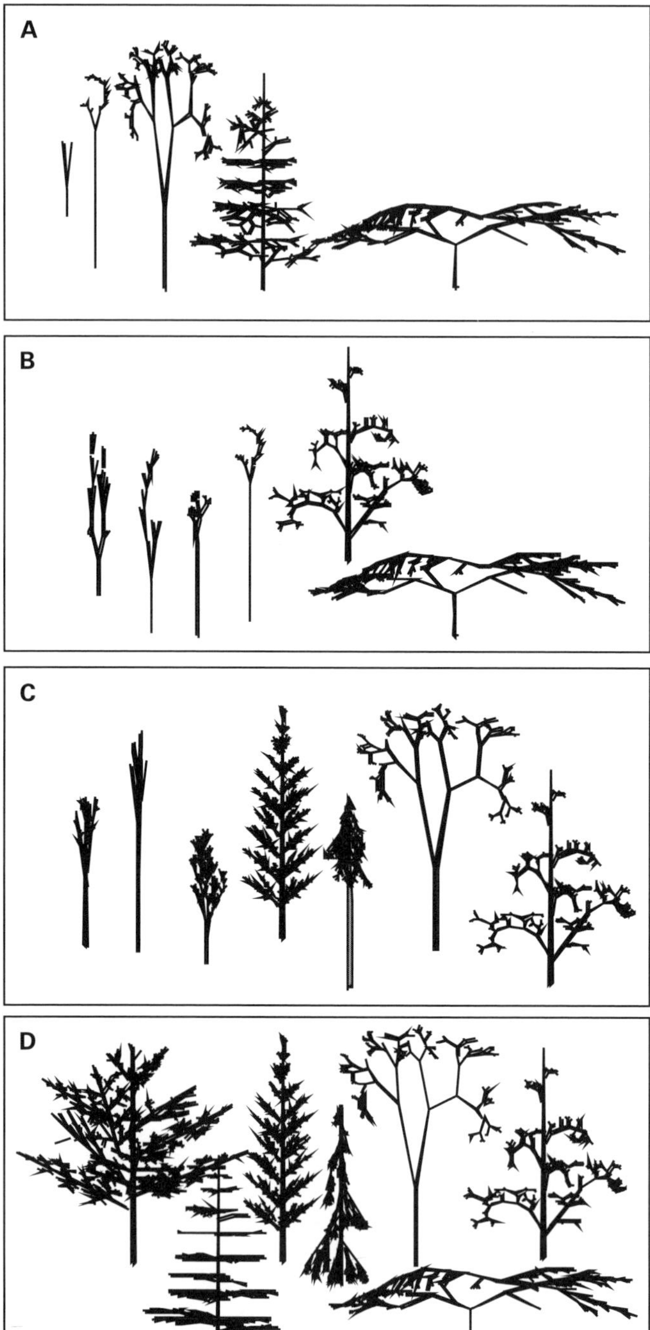

Figure 14.8
Phenotypes identified by adaptive walks on three-task landscapes capable of optimizing light interception, mechanical stability, and water conservation (A); light interception, spore dispersal, and water conservation (B); mechanical stability, spore dispersal, and water conservation (C); and light interception, mechanical stability, and spore dispersal (D).

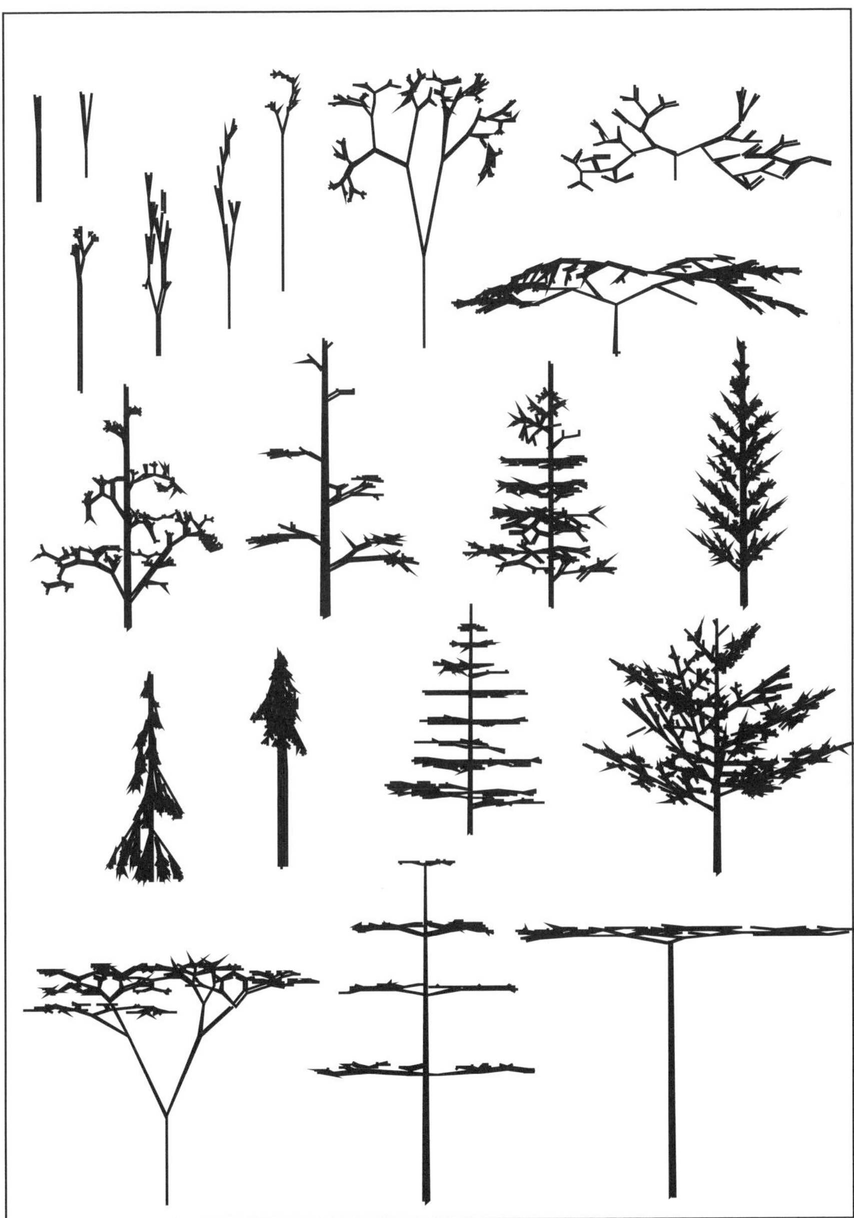

Figure 14.9
Phenotypes, identified by adaptive walks on the four-task landscape, that are capable of optimizing mechanical stability, water conservation, light interception, and spore dispersal.

three-task landscapes (figure 14.8). When all four tasks are considered simultaneously, a total of twenty equally efficient morphologies is reached by an adaptive walk (figure 14.9).

Although the number of morphologies reached by adaptive walks increases as the complexity of landscapes increases, the relative fitness of functionally optimal morphologies, on average, decreases (table 14.1). It must be noted that the "currency" by which fitness is measured differs across the various fitness landscapes (e.g., amount of light harvested, spore number and dispersal range, and bending moment at the base of each morphology). However, comparisons of relative fitness across the different fitness landscapes are possible provided the fitness of the ancestral *Cooksonia*-like morphology that initiates each adaptive walk on any landscape is used to normalize the fitness of all other morphological variants for each landscape. With this protocol, the average relative fitness of morphologies identified as optimizing two or four tasks is 11.6 and 2.4, respectively (table 14.1). In contrast, the average relative fitness of those morphologies capable of maximizing the performance of one task is 35.3. Therefore, as the number of tasks used to define relative fitness increases, the capacity to perform these tasks decreases.

Unstable Landscapes

Environments can change, often dramatically, even over ecological time scales (10^2 to 10^4 yr). It is therefore naïve to believe that selection persistently acts on the performance of one task or any particular combination of tasks, especially over time scales relevant to evolutionary or geological history (10^5 to 10^6 yr). As noted, the consequences of shifting selection can be modeled by initiating an adaptive walk on one landscape and subsequently changing the landscape one or more times. In this way, the criteria used to quantify relative fitness change as a walk proceeds through a morphospace. Unfortunately, there are no a priori rules for how or when a particular landscape changes. Therefore, the number of permutations of shifting landscapes is literally astronomically large.

Nevertheless, the fossil record and the preceding simulations provide some guidance. As noted, the oldest known vascular land plant fossils are *Cooksonia*-like in their general appearance. Computer simulations also indicate that these morphologies were capable of maximizing water conservation (see figure 14.6A). To explore the consequences of shifting selection, it is therefore reasonable to initiate an adaptive walk on a fitness landscape defined by conserving water and to subsequently shift this landscape into one or more of any of the eleven multitask landscapes as the walk proceeds through the morphospace.

Two such simulations are shown in figure 14.10. Each simulation begins on the same landscape (A), then enters two different two-task landscapes, after which each enters the same three-task landscape. Both walks come to an end in the single four-

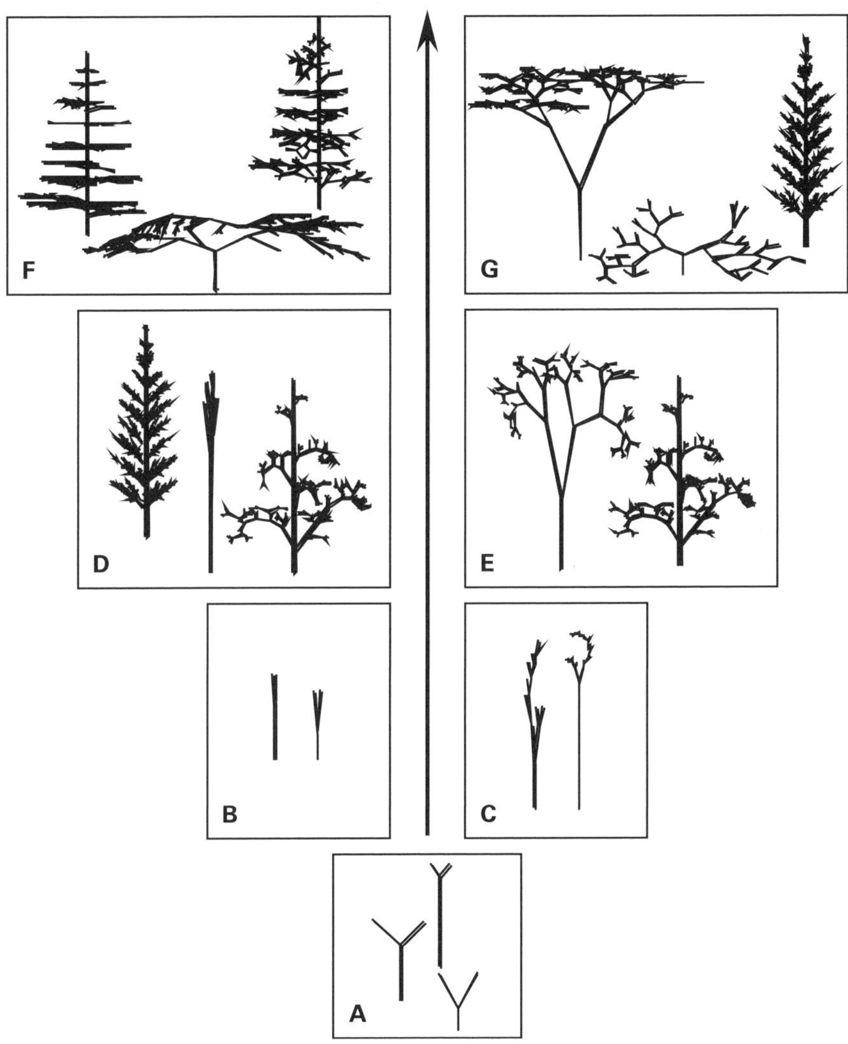

Figure 14.10
Phenotypes identified by two different adaptive walks on shifting fitness landscapes. Following the vertical arrow, the two walks begin with the phenotypes identified as maximizing water conservation (A); bifurcate into two different two-task landscapes (water conservation and mechanical stability (B), and water conservation and spore dispersal (C); enter the same three-task landscape (mechanical stability, spore dispersal, and water conservation (D–E); and come to closure by reaching accessible morphological optima on the four-task landscape (F–G).

task landscape. These two simulations illustrate an important feature that resurfaces in all similar walks through unstable landscapes. Even though adaptive walks enter the same fitness landscape, they locate different morphological optima, depending on how relative fitness was defined in previous portions of walks. For example, among the five optimal morphologies reached by the two adaptive walks entering the two three-task landscapes, only one is the same, whereas none of the morphological optima reached by the two walks on the four-task landscape is the same (figure 14.10).

Taken at face value, simulations such as these indicate that the morphologies with the highest relative fitness at any stage in the adaptive evolution of a lineage depend in part on prior selection regimes, since these regimes define the range of phenotypes that are available for the next round of selection (i.e., morphological optimization is historically contingent). If there is a lesson to be learned, it is that we should not expect the mechanisms of adaptive evolution to achieve the best *conceivable* morphologies, but only those that are the best relative to what is available based on past history.

Developmental Constraints

Thus far, it has been assumed that all adaptive walks are developmentally "unfettered." That is, changing the numerical values of one parameter is assumed to have no direct effect on the numerical values of other parameters. Likewise, all transformations are assumed to be equiprobable. But it is clear that development is a highly coordinated process such that transformations in one morphological feature are dependent on changes in other features. Additionally, all transformations are not equally possible because all possible genomic variants are not achieved by even very large populations of sexually reproducing organisms. Even though plants may be more phenotypically plastic compared with animals (see Sultan, 1987, 1992; Scharloo, 1991), the importance of developmental constraints in understanding adaptation cannot be neglected (Gould, 1980; Thomas and Reif, 1993).

The role played by developmental constraints in plant evolution is far too complex to address here. But it can be evaluated, albeit crudely, by restricting adaptive walks from entering designated portions of the morphospace for early vascular plants, because each step in a walk entails finding neighboring phenotypes with equal or higher fitness than those identified by previous steps and because each step requires a transformation in the branching angle, the rotation angle, or the probability of branching (each of which mimics a developmental change). The simplest mathematical way to model this is to limit walks to phenotypes that have only equal branching ($p_1 = p_2$), or nonovertopped branches ($\phi_1 = \phi_2$), or nonplanated branches ($\gamma_1 = \gamma_2$) such that phenotypes with $p_1 \neq p_2$, $\phi_1 \neq \phi_2$, and $\gamma_1 \neq \gamma_2$ are developmentally impossible. At issue is whether walks that are thus restricted reach more, fewer, or equal numbers

Table 14.2
Comparisons of mean number (±SE) of maximal (single-task) and optimal (multitask) phenotypes reached by unrestricted adaptive walks (see table 14.1) and those restricted to morphologies with equal branching ($p_1 = p_2$), nonovertopped branches ($\phi_1 = \phi_2$), and nonplanated branches ($\gamma_1 = \gamma_2$)

Number of Tasks	Unrestricted Walks	Restricted Walks		
		($p_1 = p_2$)	($\phi_1 = \phi_2$)	($\gamma_1 = \gamma_2$)
1	2.50 ± 1.0	2.25 ± 1.0	2.25 ± 1.5	**2.75 ± 2.1**
2	3.33 ± 1.6	2.50 ± 0.8	**3.83 ± 0.4**	2.83 ± 1.2
3	6.25 ± 0.5	6.25 ± 0.5	**8.25 ± 1.0**	**8.25 ± 1.0**
4	20	18	18	19

Note: Numbers in bold type indicate that restricted walks reach more maximal or optimal phenotypes than unrestricted walks.

of phenotypes capable of maximizing the performance of one task or optimizing the performance of two or more tasks (i.e., the number of phenotypes reached by unfettered walks serves as a null hypothesis).

The results of this approach are summarized in table 14.2. With the exception of the adaptive walks on the four-task landscape, developmentally fettered walks reach an equal or a larger number of phenotypes as unrestricted walks do. For example, walks on the four single-task landscapes confined to those morphologies with $\gamma_1 = \gamma_2$ reach, on average, 2.75 phenotypes per landscape, whereas developmentally unfettered walks on the same landscapes achieve, on average, 2.50 phenotypes. Similar comparisons show that restricting the access of adaptive walks by imposing developmental constraints has little or no effect on the number of phenotypic variants occupying adaptive peaks.

However, the relative fitness of the phenotypes reached by constrained walks is significantly less than that of the phenotypes reached by unconfined walks. For example, the mean relative fitness of the variants reached by unfettered walks on single-task landscapes is 35.3 (see table 14.1), whereas the mean relative fitness of the variants reached by walks on the same landscapes but confined to morphologies with planated lateral branches (i.e., $\gamma_1 = \gamma_2$) is 1.61. Simulations such as these indicate that the imposition of a developmental constraint does not necessarily reduce the number of phenotypes that can be reached by adaptive walks, but it can significantly reduce the repertoire of these morphologies, and thus the relative fitness of morphologies occupying adaptive peaks.

Adaptive Trends and the Telome Theory

All of the computer simulations presented so far bear on one of the most pervasive and far-reaching theories to explain early land plant morphological evolution. First proposed by Zimmerman (1930, 1953, 1965) and later expounded with vigor by others (see Stewart and Rothwell, 1993), this theory, called the telome theory, argues that the most ancient tracheophytes consisted of simple paired distal axes (called telomes) subtended by unbranched axes (called mesomes) (see figure 14.4A). These telomes and mesomes are envisioned to have been subsequently modified by one or more developmental processes—planation, overtopping, reduction, recurvature, and webbing—to obtain all known extant or extinct plant morphologies.

Each of these processes is easily envisioned, although the developmental mechanisms by which they are achieved are as undoubtedly complex as they are currently unknown. Planation occurs when neighboring axes become oriented in a single plane with respect to the horizontal; overtopping results from the differential growth in the lengths of interconnected axes such that some become longer or shorter than their companions; reduction in the size of some telomes can aggregate or positionally subordinate some axes with respect to others; recurvature requires the differential expansion of one side of an axis with respect to the opposing side; and webbing is the developmental introduction of tissues between adjoining (and presumably planated) axes.

Zimmerman argued that all of the morphological transformations attending early land plant evolution can be explained by the operation of two or more of these five processes. For example, fern and seed plant leaves (traditionally called megaphylls) are believed to have evolved by reduction and overtopping (to yield plant architectures with main vertical stems bearing lateral branching systems) and by planation and webbing of lateral branches (to give rise to megaphylls) (figure 14.11A). Likewise, the intricate reproductive organs of modern-day horsetails are postulated to have evolved by means of reduction, recurvature, and some degree of fusion or webbing.

The telome theory has been criticized, and rightly so, for a variety of reasons (Niklas, 2000; Kaplan, 2001). One obvious problem with the theory is its vagueness regarding the developmental mechanisms responsible for overtopping, planation, and the other processes. Indeed, these terms are descriptive rather than explicative in nature. Another criticism is that the telome theory never explains why certain morphological transformations occur rather than others, nor does it stipulate the sequence of processes foreshadowing the appearance of a particular morphology. Why should planated and webbed lateral branch systems evolve? Are the leaves of ferns or seed plants functionally adaptive in terms of light interception or some other biological requirement? Did these megaphylls evolve as the result of the simultaneous operation of reduction, overtopping, planation, and webbing, or did planation and webbing occur after reduction and overtopping? Questions such as these can be answered retrospectively (and only in small part) by examining the fossil record, but the telome theory sheds little light on them.

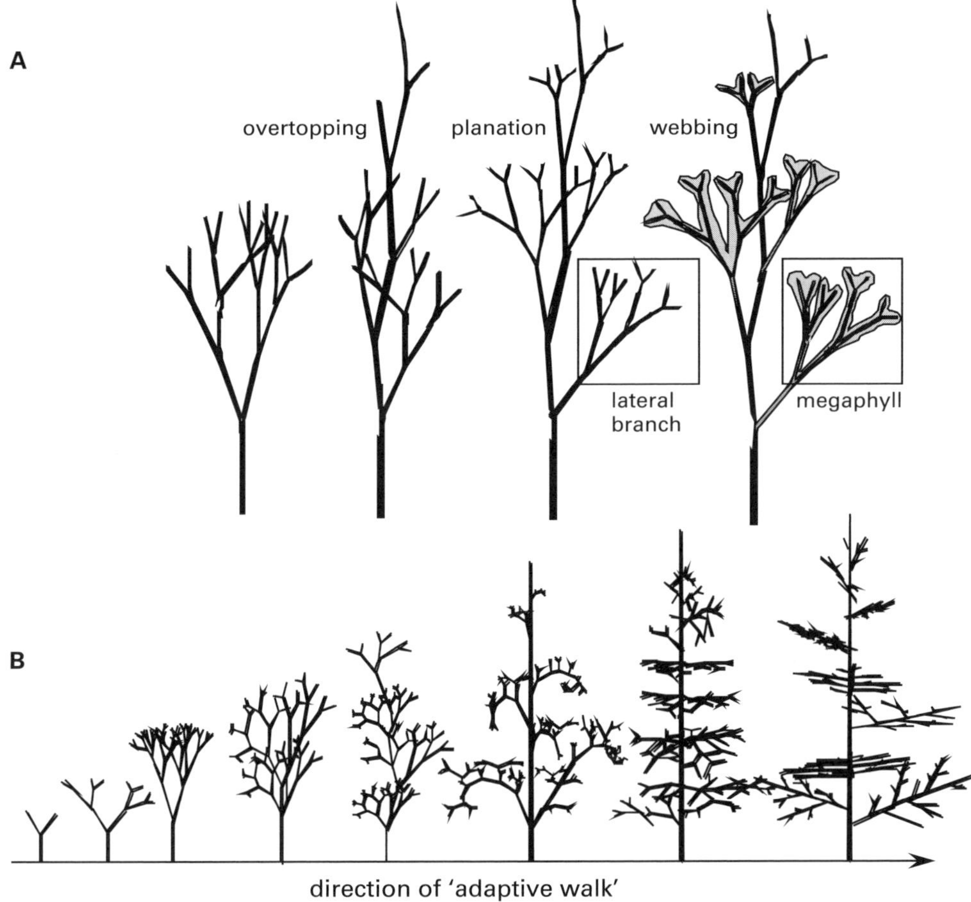

Figure 14.11
Hypothetical transformation series (involving overtopping, planation, and webbing) leading to the evolution of a "megaphyll" as envisioned by the telome theory (A) and as simulated by an adaptive walk through the morphospace for early vascular land plants (B) optimizing light interception and mechanical stability (see fig. 14.7E). Successive transformations (arranged from left to right) involve overtopping and planation (as described by the telome theory).

Zimmerman's ideas are nevertheless useful because they provide a lexicon of terms for the morphological transformations observed in the fossil record and for those identified by the computer simulations presented here. In turn, these simulations suggest the adaptive significance of the transformations envisioned by the telome theory. For example, adaptive walks on one-task landscapes identify overtopped morphologies with planated lateral branching systems as those capable of maximizing spore dispersal, light interception, or mechanical stability (see figure 14.6B–D). Similar morphologies are reached by the adaptive walks on multiple task landscapes (see figures 14.7–14.9). This convergence indirectly supports the supposition that overtopping and planation, which have occurred independently in a number of plant lineages, may be functionally adaptive.

This supposition is reinforced when we examine the sequences of morphologies identified by different adaptive walks. For example, starting with a simple *Cooksonia*-like phenotype, the adaptive walk on the fitness landscape defined by light interception and mechanical stability identifies a larger, more branched variant (figure 14.11B). As this particular walk progresses, it identifies larger and more overtopped phenotypes bearing lateral branching systems that become more planated. Although webbing is not within the repertoire of the morphospace used in this simulation, more complex computer simulations indicate that the capacity to intercept sunlight is dramatically improved if some or all of the distalmost axes (telomes) of phenotypes are interconnected by photosynthetic tissues. Finally, the sequence of morphologies shown in figure 14.12 gives credence to the notion that overtopping likely preceded planation. In turn, planation likely preceded the appearance of webbed leaves during the evolution of vascular land plant lineages (i.e., computer simulations provide circumstantial evidence that overtopping, planation, and webbing are functionally adaptive).

Summary

Computer models such as the ones presented here are heuristic tools. They provide an opportunity to test assumptions about how a particular biological or physical system operates or behaves. Their validity can be evaluated by comparing predicted behavior with observed behavior. When observation and prediction disagree, the assumptions upon which a model rests are either incorrect or incomplete. However, the obverse is not true. When predicted and observed behaviors agree, the assumptions upon which a model rests cannot be said to be sufficient and necessary. The reason is simple—a model can describe the behavior of a system for the wrong reasons. This caveat is important, because the only rigorous test of a computer model is to experimentally manipulate the system it purports to describe and see if the model predicts the outcome for each manipulation.

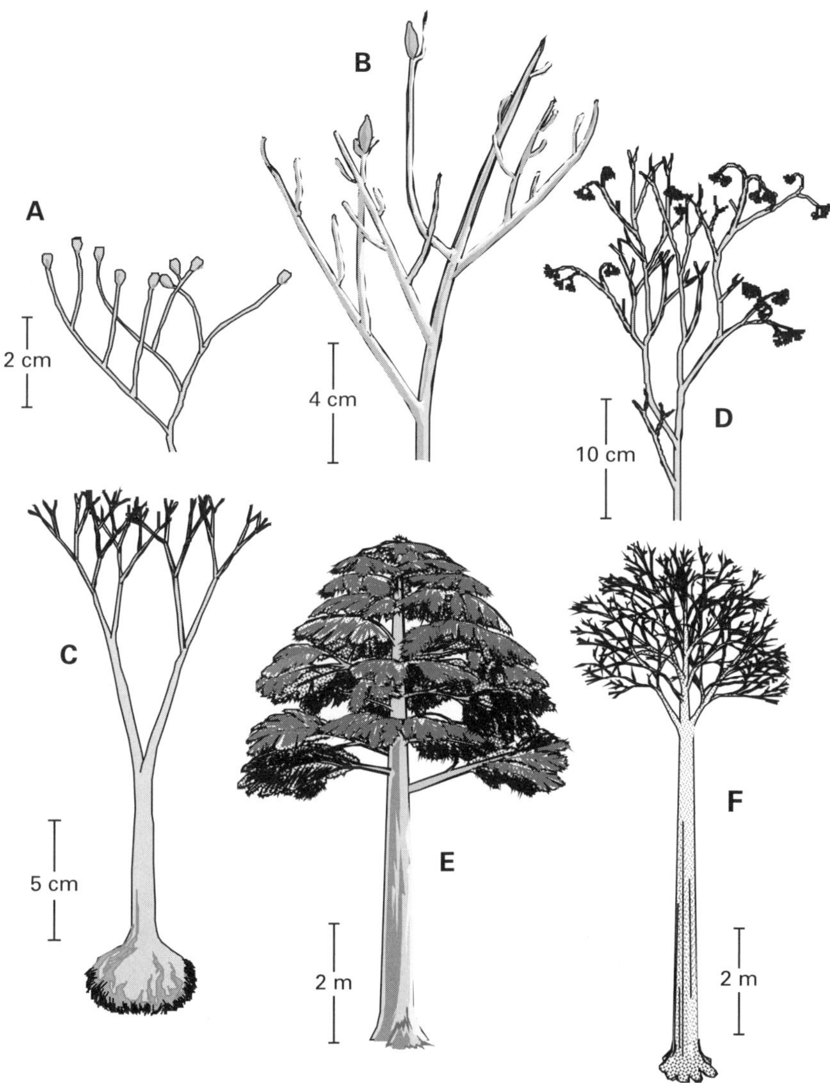

Figure 14.12
Reconstructions of representative early Paleozoic fossil plants. (A) *Steganotheca*. (B) *Rhynia*. (C) *Horneophyton*. (D) *Psilophyton*. (E) *Archaeopteris*. (F) *Lepidodendron*.

Unfortunately, we cannot experiment with history. We can only observe it. For this reason, the most conservative interpretation of the simulations presented here is that six general properties emerge logically (mathematically) from the assumptions made about early vascular plant evolution. These properties are as follows:

1. The number of equally fit morphological variants is predicted to increase as the number of functional tasks subject to selection increases.
2. The relative fitness of these phenotypes decreases as the number of tasks increases.
3. Therefore, morphological diversification is easier on complex as opposed to simple fitness landscapes.
4. Constraints on how morphology can be developmentally altered do not a priori limit the number of equally fit variants that can be reached by adaptive walks.
5. However, the relative fitness of these variants is significantly lower than the phenotypic optima that can be reached by unfettered adaptive walks.
6. Adaptive walks on shifting fitness landscapes (used to mimic changes in the focus of selection) identify morphological optima that often differ significantly from those on stable fitness landscapes (used to mimic constant selection).

Currently, the only criteria for evaluating whether these properties pertain to real land plant evolution is the extent to which the morphologies identified as occupying adaptive peaks coincide with those observed in the fossil record and the extent to which each of these six properties makes biological sense. In terms of the first criterion, the walks on the landscape for water conservation—a biological task undoubtedly required for surviving on land—identify phenotypes that are strikingly similar to those of the most ancient vascular land plants (e.g., *Cooksonia*, *Steganotheca*, *Rhynia*, and *Horneophyton*) (figure 14.12A–14.12C). As more tasks are added to construct more functionally demanding landscapes, other more complex morphologies are identified as occupying adaptive peaks. Some of these variants are remarkably similar in general appearance to more evolutionarily derived vascular plant fossils (e.g., *Psilophyton*, *Archaeopteris*, *Lepidodendron*) (figure 14.12D–14.12F). Finally, the morphological transformations attending some walks coincide with those proposed by the telome theory, and many of these transformations reoccur in the fossil records of different land plant lineages, suggesting that they are functionally adaptive.

In terms of the second criterion, each of the six properties makes biological sense. For example, the rapid rate of morphological diversification observed after land plants evolved mechanisms to cope with dehydration (see Niklas et al., 1984; Knoll et al., 1984) is consistent with the prediction that morphological diversification becomes easier on complex as opposed to simple fitness landscapes. Likewise, it is biologically reasonable to suppose that the morphological diversity manifested by extant species occupying similar or identical habitats reflects the fact that very different phenotypes can have equivalent capacities for growth, survival, or reproductive success.

Computer simulations of morphological evolution are still very much in their infancy, especially in terms of constructing morphospaces and understanding the developmental mechanisms that permit or confine phenotypic transformations in them (see Thomas and Reif, 1993; McGhee, 1999; Niklas, 2003). However, as conceptual tools, they provide opportunities to explore the logical consequences of popular metaphors for evolution, such as Sewall Wright's "adaptive walks on fitness landscapes," and by so doing, they quantify the possible biological structure and dynamics of opportunistic historical events that distinguish some evolutionary episodes as more adaptive than others.

References

Berner RA, Beerling DJ, Dudley R, Robinson JM, Wildman RA Jr. (2003) Phanerozoic atmospheric oxygen. Ann Rev Earth Planet Sci 31: 105–134.

Chaloner WG, Sheerin A (1979) Devonian macrofloras. In: The Devonian System (House MR, Scrutton CT, Bassett MG, eds), 145–161. Special Papers in Palaeontology no. 23. London: Palaeontological Association.

Gates DM (1965) Energy, plants, and ecology. Ecology 46: 1–16.

Gensel PG, Andrews HN (1984) Plant Life in the Devonian. New York: Praeger.

Gould SJ (1980) The evolutionary biology of constraint. Daedalus 109: 39–52.

Graham LE (1993) Origin of Land Plants. New York: Wiley.

Horn HS (1979) Adaptation from the perspective of optimality. In: Topics in Plant Population Biology (Solbrig OT, Jain S, Johnson GB, Raven PH, eds), 48–61. New York: Columbia University Press.

Kaplan DR (2001) The science of plant morphology: Definition, history, and role in modern biology. Amer J Bot 188: 1711–1741.

Kauffman SA, Levin SA (1987) Towards a general theory of adaptive walks on rugged landscapes. J Theor Biol 128: 11–45.

Knoll AH, Niklas KJ, Gensel PG, Tiffney BH (1984) Character diversification and patterns of evolution in early vascular plants. Paleobiology 10: 34–47.

McGhee GR Jr. (1999) Theoretical Morphology: The Concept and Its Applications. New York: Columbia University Press.

Niklas KJ (1992) Plant Biomechanics: An Engineering Approach to Plant Form and Function. Chicago: University of Chicago Press.

Niklas KJ (1994) Morphological evolution through complex domains of fitness. Proc Natl Acad Sci USA 91: 6772–6779.

Niklas KJ (1997a) The Evolutionary Biology of Plants. Chicago: University of Chicago Press.

Niklas KJ (1997b) Adaptive walks through fitness landscapes for early vascular land plants. Amer J Bot 84: 16–25.

Niklas KJ (1997c) Effects of hypothetical developmental barriers and abrupt environmental changes on adaptive walks in a computer-generated domain for early vascular land plants. Paleobiology 23: 63–76.

Niklas KJ (2000) The evolution of leaf form and function. In: Leaf Development and Canopy Structure (Marshall B, Roberts JA, eds), 1–36. Sheffield, UK: Sheffield Academic Press.

Niklas KJ (2003) The bio-logic and machinery of plant morphogenesis. Amer J Bot 90: 515–525.

Niklas KJ, Kerchner V (1984) Mechanical photosynthetic constraints on the evolution of plant shape. Paleobiology 10: 79–101.

Niklas KJ, Tiffney BH, Knoll AH (1984) Apparent changes in the diversity of fossil plants: A preliminary assessment. Evol Biol 12: 1–89.

Nobel PS (1983) Biophysical Plant Physiology and Ecology. San Francisco: W. H. Freeman.

Okubo A, Levin SA (1989) A theoretical framework for data analysis of wind dispersal of seeds and pollen. Ecology 70: 329–338.

Raup DM (1961) The geometry of coiling gastropods. Proc Natl Acad Sci USA 47: 602–609.

Raup DM (1962) Computer as aid in describing form in gastropod shells. Science 138: 150–152.

Raup DM, Michelson A (1965) Theoretical morphology of the coiled shell. Science 147: 1294–1295.

Raven JA (1984) Physiological correlates of the morphology of early vascular plants. Bot J Linn Soc 88: 105–126.

Raven JA (1985) Comparative physiology of plant and arthropod land adaptation. Philos Trans Roy Soc Lond B309: 272–388.

Scharloo W (1991) Canalization, genetic and developmental aspects. Ann Rev Ecol Syst 22: 265–294.

Stewart WN, Rothwell GW (1993) Paleobotany and the Evolution of Plants. Cambridge: Cambridge University Press.

Sultan SE (1987) Evolutionary implications of phenotypic plasticity in plants. Evol Ecol 21: 127–178.

Sultan SE (1992) Phenotypic plasticity and neo-Darwinian legacy. Evol Trends Plts 6: 61–71.

Taiz L, Zeiger E (2002) Plant Physiology (3rd ed). Sunderland, Mass.: Sinauer.

Taylor TN, Taylor EL (1993) The Biology and Evolution of Fossil Plants. Englewood Cliffs, N.J.: Prentice-Hall.

Thomas RDK, Reif W-E (1993) The skeleton space: A finite set of organic designs. Evolution 47: 341–360.

Willis KJ, McElwain JC (2002) The Evolution of Plants. New York: Oxford University Press.

Wright S (1931) Evolution in Mendelian populations. Genetics 16: 97–159.

Wright S (1932) The roles of mutation, inbreeding, crossbreeding and selection in evolution. Proceedings of the Sixth International Congress of Geneticists 1: 356–366.

Zimmerman W (1930) Die Phylogenie der Pflanzen. Jena: Gustav Fischer.

Zimmerman W (1953) Evolution: Die Geschichte ihrer Probleme und Erkenntnisse. Freiburg: Karl Alber.

Zimmerman W (1965) Die Telomtheorie. Stuttgart: Gustav Fischer.

15 Computer Modeling in Behavioral and Evolutionary Ecology: Whys and Wherefores

James A. R. Marshall and Nigel R. Franks

The idea of modeling biological systems is by no means a new one. A textbook early example is the modeling of cyclic variations in the number of Canadian lynx and snowshoe hares trapped by the Hudson's Bay Company. Company records showed that the numbers of lynxes and hares trapped varied cyclically over the years, but the reason for this was unknown. The mathematicians Vito Volterra and Alfred Lotka independently developed a simple mathematical model of change in such predator and prey populations using coupled differential equations, which can be applied to provide a satisfying first approximation of the Hudson's Bay Company's trading records, and is in fact a special case of a more general evolutionary model,[1] the replicator equation (Schuster and Sigmund, 1983).

Such models can typically be approached with only a pen, paper, and the requisite skills; indeed, these were the only resources available to Lotka and Volterra in the early twentieth century. But, of course, the mid-twentieth century saw the arrival of something completely new, with the potential to revolutionize many aspects of life, including mathematical biology: the electronic computer. The revolution began somewhat slowly, however. The first applications of computers to mathematical problems were typically numerical solutions of more complex mathematical models; the U.S. military used them to calculate firing tables using mathematical models of ballistics (Weik, 1961), and the mathematician Stanislaw Ulam developed the Monte Carlo method for numerical solution of equations (Metropolis and Ulam, 1949). The nature of these early applications is perhaps unsurprising; such mathematical models were meat and drink to the pioneers of the computer, and indeed the naming of two of the very first automated computers, Babbage's Difference and Analytical Engines, is indicative of the kind of problems they were designed to solve. Nevertheless, even this early pioneer recognized the potential for implementing simple models (Babbage, 1837).

The advent of the electronic computer, however, also had the potential for a radically different kind of modeling. This kind of modeling made use of the computer's ability to follow simple rules quickly and endlessly, and to determine their consequences. This allows a model to be specified in terms of the rules governing its components and their interactions, rather than describing the model in terms of its high-level behavior. John von Neumann is frequently credited with one of the first examples of this new approach when he began using cellular automata to study models of self-reproduction (von Neumann, 1966). Cellular automata models specify a grid of cells, each with a state, and simple state transition rules for the cells that take account of the states of neighboring cells. The only general way to determine the result of specifying a certain configuration of cell states and rules is to iteratively update each cell's state according to those rules (Smith, 1971), a job that the computer is suited to perfectly. Many similar models were developed over the years, the first examples of what today we would term individual-based models. For the remainder of this chapter we shall refer to such individual-based modeling as computer modeling, rather than the application of computers to provide numerical solutions of systems of equations.

Why Use Computer Modeling?

At the start of the twenty-first century, the prevalence of information technology and widespread familiarity with computer programming in the scientific community makes a strong case for computer modeling simply through ease of use. A competent computer programmer can frequently implement a simple computer model very rapidly, and start generating results almost immediately. Development of computer models has been facilitated still further by the arrival of new paradigms in computer science, notably object-oriented design and programming, and by the recent introduction of software development kits for the creation of such models. In fact, computer modeling tool kits are now available that allow even nonprogrammers to develop models within a particular architecture or paradigm.

Scientific considerations also motivate the use of computer modeling. Frequently the formulation of a computer model can be more intuitive than comparable equation-based models; it can be easier to imagine rules governing behavior of individuals, or even to elucidate them through experimental observation of a biological system, than to try to specify the rules governing the population-level behavior of the system (Taylor and Jefferson, 1994). It is arguably more natural to specify component-level behaviors and determine the macrolevel behaviors they give rise to through iteration than to attempt to specify these macrolevel behaviors in the first instance. Indeed, in attempting to specify macrolevel behavior there is a real danger of missing some crucial aspect of the interaction of individuals in the system that has a significant impact on the high-level behavior of the system (e.g., Marshall and Rowe, 2003).

The key concept here is "emergence," an overused and often controversial term that is typically used to mean the way in which unforeseen results can arise from the nonlinear interactions of a system's components (Schelling, 1980; Holland, 1996). Further support for the use of computer modeling comes from the comparative ease with which it deals with features frequently encountered in real biological systems, such as randomness, spatiotemporal effects, and finite populations (Taylor and Jefferson, 1994). While some such features can be dealt with in equation-based models, this is typically a much more technical process than in a computer model.

Caveats of Computer Modeling

The arguments for computer modeling are not entirely one-sided, however. Caveats that are unique to computer modeling concern the scope for error that exists in engineering a piece of software of any significant complexity. Logical errors are perhaps the most common form of such errors; these are mistakes in the implementation of the logic of a model that mean the rules it follows are not those intended by the modeler. For this reason, Ropella et al. (2002) have argued for multiple independent software implementations of the same model, and multiple representations of the model as software code, formal, textual, and diagrammatic representations. Other technical issues include the use of synchronous or asynchronous update mechanisms (Huberman and Glance, 1993), the careful use of pseudo-random number generators (Ropella et al., 2002), and care in the use of floating-point arithmetic (Polhill et al., 2004).

While some of the above problems may also be encountered in numerical solution of mathematical models, there is a fundamental issue of reproducibility in computer modeling. The sometimes extreme sensitivity to these kinds of errors, coupled with the impracticality of including a complete description of a computer model that includes source code in a published paper, can lead many scientists justifiably to view results collected from computer models with suspicion. By contrast, it is standard practice to include a full formal description of a mathematical model in a published paper. This enables reviewers and readers to verify the model and reproduce its results, makes transparent its assumptions, and hence leads to greater confidence in the conclusions drawn from it.

In theory there is no reason why computer models should not be made just as open to scrutiny; there is nothing to prevent authors of a paper from making their model and data available online for readers and reviewers. Unfortunately, publishing models online in this way is far from standard practice, and frequently the inadequate level of description of the model that is included in a paper can lead to the publication of work that is completely irreproducible. In practice, even if source code for a computer model is made available, it can often be beyond the resources of other researchers to fully validate the model. An interesting alternative approach is "model docking," in which an attempt is made to "align" two existing but different models to see if one model can generate the same results as the other if correctly parameterized (Axtell et al., 1996). "Model docking" can lead to greater confidence in the conclusions drawn from computer models if successful, and illuminating insights if unsuccessful.

Using Computer Models in Behavioral and Evolutionary Ecology

Studies in behavioral and evolutionary ecology can have different objectives than those in theoretical biology. Theoretical biology is frequently engaged in a search for general principles, whereas while both behavioral and evolutionary ecology can also be concerned with such general principles, in many cases they are concerned principally with understanding a particular organism or system.

Our approach to using computer modeling in behavioral biology is based on the concept of a "design space." Dawkins (1986) discussed this idea in terms of his famous artificial creatures, known as biomorphs, when he described "Biomorph Land" thus:

> There is a definite set of biomorphs, each permanently sitting in its own unique place in a mathematical space. It is permanently sitting there in the sense that, if only you knew its genetic formula, you could instantly find it; moreover its neighbours in this special kind of space are the biomorphs that differ from it by only one gene.... When you first evolve a new creature by artificial selection in the computer model, it feels like a creative process. So it is, indeed. But what you are really doing is *finding* the creature, for it is, in a mathematical sense, already sitting in its own place in the genetic space of Biomorph Land. (Dawkins, 1986:80)

Dawkins's idea is of a design space of different morphologies, or body plans, and it is this that a population explores as it undergoes evolution. That is to say, evolution works within a very large and high-dimensional design space of different possible morphologies, of which only a very small proportion is ever realized in the population.

In fact, Dawkins's idea is related to an earlier idea by Raup and Michelson (1965), applied elsewhere in this volume to morphospaces of brachiopods and bryozoa (see chapter 5) and land plants (chapter 14). However, there is no reason why the concept of a design space should be applied only to morphology. There are many other aspects of phenotype than physical body plan, including, in particular, behavior. It follows that in the same way that only a tiny fraction of possible morphologies is ever realized in a population, a particular observable behavior with a given utility will be only a single instance of a much larger set of possible alternative behaviors, some capable of achieving the same or even better utility for the organism exhibiting it.

Assuming that the behavior in question is under some measure of genetic control, and hence subject to evolutionary pressures, we would again expect a population undergoing evolution to describe a trajectory through this design space. It then becomes natural, and interesting, to ask what alternative behaviors exist that would seem to be equally adaptive; if any behaviors exist that would seem to be more adaptive; and ultimately why evolution has led to the observed behavior rather than any equally good, or better, alternative. Attempting to answer this question would begin to address a fundamental question in biology, the role of contingency in evolution (Gould, 1989).

It is precisely to explore such design spaces that we propose the use of computer modeling. The incredible advances in computational power of the previous decades, and the ability of computers, given a correct program, to infallibly execute the billions of calculations necessary, make them particularly suited to the systematic exploration of design spaces.

Of course, if we were to consider a more sophisticated behavior, such as a complex human or primate behavior, identifying the controlling factors, and hence the size and shape of the design space, would be exceptionally difficult. It is for this reason that we consider collective behavior, such as that exhibited by social insects, to be more amenable to the design space approach. Social insect societies are a fine example of the kind of complex systems described earlier; their collective behaviors arise in a nonlinear fashion from the local interactions between their components. While similar complex systems doubtless control behaviors in, for example, the primate brain, social insects have the great advantage of being observable and manipulable, and hence the rules governing their local interactions can be captured with a reasonable degree of confidence through careful experimentation.

Thus computer modeling comes into its own in two respects when modeling the collective behavior of social insects. First, computer modeling is characterized by describing the component behaviors of a system, then iterating these behaviors to give rise to higher-level behavior. Second, the component-level behaviors required are comparatively accessible to observation. The control of a collective behavior by local interactions between components provides the final aspect of our design space approach; the component-level interactions of a complex system can often be specified using a small number of easily identifiable parameters. Hence the size and shape of the design space can be determined and, furthermore, the position of the existing behavior within this design space can be approximated through experimental measurement of these parameters in the biological system. However, having identified the parameters, we are able to explore the parameter space or some subspace of it that interests us. In doing so, we can go beyond the one exemplar of a particular system that nature has provided us with and systematically examine alternatives, comparing their characteristics against the instance we have available to us.

We shall demonstrate the design space approach to computer modeling in behavioral biology through two examples. The first examines the nest site assessment behavior of the ant *Temnothorax albipennis* (formerly *Leptothorax albipennis*), and the second investigates the collective decision-making process underlying emigration in *T. albipennis*.

Nest Area Assessment Using Buffon's Needle

When an ant colony needs to emigrate from its current nest site—due to destruction of that nest site, for example—ant scouts explore the surrounding area to find and evaluate potential new nest sites in rock crevices and the like. The ants take many criteria into account (Franks et al., 2003b), one of the more important of which is the area of a potential new nest site. Assessing the size of a potential nest site is a hard problem for a scout ant, yet the decision of an individual scout about whether a site is suitable to house the whole colony can have a significant effect on the colony's survival. Size assessment is difficult because scouts cannot simply measure diameter, since nests are often irregularly shaped. In addition, they are usually dark, and the ants' vision does not permit them to correctly estimate area. The evaluation method used by ant scouts of *Temnothorax albipennis* is therefore an interesting area of study.

After experimentally discounting heuristics such as perimeter length or mean free-path length, "Buffon's Needle" has been suggested instead (Mallon and Franks, 2000; Mugford et al., 2001). Proposed by the eighteenth-century naturalist and mathematician Comte Georges-Louis Leclerc de Buffon, Buffon's Needle states that a needle of known length, repeatedly dropped at random onto a plane covered with equidistant parallel lines, can give a good estimate of pi based on the frequency with which it intersects those lines. Buffon's Needle can be adapted to estimate area on a plane (Newman, 1966; Franks, 1982); for the ant scouts, it works on the principle that a scout lays a pheromone trail of defined length while wandering throughout a potential nest site, then evaluates the approximate size of the nest site by wandering again within the site while assessing the frequency with which it crosses its previously laid trail (figure 15.1). As seen from figure 15.1, the scout also assesses the integrity of the site by following the walls at its perimeter. All evidence suggests that *Temnothorax* ants assess sites by working alone and that they do this by deploying individual specific trail pheromones (Mallon and Franks, 2000).

A computer model of this nest assessment behavior has been developed by Şahin and Franks (2002). This model demonstrates that the proposed assessment mechanism of ant scouts, using Buffon's Needle, works in practice, as well as a trade-off

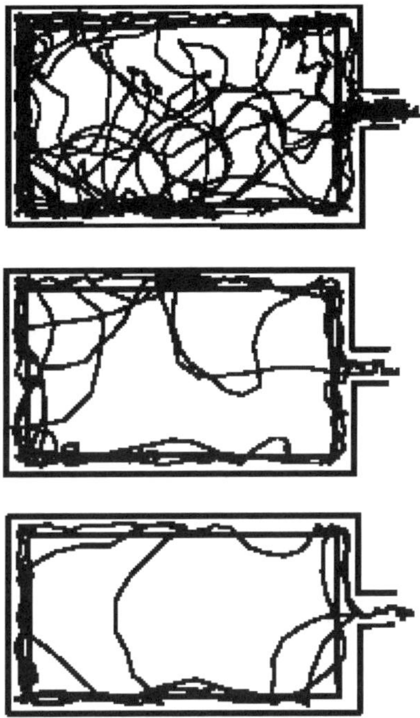

Figure 15.1
The path of a single scout (thin black line) on each of its three successive visits to the same potential nest site. The ant appears to spend a considerable part of its visit near the internal perimeter of the new nest. Nevertheless, in general, the number of intersections between second and first visit paths in the central region of the nest (within the inner box) and the edge region of the nest (between the two boxes) is similar. The first visit path in the central region of the nest line is reasonably uniformly distributed. This should ensure that the Buffon's needle algorithm gives a reasonably accurate estimation of the nest area. (Reproduced from Mallon and Franks, 2000.)

between wall-following behavior and exploration of the central nest area in effective nest site assessment. The continuum between pure wall-following behavior and pure exploration behavior explored by Şahin and Franks's (2002) model is just one part of the design space associated with that behavior. Since the behavior is quite sophisticated, this design space is in fact extremely large.

Another interesting region of the design space relates to the use of a "two-pass" strategy by the scouts; they are believed to lay a pheromone trail during their first visit to a new potential nest site, then assess frequency of crossing this trail during their subsequent visits. A particularly interesting question is why such a two-pass strategy is used. Time is of the essence during an emigration when the colony is in danger—if the colony's original nest site has been destroyed for example—and in any case we would expect animals to minimize their energy expenditure where possible, such as by minimizing distance traveled between potential nest sites.

In principle, then, it would seem more efficient to compress the two stages into one. Hence we are interested in whether the two-pass strategy employed by the real ants is algorithmically superior to a one-pass strategy in which the ant wanders within the nest site, simultaneously laying pheromone and assessing the frequency of intersection with its own trail. Is there an algorithmic reason why a one-pass strategy is not suitable? For instance, does a one-pass strategy bias area estimation in a way which an ant cannot compensate for when classifying a nest's size? Such a bias might occur because the ant is laying pheromone and detecting it in adjacent locations, which could result in a different frequency of path crossings compared with the two unrelated random walks used with the two-pass strategy.

Marshall et al. (2003) implemented a computer model to study this question, and concluded from its results that in the simplest case, there was no apparent algorithmic reason why the two-pass strategy should be favored over a one-pass strategy. Therefore, the conclusion was that the use of a two-pass strategy might in the end be due to a lack of selective influence on the behavior, physical constraint preventing ants from laying and detecting pheromone at the same time, or evolutionary constraint preventing evolution from discovering the one-pass strategy. This work is being continued to study the assessment behavior more exhaustively and in more realistic situations (figure 15.2).

Interesting as the question of different possible strategies for nest area assessment is, other behaviors can be found that are even more suited to the design space approach using computer modeling. One unsatisfactory aspect of the nest assessment model in this regard is that the hypothetical alternative behavior, the one-pass strategy, was generated a priori by the human beings conducting the investigation. It would be nice if the computer model could by itself explore the design space with minimal prior guidance from the investigators, but in the case of the nest assessment behavior it is likely that the complexity of the behavior would render this impractical. Thus the alternative behaviors to be considered must be proposed by the investigators, with the computer model limited to their evaluation. Additionally, one of our original criteria for the suitability of computer modeling for a particular system was that the system be a complex one; in other words, a system in which higher-level behaviors emerge from the nonlinear local interactions of its components.

The nest assessment model investigates an individual-level behavior; the main advantage of the computer model comes from the complicated nature of that behavior's interaction with the environment. In principle the question posed is geometrical in nature, and a sufficiently proficient geometer might come up with an answer. In practice, however, the problem is probably exceptionally difficult to tackle analytically, and the benefits from using a computer model are in the simplicity of the way in which the problem is addressed and the ease with which answers can be obtained.

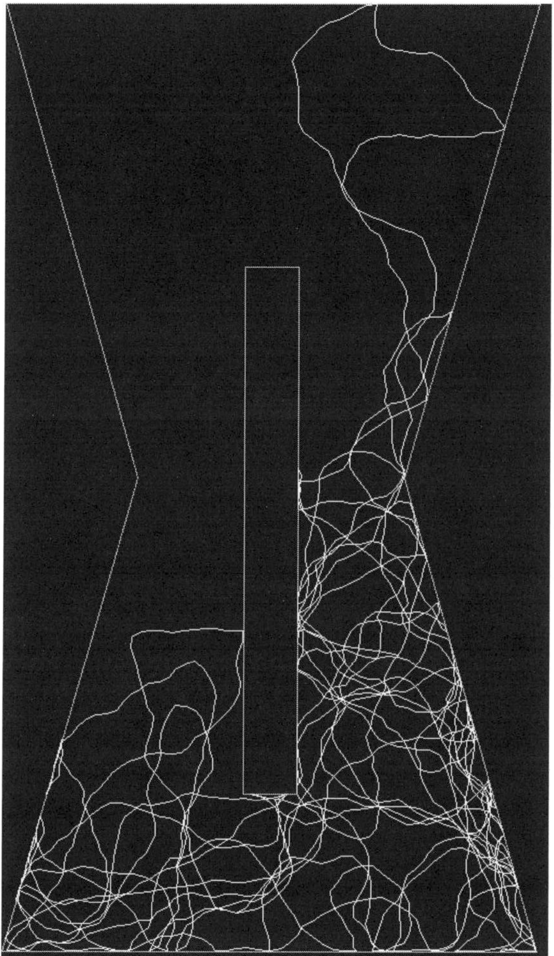

Figure 15.2
The pheromone trail of a simulated ant scout assessing the area of a virtual nest site.

Collective Decision-Making During Emigration

A behavior that more closely meets our criteria for the applicability of computer modeling is the house-hunting behavior of *T. albipennis*. This is because this behavior is one in which a high-level decision arises from the nonlinear interactions of the ants; in other words, the colony's collective decision-making system is a complex system. As will be seen below, this has the added advantage of making the design space for the system easier to delineate, and hence suitable for automatic exploration by the computer model.

During its lifetime a colony may need to emigrate from its current nest site to a new one. Such a necessity may arise due to the destruction of the original nest site, but colonies are also able to move opportunistically to improve the quality of their housing (Dornhaus et al., 2004). These different circumstances place different requirements on the colony. For example, an emergency emigration after destruction of the original nest must be concluded as quickly as possible to remove the colony from danger, even at the risk of making a poor choice of new nest site, whereas a colony looking to trade up from its existing home has more time to make the right choice. Furthermore, there may be environmental cues—indicating the presence of a predator, for example—that lend urgency to the emigration process.

The ants are able to sense these different circumstances and vary the speed of their decision-making accordingly, at the expense of always making the best choice the first time: the ants' decision rule exploits a speed-accuracy trade-off in the problem being addressed (Franks et al., 2003a). The collective decision arrived at by the colony is, furthermore, achieved with a minimal amount of direct comparison of alternatives by individual ant scouts. It is a truly decentralized approach to decision-making, with no one individual acting as controller. However, despite this decentralization, colonywide agreement about the best choice is crucial to prevent the colony from splitting during emigration.

The decision-making process involved in colony emigration in *T. albipennis* revolves, as many social insect behaviors do, around recruitment. Unlike the mass recruitment methods of pheromone trail-laying in ants (Holldobler and Wilson, 1990) or, to a lesser extent, the waggle-dance in honeybees (von Frisch, 1954), the recruitment behavior in this case is individual. Ant scouts each discover and evaluate a nearby nest site, and begin recruiting other ants to their choice after a time delay that is inversely proportional to the perceived quality of the site. The decision-making process is a two-stage one, in that two recruitment methods are used. The first, tandem running (figure 15.3A), is slow and involves an ant scout leading another ant to its preferred nest site; tandem running reflects a provisional acceptance of a nest site. The second, social carrying (figure 15.3B), is approximately three times quicker than tandem running and involves an ant scout picking up and carrying a passive ant or a brood item and transporting it to the new nest site; social carrying reflects a full acceptance of a nest site. Social carrying to a nest site begins when a quorum threshold is exceeded for that site; each time a scout visits its preferred site, it estimates the number of other ants already there, and if this exceeds some threshold, the scout begins social carrying (Pratt et al., 2002; Franks et al., 2002).

What is particularly interesting is that, as already mentioned, the ant scouts are able to vary the quorum threshold they use according to the circumstances of the emigration; a high quorum threshold results in a slower, less error-prone decision, and is typically used during emigrations in benign environmental conditions when the col-

Figure 15.3
(a) Recruitment via tandem running in the ant *Temnothorax albipennis*. The worker at the front is leading the tandem run, and the follower is about to signal its presence by tapping with its antennae on the gaster of the leader. (b) Recruitment by transport in *T. albipennis*. One worker is carrying another to the new nest site. (Both photographs by S. C. Pratt; reproduced from Franks et al., 2002.)

ony is not in danger at its original nest site; a low quorum threshold, on the other hand, leads to faster but more error-prone decisions, and is typically used when the environment is hostile or the colony's original nest site has been destroyed (Franks et al., 2003a).

There are many parameters associated with the collective decision-making process described above. For example, the rate at which ants in the original nest site go out to search for new nest sites, the difference in the recruitment delay induced by a mediocre and a superior nest site, and the rate at which ants recruit other ants, among others. All of these parameters affect the performance of the collective behavior in question. It is then a natural question to ask exactly what effect changing these parameters has on performance. For the house-hunting behavior, natural performance measures are the speed and error rate involved in using the minimal quorum size (i.e., using purely individualistic decision-making) and the rates at which the error rate improves, and the decision time increases, as the quorum size is increased. With a clearly delineated parameter space we are able to automatically explore how performance changes in different regions of the space, and how the performance of the system realized by nature compares with possible alternatives.

If the parameter space is sufficiently small, we may be able to explore it systematically. The ants' house-hunting behavior can be described by six or seven parameters, but by selecting three relevant ones and systematically plotting the system's performance in this subspace of the parameter space, we can quickly see the effect of these parameters on the system's performance (figure 15.4).

This approach has been used by Marshall et al. (2006) to examine the optimality of the decision-making system used by *T. albipennis* during emigration. Using their computer model, Marshall et al. conclude that the speed-accuracy trade-off exhibited during decision-making by the ants is not entirely explained by current models, and that the ants should always be able to make faster, completely accurate decisions.

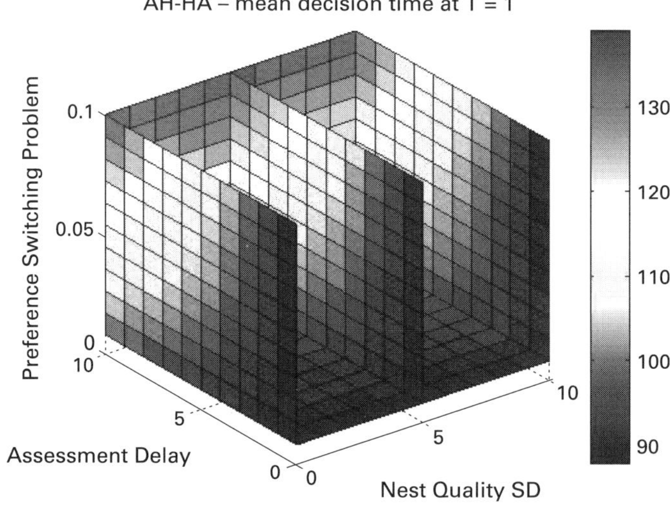

Figure 15.4
A visualization in three dimensions of the performance of the ant house-hunting algorithm, varying with preference switching rate, assessment noise, and cost. The performance measure used in this particular example is the average number of colony members choosing the best nest site for the minimum quorum threshold of 1. This visualization shows how increasing the time required to perform nest site assessments affects the relationship between the willingness of scouts to evaluate alternative nest sites, and the time taken to reach collective decisions. Such visualizations can provide a quick and intuitive entry point to the exploration of a design space, but spaces of higher dimension may require more advanced techniques, such as data-mining. (Reproduced from Marshall et al., 2006.)

This conclusion is supported by appropriate parameterization of an equation-based model of the same system (Pratt et al., 2002). The strength of the computer modeling approach lies in the level of detail with which the model can be constructed. Frequently, for complex systems the devil is indeed in the details, and where a rate-governing constant in a differential equation model may be arbitrarily limited to give rise to a given phenomenon, in a computer model the underlying reasons for this limit may be more thoroughly explored. Thus, in the emigration system, Marshall et al.'s computer model showed the importance of both the noise and the time cost inherent in nest assessment for the optimality of the decision-making system.

Sensitivity analysis of a system varying three or fewer parameters allows us to develop our understanding of the system's behavior using simple visualization techniques (figure 15.4). However, to consider the interactions of all the parameters, we must identify another method, because we cannot easily visualize higher-dimensional spaces. The application of machine learning techniques to mine this high-dimensional data set and try to discover rules to classify the system's performance in different regions is a promising approach. We are currently working on this analysis of the performance of possible different configurations of the system, and their comparison with the configuration realized in *T. albipennis*.

Our motivation for studying the decision-making system of *T. albipennis* is twofold. First, we are interested in increasing our understanding of the underlying biology. Additionally, we are interested in designing a new decentralized decision algorithm for problems in computer science. Since Babbage, computer science has concentrated on centralized computing paradigms. The massively parallel computing environment realized by the Internet presents modern computer scientists with an environment much closer to that experienced by organisms such as *T. albipennis*, in which noiseless, global information is impossible to obtain. Accordingly, computer scientists are increasingly seeking inspiration from nature in tackling the difficult control and optimization problems they face (e.g., Bonabeau et al., 1999). In so doing, it is important to verify that the behavior observed in nature is indeed optimal, and under what circumstances—a goal which the design space approach to computer modeling provides a positive step toward.

Views on the Scientific Status of Computer Models

In this chapter we have argued for a design space approach to using computer models in the course of scientific research. In recent years debate has focused on the scientific status of such computer models. This debate largely originates in the artificial life community, which frequently deals with questions of theoretical biology. In particular, there are two distinct views of the status of computer models: the first is that they are *simulations* of life, the second is that they are *instantiations* of life.

The proponents of the "instantiations of life" viewpoint argue that the study of different kinds of life in media different from those of conventional biology, such as computer processors and memory (Ray, 1994), can help elucidate general principles underlying life and its origins without being distracted by artifacts that are introduced through considering only one particular medium, such as carbon-based biology. This viewpoint is controversial, largely because a generally agreed definition of life is problematic.

The controversy can perhaps best be avoided by taking the "simulations of life" viewpoint, in which computer models are precisely that: scientific models that are tools to gain understanding of, and make testable predictions about, a system. This viewpoint is exemplified in the "physics model" approach of Kitano et al. (1997), in which a model implementing a particular theory is validated against empirical data, then used to generate results that lead to new hypotheses. As its name suggests, this approach is close to that used in the physical sciences. The main problem encountered in this approach is the validation stage; if a model fails to reproduce empirical results, is the model incorrect, or is the underlying theory incorrect?

Bedau (1999), on the other hand, proposed that computer models be viewed as "emergent thought experiments." This approach treats computer simulations as a means of extending traditional thought experiments into areas, such as complex systems, in which the human mind cannot apprehend the implications of the premises of the original thought experiment. As Dennett (1994) said, "Artificial Life . . . can be conceived of as a *sort* of philosophy—the creation *and testing* of elaborate thought experiments, kept honest by requirements that could never be imposed on the naked mind of a human thinker acting alone" (291, author's emphasis).

While in a traditional thought experiment the conclusions follow intuitively from the premises, in an emergent thought experiment the conclusions "emerge" nonintuitively from the premises. This is in agreement with our original motivation for using computer modeling, that the systems being modeled are highly nonlinear. Bedau went on to assert that such experiments "can explain how some phenomenon occurs only if [they produce] actual examples of the phenomena in question, it is not sufficient to produce something that represents the phenomenon but lacks its essential properties" (21). This has parallels with the warning by Bonabeau and Theraulaz (1994) that the ability to represent the behaviors of two different systems using the same model should not lead us to believe that the systems are necessarily the same at some fundamental level.

Di Paolo et al. (2000), critical of both the physics model approach of Kitano et al. (1997) and the emergent thought experiment approach of Bedau (1999), proposed that computer models be considered "opaque thought experiments." They criticized the physics model approach because of the validation problem identified above. For the emergent thought experiment approach they argued that such experiments either could not be thought experiments because thought experiments are by definition not sources of empirical data, or, if they were not sources of empirical data, then they could not produce actual examples of the phenomenon in question.

Di Paolo et al.'s opaque thought experiment is similar to Bedau's emergent thought experiment in that it considers a defining aspect to be the nonintuitive emergence of results from premises, but the use to which the model is put is different. Specifically, the opaque thought experiment is designed to implement a particular theory; then observations are collected from it during what is known as the "exploratory phase." What follows in the "experimental phase" is the formulation of hypotheses that organize these observations and are tested through experimentation with the model. This results in an explanation of how the model gives rise to these observations, and it is in the "explanatory phase" that the traditional loop of model-experiment-revise is closed; the explanation of the observations in terms of the model is compared with the theory implemented, possibly resulting in revision of that theory if necessary.

We consider that our approach to computer modeling in behavioral ecology most closely coincides with the "opaque thought experiment" of Di Paolo et al.; the computer model allows us both to determine that the individual behaviors we have observed experimentally in the laboratory are sufficient to give rise to the observed macrobehavior, and to go beyond this by exploring their necessity through examining alternative parameterizations and configurations of the system.

Conclusion

Since the advent of the electronic computer, a new approach to modeling biological systems has been opened up. Computer modeling relies on this newly acquired computational power to model a system through representing its components and their interactions, and iterating them. Frequently, traditional equation-based approaches to modeling can have difficulty in dealing with the nonlinear and finite natures of the components and interactions found in real biological systems. This new approach to modeling, however, is not without its drawbacks. While the relaxation of the kinds of simplifying assumptions used in analytical approaches allows systems to be modeled more realistically, there is a price to pay in the increased difficulty of interpreting the results of computer models. Various approaches to interpreting computer models have been proposed for the field of theoretical biology, one of the more interesting of which is the "opaque thought experiment" approach, in which explanation of the model's unexpected results in terms of its implementation feeds back into comparison with, and possible revision of, the underlying theory the model is intended to investigate.

For behavioral biology we propose a design space approach to using computer modeling. The idea underlying this approach is that many behaviors exhibited by organisms or biological systems are single instantiations of a much larger range of alternative behaviors that might achieve the same end, the so-called design space. By using computer modeling we are able to explore this design space in a way that conventional biology does not permit. We propose that the behaviors of social insects are a particularly useful set of systems to apply this approach to. First, they are observable and manipulable in a way that, for example, the brain is not. Second, their collective behaviors in particular are examples of complex systems that are most naturally represented using computer models, and that provide a clearly delineated design space of alternatives to explore. We are not unique in using this approach, which has also been successfully applied to, for example, fish shoals (Couzin et al., 2002) and honeybee swarms (Couzin et al., 2005).

We hope that extending the design space approach may enable us to begin to address questions on the evolutionary history of specific behaviors. If we take the emigration behavior of *T. albipennis* as an example, we might propose an explanation for the evolution of this complex behavior as follows. Initially ants were solitary decision-makers, and simply started carrying their nest mates to any reasonable nest they discovered. Selection, however, favored colonies with slower, more deliberate workers, who took their time over the decision and showed nest mates the way to newly discovered sites. In doing so, these ants took into account the opinions of their nest mates, but also retained the ability to switch to their earlier, individualistic decision-making when faced with adverse conditions.

With the design space approach such explanations might be rescued from simply being labeled as just-so stories. For example, it may be possible to use an adaptive walk approach, exemplified by Niklas's work on land plant optimization (see chapter 14), which might allow us to demonstrate the adaptive value of putative steps in the evolution of complex behaviors. Of course Niklas uses fossil forms to initialize his adaptive walks, and while fossils of morphology are comparatively easy to obtain, "behavioral fossils" are somewhat harder to find, although developments in the study of trace fossils are very interesting (e.g., Miller, 2003). However, it is also frequently possible to observe more "primitive" behaviors in relatives of a species, and these kinds of "behavioral fossils" may serve the same purpose when considering the evolution of a behavior.

In using computer modeling, we are motivated not just by the potential to gain deeper insights into our biological systems by exploring alternative possible behaviors. We are also interested in deriving useful algorithms that can be applied to general problems in engineering and computer science. In designing such algorithms it is important not to assume that nature has already discovered the best possible solution and slavishly follow it but, as discussed by Bonabeau et al. (1999), to take inspiration from nature and use computer modeling to seek to understand and go beyond what nature has achieved.

Acknowledgments

We gratefully acknowledge Tim Kovacs and Anna Dornhaus for collaboration on the nest assessment and house-hunting models, Anna Dornhaus and Ana Sendova-Franks for comments on drafts of this chapter, Nic Minter for discussions on trace fossils, and members of the Ant Lab at the University of Bristol for always helpful discussions. We also thank George McGhee and Karl Niklas for drafts of their chapters, and Werner Callebaut for bringing Giorgio Israel's account of the history of the Lotka-Volterra equation to our attention. NRF acknowledges the support of the BBSRC (Grant no. E19832). JARM and NRF acknowledge the support of the EPSRC (Grant no. GR/578674/01).

Note

1. Interestingly, Lotka and Volterra apparently would not have been excited by the applicability of a general form of their equations to population genetics, hypercycles, and animal behavior. Indeed, both Lotka and Volterra apparently considered it unimportant that Lotka had in fact discovered an analogous form of the predator-prey equations five years previously when considering a hypothetical chemical oscillator reaction. Rather, their main concern over priority seems to have been regarding the application of the equations to the specific problem of predator-prey interactions (Israel, 1996).

References

Axtell R, Axelrod R, Epstein J, Cohen MD (1996) Aligning simulation models: A case study and results. Comp Math Org Theo 1: 123–141.

Babbage C (1837) Ninth Bridgewater Treatise: A Fragment (2nd ed). London: John Murray.

Bedau MA (1999) Can unrealistic computer models illuminate theoretical biology? In: Proceedings of the 1999 Genetic and Evolutionary Computation Conference Workshop Programme (Wu AS, ed), 20–23. San Francisco: Morgan Kaufmann.

Bonabeau E, Dorigo M, Theraulaz G (1999) Swarm Intelligence: From Natural to Artificial Systems. New York: Oxford University Press.

Bonabeau E, Theraulaz G (1994) Why do we need artificial life? Art Life 1(3): 303–325.

Couzin ID, Krause J, Franks NR, Levin SA (2005) Effective leadership and decision making in animal groups on the move. Nature 433: 513–516.

Couzin ID, Krause, J, James R, Ruxton GD, Franks NR (2002) Collective memory and spatial sorting in animal groups. J Theoret Biol 218: 1–11.

Dawkins R (1986) The Blind Watchmaker. Harlow U.K.: Longman.

Dennett D (1994) Artificial life as philosophy. Art Life 1(3): 291–292.

Di Paolo EA, Nobel J, Bullock S (2000) Simulation models as opaque thought experiments. In: Artificial Life VII: Proceedings of the Seventh International Conference on Artificial Life (Bedau MA, McCaskill JS, Packard NH, Rasmussen S, eds), 497–506. Cambridge, Mass.: MIT Press.

Dornhaus A, Franks NR, Hawkins RM, Shere HNS (2004) Ants move to improve: Colonies of *Leptothorax albipennis* emigrate whenever they find a superior nest site. Anim Behav 67(5): 959–963.

Franks NR (1982) A new method for censusing animal populations: The number of *Eciton burchelli* army ant colonies on Barro Collarado Island, Panama. Oecologia 52: 266–268.

Franks NR, Dornhaus A, Fitzsimmons JP, Stevens M (2003a) Speed versus accuracy in collective decision making. Proc Roy Soc Lond B270(1532): 2457–2463.

Franks NR, Mallon EB, Bray HE, Hamilton MJ, Mischler TC (2003b) Strategies for choosing between alternatives with different attributes: Exemplified by house-hunting ants. Anim Behav 65(1): 215–223.

Franks NR, Pratt SC, Mallon EB, Britton NF, Sumpter DJT (2002) Information flow, opinion polling and collective intelligence in house-hunting social insects. Phil Trans Roy Soc Lond B357(1427): 1567–1583.

Gould SJ (1989) Wonderful Life: The Burgess Shale and the Nature of History. New York: Norton.

Holland JH (1996) Hidden Order: How Adaptation Builds Complexity. Reading, Mass.: Perseus.

Hölldobler B, Wilson EO (1990) The Ants. Cambridge, Mass.: Belknap Press of Harvard University Press.

Huberman B, Glance N (1993) Evolutionary games and computer simulations. Proc Natl Acad Sci USA 90(16): 7715–7718.

Israel G (1996) La Mathématisation du Réel. Paris: Seuil.

Kitano HS, Hamahashi S, Kitazawa J, Takao K, Imai S (1997) The virtual biology laboratories: A new approach of computational biology. In: Proceedings of the Fourth European Conference on Artificial Life (Husbands P, Harvey I, eds), 274–283. Cambridge, Mass.: MIT Press.

Mallon EB, Franks NR (2000) Ants estimate area using Buffon's needle. Proc Roy Soc Lond B267: 765–770.

Marshall JAR, Dornhaus A, Franks NR, Kovacs T (2006) Noise, cost and speed-accuracy trade-offs: Decision making in decentralised systems. J Roy Soc Interface 3: 243–254.

Marshall JAR, Kovacs T, Dornhaus A, Franks NR (2003) Simulating the evolution of ant behaviour in evaluating nest sites. In: ECAL 2003: Lecture Notes in Artificial Intelligence 2801 (Banzhaf W, Christaller T, Dittrich P, Kim JT, Ziegler J, eds), 643–650. Berlin: Springer-Verlag.

Marshall JAR, Rowe JE (2003) Viscous populations and their support for reciprocal cooperation. Art Life 9(3): 327–334.

Metropolis N, Ulam S (1949) The Monte Carlo method. J Amer Stat Assn 44: 335–341.

Miller W (ed) (2003) Palaeogeog., Palaeoclimat., Palaeoecol. 192. Special Issue on New Interpretations of Complex Trace Fossils.

Mugford ST, Mallon EB, Franks NR (2001) The accuracy of Buffon's needle: A rule of thumb used by ants to estimate area. Behav Ecol 12: 655–658.

Newman EI (1966) A method of estimating the total length of root in a sample. J Appl Ecol 3: 139–145.

Polhill JG, Izquierdo LR, Gotts NM (2006) What every agent based modeller should know about floating point arithmetic. Environ Model Softw 21: 283–309.

Pratt SC, Mallon EB, Sumpter DJT, Franks NR (2002) Quorum sensing, recruitment, and collective decision-making during colony emigration by the ant *Leptothorax albipennis*. Behav Ecol Sociobiol 52(2): 117–127.

Raup DM, Michelson A (1965) Theoretical morphology of the coiled shell. Science 147: 1294–1295.

Ray TS (1994) An evolutionary approach to synthetic biology: Zen and the art of creating life. Art Life 1(1): 179–210.

Ropella GEP, Railsback SF, Jackson SK (2002) Software engineering considerations for individual-based models. Nat Resour Model 15(1): 5–22.

Şahin E, Franks NR (2002) Simulation of nest assessment behaviour by ant scouts. In: ANTS 2002: Lecture Notes in Computer Science 2463 (Dorigo M, Di Caro G, Sampels M, eds), 274–281. Berlin: Springer-Verlag.

Schelling TC (1980) Micromotives and Macrobehavior. New York: Norton.

Schuster P, Sigmund K (1983) Replicator dynamics. J Theor Biol 100: 533–538.

Smith AR III (1971) Simple computation-universal cellular spaces. J Assn Comput Mach 18: 339–353.

Taylor CE, Jefferson D (1994) Artificial life as a tool for biological inquiry. Art Life 1(1): 1–14.

von Frisch K (1954) The Dancing Bees: An Account of the Life and Senses of the Honey Bee (Ilse D, trans). London: Methuen.

von Neumann J (1966) The Theory of Self-Reproducing Automata (Burks AW, ed). Champaign: University of Illinois Press.

Weik, MH (1961) The ENIAC story. J Amer Ordnance Assn (Jan.–Feb.): 3–7.

16 Modeling in EvoDevo: How to Integrate Development, Evolution, and Ecology

James P. Collins, Scott Gilbert, Manfred D. Laubichler, and Gerd B. Müller

Evolutionary developmental biology, or EvoDevo, integrates perspectives from evolutionary and developmental biology, and increasingly also from ecology, to understand patterns and processes of phenotypic evolution. Initially progress in EvoDevo was driven mainly by conceptual approaches, beginning with such ideas as "burden" and "developmental constraint," designed to account for the specific and limited trajectories of morphological change. Then the discovery of a small number of highly conserved developmental genes, foremost the *Hox* genes, gave rise to notions such as the "genetic toolkit of development" (Carroll et al., 2005). These concepts were generally well received, as evidenced by scientific and popular pronouncements of "success" and of a "new" or a "completed synthesis." But in the meantime an increasingly vocal chorus of critical voices emerged, questioning to what extent the reality of EvoDevo has lived up to its promise, and even whether such a synthesis of evolutionary and developmental biology is at all possible. Of greatest concern has been the lack of genuine models for EvoDevo, abstract and organismal, that integrate diverse perspectives in ways that could make them a reference point for EvoDevo explanations.

In contrast, transmission genetics had Mendel's rules as an abstract formalism and *Drosophila* as its iconic model system. The Modern Synthesis had the theory of allopatric speciation and local adaptation represented by, for instance, Sewall Wright's adaptive landscapes and any number of examples of adaptive radiation, such as Darwin's finches and the cichlids of the East African lakes. There are also examples of close links between empirical case studies and formal models relating to both evolutionary and developmental biology, such as the Hymenoptera for inclusive fitness, or *Hydra* for reaction-diffusion models and gradients of determining morphogenetic factors. None of the current concepts and research programs in EvoDevo, however, have reached the degree of cohesion that is characteristic of representative models in other domains. In this chapter we first explore the history of and the conceptual difficulties in modeling development *and* evolution, paying special attention to the kinds of model systems used in different explanatory contexts. Next, we address the question of modeling EvoDevo, both in its conceptual approaches and in its potential new model systems.

Modeling Evolution *and* Development

Evolutionary ideas have been closely linked to embryological phenomena ever since both captured the imagination of researchers in the early nineteenth century. Since these temporal processes were difficult to observe, models—both theoretical and material—were crucial to the development of these scientific enterprises. In embryology, practical demands led to the study of a small number of organisms, such as the frog, the chick, and, after the establishment of marine stations in the later decades of the century, marine invertebrates. Painstaking observations allowed the reconstruction of embryological stages, and comparative studies led to the formulation of embryological "laws" and the theory of recapitulation. The latter took on a particular meaning after Darwin's work paved the way for a general acceptance of the evolutionary history of organisms.

A phylogenetic perspective provided the first context for modeling development and evolution when developmental sequences were interpreted as a window into the evolutionary past of organisms. Embryological observations became the basis for reconstructing genealogical relationships. Homologies between characters were modeled as a sequence of morphological transformations, while embryological and comparative data provided the evidence supporting these reconstructions.

These attempts at reconstructing phylogeny left many dissatisfied. A younger generation of researchers was especially concerned with the high degree of ambiguity in those fundamentally historical models. Headed by Hans Driesch and Wilhelm Roux, they had as their alternative goal understanding development mechanistically as a way to uncover the causal connections between different embryological stages and structures. The resulting new science of *Entwicklungsmechanik* provided a different context for modeling development. The first models, such as those of Wilhelm His, were inspired by mechanical principles of folding, bending, and differential growth of tissues and cell layers (Hopwood, 2000). These models became more physiological when phenomena such as regulation and differentiation (specialization, division of labor, etc.) rose in prominence.

The physiological paradigm, in contrast to the historical perspective, also emphasized experimental intervention, which required cultivating model organisms suitable for these tasks. The chick, amphibians (frogs and salamanders), flatworms, and, to some extent, sea urchins were especially suited for experimental manipulations such as transplantation and selective destruction of embryological tissues. Experiments in transplantation gave rise to a biochemical approach when researchers tried to uncover the chemical nature of signals that had the power to shape the organism, such as the chemical nature of the organizer. The biochemical orientation and the new technologies that came with it again changed the ways development was modeled. Tissue and cell cultures greatly aided in the search for the chemical determinants of development and became a new model for these processes.

While a combination of developmental mechanics and physiology was used to model the development of individual organisms, new trends in evolutionary biology were focusing on variation and its genetic basis. In the study of phenotypic variation the emphasis shifted very early to the behavior of those (abstract) factors—genes—that could be correlated with the observed patterns of phenotypic inheritance. Even though development as well as the environment played a crucial role in conceptualizing genes—as in Johannsen's definitions of genotype and phenotype (Johannsen, 1911) or in Woltereck's idea of a reaction norm (Woltereck, 1909)—the *Drosophila* model system soon privileged a genetic account of phenotypic variation. It is widely known how this model then became the basis for the Modern Synthesis in evolutionary biology and how it was transformed by the molecular revolution in biology after World War II.

The privileging of genetics can also be seen in the standard seven model organisms of developmental biology: the fruit fly *Drosophila melanogaster*; the nematode *Caenorhabditis elegans*, the mouse *Mus musculis*, the frog *Xenopus laevis*, the zebra fish *Danio rerio*, the chick *Gallus gallus*, and the mustard *Arabidopsis thaliana*. Except for the chick, all of these systems are especially suitable for genetic approaches.

But these trends are only one side of the history of modeling development and evolution. Important parallel developments included the physiological genetics of Richard Goldschmidt, who conceived of the gene as something far less corpuscular and more dynamic (i.e., physiological; Goldschmidt, 1927), and Alfred Kühn's physiological developmental genetics (Kühn, 1941; Laubichler and Rheinberger, 2004). The latter was based on a concept of interlocking genetic and physiological systems that we would now call gene cascades, reaction chains, and substrate chains that interact to produce a phenotypic effect. Goldschmidt and Kühn worked with different model systems—*Lymanntria* and *Ephestia*, respectively—that were less suited for genetic analysis but much better for physiological and biochemical studies.

Besides Kühn and Goldschmidt, Hans Przibram at the Vienna Vivarium emphasized an integrated approach to modeling development and evolution in the early twentieth century. Using a variety of organisms, the work at the Vivarium Institute headed by Przibram emphasized the study of the whole life cycle of organisms and organism-environment interactions. This included studies of regeneration as a model for normal development, endocrinology, experimental evolution, and the study of reaction norms. But—and this is particularly interesting in the context of the present volume—Prizbram was also keen on the integration of experimental and theoretical work (e.g., Przibram, 1923). The latter included attempts to mathematically model developmental processes and a major push for a quantitative orientation within biology. Paul Weiss and Ludwig von Bertalanffy, who later championed theoretical and mathematical approaches in biology, such as the field concept and the idea of a general systems theory, were connected to Przibram and the Vienna Vivarium.

It seems, then, that one reason for the small number of model systems in current developmental biology has to do with the alliance between developmental biology and genetics. When developmental biology was more of a "physiological science," it had different models—newts, sea urchins, ranid frogs, ambystomid salamanders, slime molds, flatworms, and chicks. As it became a "genetic science," the environment was relegated to a smaller role. Part of the explanation for the difference lies in the fact that a predictable, controlled genetic analysis depends on a model organism whose phenotype is not significantly controlled by the environment.

As Sonia Sultan (2003) points out, "Neo-Darwinian botanists were often quite frustrated in their attempts to discern genetically based local adaptations through this 'environmental noise,'" and this led them to overlook the adaptive nature of developmental plasticity. On the zoological side of developmental biology, the desire to link developmental biology with genetics (and the desire to breed the animals easily) led to adoption of a limited number of model species selected for the lack of significant environmental contributions to the phenotype (Bolker, 1995).

In this context it is interesting to read the preface to volume 1 of *Current Topics in Developmental Biology* (1966), which was written by Joshua Lederberg, who was a geneticist, not an embryologist. He proposed that if developmental biology were going to make progress, it required a model organism such as *E. coli* B. He suggested two such models: the mouse (as a surrogate for humans) and "some very simple system like a rotifer or nematode." Lederberg was prophetic; most of today's model organisms in developmental biology are favored for reasons similar to those he suggested.

Today EvoDevo and its subdiscipline, ecological developmental biology, both assert that the canonical model systems are starting points for evolutionary and developmental investigations, but that they may give a biased view of nature. They are good starting points because countless efforts have led to a detailed understanding of these systems; they may be biased because they often represent derived and specialized lineages that may not be suitable for questions about the *evolution* of morphological novelties and patterns of phenotypic evolution. First, these animals can give the erroneous impression that everything needed to form the embryo is within the fertilized egg. Second, in the laboratory these animals may not provide adequate explanations for the way animals develop in the wild. Tadpoles in the wild, for example, may look different from tadpoles reared in aquaria because their phenotype develops, in part, from cues emanating from competitors and predators absent in the laboratory Environmental chemicals that are harmless in the laboratory may be dangerous to developing organisms in the wild (Colburn et al., 1996; Hayes et al., 2003; Relyea and Mills, 2003). Third, as Neff and Rine (2006) noted, "Model organisms have become a 'comfort zone' for biologists, luring them away from investigating questions that cannot be answered with any of the existing models." And fourth, the organisms used for modeling a particular phenomenon may be idiosyncratic.

Species can be defined as those organisms which develop in a particular way, using particular molecules and processes. Thus, the development of a single organism, by this definition, cannot circumscribe the development of its clade. Most arthropods probably do not use a gradient of Bicoid to form their head, even though this gradient is remarkably important for *Drosophila*. Most amphibians probably do not form their mesoderm as *Xenopus* does, even though *Xenopus* is a model for mesoderm formation. The mouse is a good starting point for studying other mammals, but mammalian development has diverged enormously, and certainly beyond what one murine species expresses (see Benirschke, 2006). David and Marilyn Kirk (2004) have spent their professional lives sorting out the ways that *Volvox carteri* distinguishes germ line from soma and have come to the conclusion that "*Volvox carteri* is an excellent model for other *Volvox carteris*," because most other *Volvox* species do this important act in different ways.

The problems of suitable model systems for EvoDevo are reflected in a workshop document published by the National Science Foundation of the United States (NSF, 2005). This booklet specifically relates to the influence of model systems in directing the course of developmental research.

> Developmental biologists have, for many years, focused their efforts to understand ontogeny by selecting a few model organisms that are genetically tractable, and that are appropriate to the study of fundamental processes of development at the genetic, molecular, and cellular levels. These efforts have led to a detailed understanding of the genetic mechanisms that are involved in the control of developmental events. Many of the findings that have emerged from this work have proven remarkably transferable among the models studied. Developmental biologists have relied on model systems with relatively little but controllable genetic variation. Consequently they have typically not studied the way developmental mechanisms differ among species, nor the variance in mechanism among individuals due to normal variation in genetic and environmental factors. Some developmental biologists have recently begun to expand their studies to include non-model species for understanding aspects of developmental processes not reflected in the models. Still others are interested in illuminating the breadth or limitations of the generalizations discovered in the model systems. Recent developments in genomic approaches have facilitated this move away from the few genetically tractable model systems.

The authors go on to contrast this with the physiological approach.

> Animal physiologists, by contrast, have been reluctant to adopt the use of a relatively small number of model species. This is in part because the physiological principles that bind the subscience cohesively, such as regulation and control of the functions required for normal operation, are known to differ between species. Thus, animal physiologists have employed a

broad array of study systems, each selected for its suitability to address a specific physiological mechanism. Interestingly, some investigators have recently advocated the adoption of model systems that are genetically tractable as a means to approach questions about the genetics and evolution of physiological mechanisms, and as a means to leverage financial support of genomic approaches, which remain costly.

The NSF then calls for an integrative developmental biology that would synthesize the methodologies, analytical tools, and conceptual approaches of these two disciplines. EvoDevo, if it is to avoid the conceptual pitfalls mentioned above, also needs to adopt a similar perspective and, indeed, has already taken the first steps in this direction. Various potential new model systems are being explored, and traditional model systems are being retooled to fit EvoDevo-type questions. In addition, a number of systematic accounts of EvoDevo call for a closer integration of mathematical models, which have traditionally been the domain of evolutionary biology, with the mechanistic perspective of developmental biology. In particular, the multiple layers of epigenetic control of development that have recently been uncovered have the potential to dramatically transform the traditional abstractions of population and quantitative genetics. They provide us with a much richer understanding of the molecular mechanisms of development, one that also transcends the traditional gene-environment dichotomy. Below we will first systematically discuss seven clusters of EvoDevo concepts, and their associated questions and research approaches, and then explore how these are reflected in new model organisms and model systems for EvoDevo.

Models in EvoDevo

Approaches

Despite recent efforts to consolidate the field, EvoDevo is still a pluralistic discipline, as is illustrated by the different approaches taken in its research programs. In part these reflect the different disciplinary origins of its practitioners (in either developmental genetics or evolutionary biology, for instance), but they are also a consequence of the number of different questions that fall within the purview of EvoDevo. Following Müller (2007), we distinguish seven types of questions, each characterized by its own set of organizing concepts, models, and explanatory strategies.

Origin of Developmental Systems The first premise of EvoDevo is that phenotypic evolution is a consequence of changes in the developmental systems of organisms. These developmental systems are, of course, themselves a product of evolution, and thus subject to evolutionary dynamics. As with phenotypic transformations more generally, we can distinguish between gradual modifications of developmental systems and the more complicated problem of their origin. The latter is tied to such questions as life cycle evolution (Bonner, 1974), the role of cell lineage competition in structuring early developmental processes and the evolution of individuality (Buss, 1987), and the origin of generic forms as a consequence of an interaction of basic physical laws with self-reproducing biological materials such as cells (Newman, 1992, 1994).

These questions point to complex interactions between physical processes and constraints, and the developmental systems that incorporate and then stabilize these processes. Conceptually, these questions are connected to the difficult problems of emergence and major transitions. Due to the likely rarity of these events, we depend on a few select model systems for theoretical and empirical studies. We also rely on heuristic models that allow us to explore possible scenarios in the origination of developmental systems (Müller and Newman, 2003; Newman et al., 2006). Explanations of the origin of developmental systems thus depend on a combination of material, heuristic, and theoretical models.

Evolution of the Developmental Repertoire Once developmental systems are established, their subsequent modification provides the foundation for further phenotypic evolution. The basic architecture and transformations of these developmental systems are thus the subjects of intense study within EvoDevo. Comparative studies reveal that developmental systems have a highly conserved architecture that is based on a small number of elements and their combinatorial transformations. The most important features of developmental systems are their modular organization (Schlosser and Wagner, 2004; Callebaut and Rasskin-Gutman, 2005), the hierarchy of regulatory pathways and networks (Wilkins, 2002; Davidson 2006), the conservation of regulatory genes and the evolution of *cis-regulatory* elements (Carroll et al., 2005; Davidson, 2006), the duplication and further deployment of regulatory genes (Holland, 1999), and the co-option of existing modules into new tasks (True and Carroll, 2002).

Taken together, these elements of the evolution of developmental systems establish a conceptual model for the genetic basis of phenotypic evolution. Within this broad vision several heuristic concepts and additional models have emerged, such as the idea of a "genetic toolkit for development," the reconstruction of the basic features of the ancestor of all bilaterally symmetrical organisms, and the "Urbilateria," or the idea of an hourglass model of development passing through a conserved "phylotypic stage." These heuristic conceptual models guide empirical research in important ways, and they also provide a starting point for theoretical models and formal treatments, such as the analysis of network properties of regulatory gene networks and general features in the evolution of signaling pathways.

Evolutionary Modification of Developmental Processes Next to evolutionary transformations of the developmental repertoire and their implications for phenotypic evolution, certain features of developmental processes also allow modifications of phenotypes. Morphologists and developmental biologists have long noticed how changes in timing of certain events during development can result in often dramatically altered phenotypes. They described these phenomena as heterochrony (Haeckel, 1866; de Beer, 1930; Gould 1977; McKinney and McNamara, 1991). Building on these classical observations and interpretations, developmental biologists have more recently analyzed the underlying developmental mechanisms and described additional elements of morphogenesis, such as morphoregulation (Edelman, 1986, 1988), ontogenetic repatterning (Wake and Roth, 1989), and dissociability (Needham, 1933; Raff, 1996). All these studies have led to a revival of morphogenesis as a topic of evolutionary research. Studies in morphogenesis also have traditionally been at the vanguard of theoretical modeling, and the recent focus on EvoDevo has initiated renewed interest in this area.

Environment-Development Interaction Phenotypic plasticity is one of the more obvious examples of how environment and genes interact to yield developmental programs with patterns of variation that can be continuous or discontinuous. The environment is often thought of as only or mainly abiotic factors, as in the way temperature can affect wing color and pattern in butterflies in a case we will describe shortly. But environment also includes other organisms, as illustrated in the way that the density and size distribution of conspecifics affects expression of cannibalism in salamanders (Collins et al., 1993). Gene-environment interactions may cause individuals to vary in physiology, morphology, or behavior, which can also yield intraspecific variation in birth and death rates that affect demography. Density, size distribution, and gene frequencies can then feed back on the development of individuals. Finally, other species acting as competitors and predators will help shape development across a range of reaction norms.

As we come to understand the complexity of the genome, it will be possible to model these systems using a vision that goes beyond a one gene-one trait perspective. Modern molecular techniques, especially the rapidity at which sequencing is now possible, are opening opportunities to study diverse model systems chosen especially for their usefulness in answering leading ecological and evolutionary questions. An especially interesting development in the last decades of the twentieth century was the increasing appreciation by evolutionary ecologists that modeling population dynamics required an incorporation of evolutionary principles. Among the factors that drove this conceptual transition was an appreciation of the convergence of ecological and evolutionary times (Collins, 1986). More recently, Hairston et al. (2005) concluded that "to understand the temporal dynamics in ecological processes it is crucial to consider the extent to which the attributes of the system under investigation are simultaneously changing as a result of rapid evolution."

All of this means that diverse model organisms that vary intraspecifically in fundamental traits, such as the capacity to metamorphose or not in salamanders, and the presence or absence of wings in insects, can be analyzed with the aim of understanding the gene-environment interactions surrounding traits indicative of key evolutionary transitions. Understanding these transitions is a central goal of EvoDevo.

Phenotypic Variation Heritable phenotypic variation is the basis for evolution. Analyzing patterns of phenotypic variation has, therefore, been a prime concern of evolutionary biologists ever since Darwin (1859) and Bateson (1894). More recently the role of developmental processes in both enabling and constraining phenotypic variation has become a major part of EvoDevo research. The main problems in this context are:

1. The observation that not all thinkable phenotypic variants are indeed possible—a fact captured by the concept of constraints, such as developmental constraints (Alberch, 1982; Maynard Smith et al., 1985) or physical, functional, and architectural constraints (see chapters 5 and 14 in this volume)
2. The idea that development can act as a buffer that filters out both genetic and environmental variation and perturbations (Katz et al., 1981)
3. The suggestion that the developmental system can bias the expression of underlying (molecular) genetic variation in such a way that the resulting phenotypic variation might appear directed—a concept termed developmental drive (Arthur, 2001)
4. The realization that the specific structures of the genetic and developmental systems are crucially important for the future capacity of species to evolve (Wagner and Altenberg, 1996; Kirschner and Gerhart, 1998).

Many of these heuristic concepts gave rise to analytical models that in different ways have become the theoretical foundation of EvoDevo.

Phenotypic Innovation Explaining phenotypic innovation is a central goal for EvoDevo (Müller and Wagner, 1991; Love, 2003). The question of how new structures and behaviors arise during evolution has been the main challenge for evolutionary biologists since Darwin. From the very beginning, development has played an important part in these explanations. Within his generalized law of recapitulation and its mechanism of terminal addition (new structures are added on at the end of developmental sequences), Haeckel already allowed for exceptions. These "caenogenetic" features represented adaptations of the developing embryo to internal and external conditions, and were therefore a consequence of development. More recently, EvoDevo researchers have focused on developmental side effects (Müller, 1990), epigenetic causation (Newman and Müller, 2000), altered *cis-regulatory* interactions

(Carroll et al., 2005; Davidson, 2006), developmental exaptation (Chipman, 2001), and environmental induction (West-Eberhard, 2003) in their efforts to develop a mechanistic model for the origin of novel phenotypes during development and evolution. So far the focus on innovation has generated heuristic and functional models (see, e.g., Müller and Newman, 2006). A main challenge for EvoDevo is therefore to embed these models within a formal and analytic account of morphological evolution.

Genetic and Epigenetic Fixation EvoDevo is based on the assumption that development is a central feature of all explanations of morphological evolution. Over the last few decades a number of heuristic and analytical models have emerged that have helped to make this general statement more concrete by showing exactly how development is reflected in specific features of the genetic and epigenetic systems. These concepts all focus on the systemic effects of the developmental, genetic, and, increasingly, environmental contexts on the expression of morphological and behavioral traits. They include Waddington and Whyte's idea of internal selection pressures in analogy to external selection (Waddington, 1953; Whyte, 1965); several models of canalization (Waddington, 1942; Wagner et al., 1997; Wilkins, 2003) that are also connected to the concept of developmental constraints, Riedl's (1978) concept of burden as a measurement for the limitations on future variability imposed by a highly integrated developmental system; Wimsatt's (1986) related formulation of generative entrenchment; Waddington's (1956) concept of genetic assimilation as an account of how selection can eventually lead to the cooption of favorable environmentally induced variation; several ideas related to the emergence of the hierarchical organization of developmental systems (Riedl, 1978; Buss, 1987; Salazar-Ciudad et al., 2001). New findings related to the molecular details of epigenetics (Stillman and Stewart, 2005) already provide the kind of experimental data that will help turn these concepts into functional and analytical models of the genetic and epigenetic bases of morphological evolution.

Model Systems and Model Organisms

An important part of modeling EvoDevo is the selection of adequate model organisms. In selecting model organisms, two sets of demands must be met—the pragmatic considerations of housing, breeding, and easy manipulation, and theoretical considerations related to whether or not a model organism is representative of the phenomenon in question. These two demands sometimes conflict. While easy manipulability facilitates experimental work and the standardization of results that ensures the quality and comparability of the experimental data, the issue of to what degree model organisms exemplify a scientific problem is tied to the question of whether it is possible to develop a more general model based on work done with one or a few selected organisms. This problem is even more acute in the context of EvoDevo (cf. Metscher and Ahlberg, 1999).

As mentioned above, the seven basic model systems of developmental biology were selected because of their easy manipulability and, except originally for the chick, also because they are well suited for research within the genetic paradigm in developmental biology. However, not all of them are particularly useful for Evo-Devo questions, since most of these organisms are highly derived and specialized, and thus not suited for modeling major phenotypic transformations or any of the other questions that are being asked in EvoDevo. It is by now almost universally accepted that addressing these problems requires new model systems. This is, in a sense, an interesting phenomenon since EvoDevo was originally characterized by its revolt against model systems (Bolker, 1995; Bolker and Raff, 1996). One of the earliest and by now well-known EvoDevo model systems was developed by Rudy Raff and coworkers, who used sea urchins of the genus *Erythrogramma* to study the differences between larval and direct development in closely related species (e.g., Raff and Wray, 1989). Now, as EvoDevo becomes stabilized around additional, specific questions, new model systems are emerging.

Each of the seven clusters of EvoDevo concepts and their associated theoretical, experimental, and empirical research strategies attracts new investigators, who introduce new model systems and organisms into EvoDevo, as well as the "repositioning" of traditional model systems such as zebra fish and *Drosophila*. These become increasingly employed for the discovery of specific differences within lineages by studying altered gene expression through evolution. The zebra fish, for instance, is seen as a useful source of genes through which the evolution of specific piscine lineages might be studied (Webb and Schilling, 2006). The dog *Canis familiaris* and the three-spined stickleback fish are also considered model systems for studying altered gene expression during evolution. The latter two are new model systems, specifically developed in the context of EvoDevo to identify genes in which small changes can make large phenotypic differences.

Other new and nontraditional organisms are the cnidarian *Nematostella*, as a model system for looking at the origins of the bilateria, and the dung beetle *Onthophagus*, which is proposed as a model system for studying the evolution and the properties of developmental plasticity. More recently there has also been a push to use social insects as a new model system for EvoDevo, especially for studying the origins of morphological, physiological, and behavioral novelties. In this regard it is important to note, as Gilbert did at a recent conference in Paris, that "model systems" in EvoDevo are not merely "model organisms." Rather, these systems include the organism plus their historical or ecological context. Here we will briefly introduce a few of these new model organisms and model systems, and discuss their significance for EvoDevo.

The Dog Model System The title of the Neff and Rine (2006) essay in *Cell* says it best: "A Fetching Model System." As recognized by Darwin (1859), artificial selection is a mode of evolution wherein harsh selective pressures imposed by human selection and mating regimes can rapidly change the appearance of an organism. There are now over 200 recognized breeds (and about 1000 local breeds) of dogs, each derived from *Canis lupus*, the wolf, starting about 135,000 years ago (Vila et al., 1997).

Dog breeding, write Neff and Rine, "has been an ongoing experiment in the rapid evolution of form and function." Moreover, the completion of the canine genome has made *Canis familiaris* "genetically tractable and poised to offer insights into evolution, development, and behavior." These authors point out that while null mutations, such as those readily produced in the mouse, can tell you about how a system breaks down, such mutations are not usually relevant for understanding natural diversity or evolution. In dogs, however, you have remarkable diversity of functioning systems. Dogs can differ fiftyfold in mass and have behaviors ranging from completely docile to overtly vicious. Moreover, these differences are heritable. There are not only breed-specific temperaments, but even dog breeds that perform the same behaviors (such as herding) differently from one another. The variation that dogs have is very different from the variation produced in the laboratory using caged mice. Studying this normal and enormous variation is critical if one wishes to study evolution or, for that matter, brain function.

The central argument for the dog as model system for EvoDevo is that it has enormous variation; that these variants are functional, not pathological; and that this variation occurs within the same species. Stockard (1941), and more recently others (Gilbert, 1991), have pointed out that differences in dog snouts represent remarkable changes in the migration and proliferation rates of cranial neural crest cells. Now that the canine genome is complete (Lindblad-Toh et al., 2005), it is hoped that comparisons can be made. There are over 50 million pedigreed dogs in the United States alone, and there should be plenty of molecular variations to map. The goal is to elucidate the genetic variations that underlie the different morphologies, embryologies, and behaviors that define each breed.

Such studies have already started. Fondon and Garner (2004) and Caburet et al. (2005) have shown that length variation of tandem repeats in protein-coding regions of developmental genes are associated with morphological changes in dog breeds. For instance, the Great Pyrenees breed is characterized by bilateral polydactyly of the first digit, which correlates with a deletion of seventeen repeats of a Pro/Gly sequence in the *Alx-4* gene. The deletion characterizes the breed, and a single Great Pyrenees dog without this polydactylous condition was homozygous for the full length of the *Alx-4* allele that characterizes all nonpolydactylous breeds. Similarly, repeat variation in the *Runx-2* gene is correlated with craniofacial depth. The gene is homologous to human *TCOF*, mutations of which cause the Treacher-Collins syndrome of facial shape anomalies, and shows variants highly associated with head depth in dogs (Haworth et al., 2001).

The *Nematostella* Model System *Nematostella vectensis*, the starlet sea anemone, is a cnidarian that represents a basal phylum. Moreover, its proponents argue that it represents two of the most fundamental transitions in animal evolution: the origin of bilateral symmetry and the origin of the mesoderm. A *Nematostella* Web site (http://www.nematostella.org) claims: "The starlet sea anemone, *Nematostella vectensis*, is becoming an increasingly important model system for the study of development, evolution, genomics, reproductive biology, and ecology." When Martindale and colleagues (2004) introduced *Nematostella* as a model system for the study of triploblasty, they proposed it for the reasons traditionally used to justify such a designation: simplicity, ability to be cultured, large number of embryos, availability of embryos all year, and rapid development of the embryo:

> *Nematostella* has many practical advantages as a developmental model, including a simple body plan and a simple life history. It is a hardy species, easy to culture (Hand and Uhlinger, 1992) and will spawn readily throughout the year under laboratory conditions (Fritzenwanker and Technau, 2002; Hand and Uhlinger, 1992). Sexes are separate and fertilized embryos develop rapidly to juvenile adults bearing four tentacles. (Martindale et al., 2004:2464)

To demonstrate the usefulness of *Nematostella* as a model organism for looking at the origins of triploblasty and bilateral symmetry, Martindale and his colleagues showed that they have the typical bilateral body plan (common to vertebrates and insects), but in a rudimentary form. Thus, the genes for dorsal-ventral polarity (*BMP* and *chordin*) are found, but they appear to be playing slightly different roles than in the more highly specialized bilaterians (Matus et al., 2006); the genes used in insects and vertebrates for germ-cell specification are found there, too, but seem to be playing more roles than expected (Extavour et al., 2005).

The finding in *Nematostella* of many of the transcription factor families known to be critical in the development of contemporary insects and vertebrates gives further reasons to look at this organism as an example of an organism that is ancestral to all the major lineages of the animal domain (Magie et al., 2005). In 2004 Finnerty and colleagues showed that *Nematostella* uses homologous genes to achieve bilateral symmetry by means of staggered *Hox* gene expression along the primary body axis. They suggested, therefore, that bilateral symmetry arose before the evolutionary split of Cnidaria and Bilateria. Thus, bilateral symmetry can first be seen in the Cnidarians; moreover, so can mesoderm. Not only are there muscle cells among the cnidarians, but these cells are expressing the "mesodermal genes" that characterize mesodermal specification in insects and vertebrates (Martindale et al., 2004). Basal metazoa, such as *Nematostella*, will thus be invaluable model systems for understanding earliest events in the evolution of higher animals (Martindale, 2005).

Martindale continues that the field needs these models to place renewed emphasis on the functional interactions of complex gene regulatory pathways in a phylogenetic context so that scientists can "unravel the legacy of morphological complexity that is seen in the animals of today." This is echoed by another laboratory that emphasizes the importance of *Nematostella* over other basal organism models:

> In recent years, a handful of model systems from the basal metazoan phylum Cnidaria have emerged to challenge long-held views on the evolution of animal complexity. The most-recent, and in many ways most-promising addition to this group is the starlet sea anemone, *Nematostella vectensis*. The remarkable amenability of this species to laboratory manipulation has already made it a productive system for exploring cnidarian development, and a proliferation of molecular and genomic tools, including the currently ongoing *Nematostella* genome project, further enhances the promise of this species. In addition, the facility with which *Nematostella* populations can be investigated within their natural ecological context suggests that this model may be profitably expanded to address important questions in molecular and evolutionary ecology. (Darling et al., 2005:211)

The Dung Beetle Model System A major change in developmental biology since the mid-1990s is the recognition that the environment plays an instructive role in producing phenotypes. Polyphenisms, norms of reaction, and developmental symbioses, long a part of ecology, are now increasingly seen as being part of developmental biology. What had been a province of exceptions is becoming the rule, as mammalian gut development has been found to be symbiotically regulated, and as evolutionarily cued epigenetic methylation is seen to alter DNA in numerous animals, including mammals (see Gilbert, 2004).

The question then becomes how best to study ecological developmental biology, or EcoDevo (Gilbert, 2001; Hall et al., 2004), and the multiple effects of environmental factors on regular development. Again, the selection of an appropriate model system proves crucial. First, one needs an animal with a readily identifiable suite of traits that change consistently with the environment. Two claimants for such an EcoDevo model system have recently come forward. The first, from Paul Brakefield's laboratory, is the Malawian butterfly *Bicyclus anynana* (see Beldade et al., 2002, 2005). In this butterfly, temperature helps determine the phenotype. During the cool, dry season the butterfly walks among the leaf litter and its cryptic brown coloration protects it. During hot, moist months, the butterfly flies, and its eyespots protect it from insect predators (Lyytinen et al., 2004). *Bicyclus* thus provides an excellent system for looking at phenotypic plasticity.

Moreover, by combining forces with Sean Carroll's developmental genetics laboratory, Brakefield's group has begun to uncover the molecular mechanism for this plasticity (Brakefield et al., 1996). A temperature rise causes an increase in the ecdysone hormone during a particular stage of larval development, and this hormone sustains expression of the *Distalless* gene in the presumptive eyespots of the imaginal wing disk. The *Distalless* protein activates a series of transcription factors that initiate color development throughout the wing spot in a concentric manner. The ability to transform the butterfly by molecular means, study its physiology, monitor its development, and analyze its ecology and evolutionary biology makes this a particularly exciting species to follow. Beldade and colleagues (2005) remarked, "This system provides the potential for a fully integrated study of the evolutionary and developmental processes underlying diversity in morphology," although it might be more cautious to restrict this claim to essentially two-dimensional color patterns.

Another eco-devo model is the dung beetle *Onthophagus* and its fascinating structural and behavioral polyphenisms. Male dung beetles can be separated into two distinct classes. The large males have head horns, while small males have rudimentary horns or are hornless. Horn length varies allometrically with body length, resulting in a bimodal distribution of horn sizes. Up to a particular body size, the males are essentially hornless. Then, after they reach this threshold, the horn grows much faster than the body. Body size is determined primarily by the amount and quality of the dung provided to the larva by its mother. When a larva runs out of food, it metamorphoses into an adult (Emlen, 1994, 1996). The regulation of horn size by food is achieved through the prepupal endocrine system, wherein ecdysone and juvenile hormone cooperate to stimulate horn growth (Emlen and Nijhout, 2001).

The hornless and horned males have very different sexual strategies (Emlen, 1997; Nijhout, 2003). The horned males defend tunnels that are dug by the females and use their horns to fight other males who want access to these females. The male with the longer horns wins. The hornless males would seem to be at a reproductive disadvantage; but not only are the horns polyphenic, so is a behavior. Instead of fighting, the hornless males either try to sneak past a defending male or dig their own tunnels into the tunnels of the females. This polyphenism results in divergent selection: large males benefit from large horns (they helps them win combats), while small males benefit from the smallest possible horns (because horns get in the way of digging and sneaking).

Thus, the polyphenism in dung beetles involves the coordination of both morphological structures and behavioral strategies by the endocrine system. But for a male dung beetle, body and attitude are due largely to the amount of dung left by the mother. Emlen (2000) sees a reciprocal relationship wherein the study of beetle development contributes to EvoDevo and EvoDevo contributes to the study of beetle development.

In principle, understanding how development affects the expression of morphological traits should explain the evolution of those traits. However, empirical studies demonstrating an immediate relevance of development for understanding evolution in natural populations are rare because most population biologists do not study the developmental mechanisms regulating the expression of their traits of interest. One trait in which to examine this question is in the horns of beetles. The behavior associated with horns, the evolution of horns, and the development of horns have been explored for the same two species; consequently, it is now possible to integrate the results from these studies and to explore how knowledge regarding the mechanism of horn development influences our understanding of beetle horn evolution. (Emlen, 2000:403)

Thus, the dung beetle might be a model system for looking at the evolution of developmental plasticity (of both form and behavior), the consequences of developmental plasticity, and the hormonal mediation by which such plasticity is regulated via environmental factors.

The Vertebrate Limb Model The amenability of embryonic limbs to experimental manipulation and the extensive fossil record of limb skeletal patterns have for a long time inspired EvoDevo perspectives on the vertebrate limb (Hinchliffe and Johnson, 1980). Mostly these models are concerned with the pattern of skeletal elements that arises in a proximodistal sequence from localized accumulations of prechondrogenic cells in embryonic limb buds. Early EvoDevo models (that would account not merely for embryonic patterning but also for its taxon-specific evolutionary modification) were based on physical processes of cell association and the macroscopic properties of growing skeletogenic tissue masses (Oster et al., 1985; Shubin and Alberch, 1986). In these types of models the evolution of skeletal patterns would occur through modulations of spatial or temporal aspects of macroscopic events of skeletogenic tissue organization (Shubin and Alberch, 1986; Müller, 1991; Hinchliffe, 2002).

The growing understanding of the molecular basis of limb development has led to the identification of some of the key genes that are associated with skeletal patterning, and consequently the models of limb evolution have shifted to relating gene expression patterns in the development of extant limbs to the evolution of the skeletal patterns (e.g., Izpisua-Belmonte et al., 1991; Tabin, 1992; Sordino and Duboule, 1996). Genetic pattern-based models of limb evolution rely explicitly or tacitly on a hierarchical developmental program notion in which gene activity affects the "positional information" provided to individual cells during the cell condensation process (Wolpert, 1969, 1989) via candidate molecules that generate a putative coordinate system along the limb's three Cartesian axes.

This spirit of positional identity also underlies recent limb models that propose a stepwise subdivision of broad initial expression domains to produce the proximodistal array of skeletal elements (Richardson et al., 2004). The limb pattern and its evolutionary modification would thus be a direct consequence of feed-foreward gene-gene interactions that specify skeletogenic patterns without relevant contributions of other developmental parameters. As a consequence, such "informational models" require a high level of regulatory intricacy to generate patterns of any complexity. Position-specific promoters (Stanojevic et al., 1991) or "smart genes" (Davidson, 1990) are invoked to explain such intricacy in other systems, and major innovations are thought to arise from shifts of gene expression domains in the early limb bud mesenchyme (Sordino et al., 1995; Wagner and Chiu, 2001).

A different class of models is based on the self-organizational properties of cell and tissue masses in a confined developmental space. Such "generic models" of limb development give priority to the capacity of precartilage mesenchymal tissues to autonomously generate regularly spaced skeletogenic accretion centers (Newman and Frisch, 1979; Newman, 1996). These kinds of patterning processes are based on a core set of self-organizing cellular and molecular processes. Gene regulatory evolution is here thought to capture, stabilize, and refine the tissue interactions that produce generic initial forms (Newman and Müller, 2005). *In silico* modeling and simulation of a minimal set of self-organizing interactions within limb budlike geometries are shown to generate patterns that correspond to the natural limb patterns (Hentschel et al., 2004). These models are generative in the sense that they include a mechanistic account for the origin of first patterns and for later skeletal innovation. Daeschler and collegues (2006) describe well-preserved pectoral appendages in recent finds of Late Devonian sarcopterygian fish that exhibit a transitional morphological and functional stage between fins and limbs, suggesting a generative rather than an informational mode of skeletogenic pattern evolution.

The predictive and heuristic roles of informational models and of generic models in limb EvoDevo differ to a great extent. Whereas the former suggest that we need to provide an ever more detailed account of all gene regulatory interactions in limb development, and hence need to continue with the genetic dissection of the limb system, the latter suggests a program that explores the rules of cell and tissue organization in limb development. These rules could explain the generation of similar patterns from developmental processes that have quite different genetic and molecular underpinnings.

Furthermore, in the informational models nearly all changes of pattern are equally possible and the continued evolutionary identity of individual skeletal elements (homology) disappears (Müller, 2003). In the generic models not all changes are equally possible, and the identity of elements is maintained. Here the history of morphogenetic structure itself is a determinant of possible evolutionary change.

The Mammalian Tooth Model In recent years, the development and the evolution of the mammalian tooth has become a major focus of convergence for paleontology and development. Since enamel is far more durable than ordinary bone, teeth often remain after all the bones have decayed. Indeed, tooth morphology has been critical to mammalian ecology and classification. At the same time, the teeth represent a circumscribed developmental module that can be studied in its molecular and morphogenetic aspects without much interference from/with other parts of the embryo. Changes in the cusp pattern of molars is seen to be especially important in allowing the radiation of mammals into new ecological niches. The question, then, is what developmental mechanism allows the mammalian molars to change their form so rapidly.

Jukka Jernvall and colleagues (Jernvall et al., 2000; Salazar-Ciudad and Jernvall, 2002) pioneered a computer-based approach to phenotype production using geographic information systems (GIS) to map gene expression patterns in incipient tooth buds. These studies have shown that specific gene expression patterns forecast the exact locations of the tooth cusps and that the differences between the molars of mice and voles are predicated on differences in gene expression.

The tooth system is particularly suited for modeling EvoDevo processes. Salazar-Ciudad and Jernvall (2004) correlated the morphogenetic kinetics with known paracrine factor properties and distributions. Small changes in a gene network, working through the interactions of the *BMP* and *Shh* proteins, are seen as crucial. *Shh* and *FGF*s (produced by the enamel knot signaling center of the developing tooth bud) inhibit *BMP* production, while *BMP* production stimulates both the production of more *BMP*s and the synthesis of its inhibitors, the *FGF*s and *Shh* proteins. In the model this generates regions of activators (*BMP*s) that block epithelial proliferation, and regions of inhibitors (*FGF*s, *Shh*) that block *BMP* synthesis and independently stimulate mesenchymal proliferation. The result is a pattern of gene activity that changes as the shape of the tooth changes, and vice versa.

Several kinds of EvoDevo predictions can be derived from such a combined molecular and morphogenetic model. Since the kinetic parameters change while the tooth is forming, this gives a great amount of flexibility to the developing system, allowing it to change at relatively rapid rates during evolution. The diverging shapes of mouse and vole molars may have resulted from very small changes in the initial molecular and topological conditions. And the model also predicts that some types of teeth are more likely to evolve in certain ways and that certain shapes are more likely to arise than others. This morphogenetic potentiality conforms significantly to the patterns observed in mammalian tooth evolution (Jernvall et al., 2000; Salazar-Ciudad and Jernvall, 2002).

Perspectives for EvoDevo Modeling

In discussing modeling strategies within EvoDevo, we painted a broad picture and touched upon several different dimensions of modeling discussed in this volume. Here, as a conclusion, we will sketch some of the trends within current EvoDevo that best illustrate the importance of modeling for the theoretical synthesis of the discipline.

A productive interaction between experimental research and heuristic models characterizes current EvoDevo research. The results of empirical and comparative studies, such as correlations of gene expression patterns with morphological transformations, give rise to heuristic models of the genetic control of development and of regulatory evolution. They also produce more theoretical models that emphasize general aspects of the evolution of developmental systems, such as the roles of gene duplication and regulatory evolution. So far, these models are mostly diagrammatic and functional; very few analytical and predictive models exist within EvoDevo. In part this is because of the disciplinary bias of current EvoDevo research, which is dominated to some degree by developmental genetics and is less guided by evolutionary theory. Insofar as evolutionary models do exist—those that deal with the problem of evolvability, the role of epistatic and epigenetic effects, canalization, or generative entrenchment—they provide a more analytical and predictive framework for what its practitioners sometimes call "developmental evolution" (Wagner, 2000)

In general, EvoDevo has the same problems as any other field of current biology—it has an overabundance of data and very few genuine theoretical principles. One function of models in EvoDevo is thus to organize and visualize large amounts of data, such as those concerning gene expression patterns in developing systems. This involves a huge amount of computational modeling and data mining, aided by hypotheses and concepts about relevant connections and links between different data sets. This theoretical biology aspect of EvoDevo will become an increasingly important tool in the future development of the discipline (Müller, 2005; Laubichler et al., 2005).

We started our analysis on a cautionary note: Unless EvoDevo develops models that integrate empirical and theoretical studies while capturing the essential features of an EvoDevo explanation, it will not live up to its promise as a new synthesis of development and evolution. Such genuine models would combine theoretical, material, and heuristic dimensions of modeling biology (cf. chapter 1 in this volume). We have argued that some developments in EvoDevo represent steps in this direction. There is now an active push for new EvoDevo model organisms, specifically chosen for the study of genuine EvoDevo questions, such as the role of genotype–environment interactions and plasticity, or the role of particular genes in morphological and behavioral differences. This is an important start. In a subsequent step the information extracted from these model organisms will need to be embedded within a framework of analytical and theoretical models that connect the specific empirical details with general processes of development and evolution. The future of EvoDevo, like that of any other discipline, will depend on the successful integration of its material and theoretical models.

References

Alberch P (1982) Developmental constraints in evolutionary processes. In: Evolution and Development (Bonner JT, ed), 313–332. Berlin: Springer-Verlag.

Arthur W (2001) Developmental drive: An important determinant of the direction of phenotypic evolution. Evol Dev 3: 271–278.

Bateson W (1894) Materials for the Study of Variation. London: Macmillan.

Beldade P, Brakefield PM, Long AD (2005) Generating phenotypic variation: Prospects from "evo-devo" research on *Bicyclus anynana* wing patterns. Evol Dev 7: 101–7.

Beldade P, Koops K, Brakefield PM (2002) Modularity, individuality, and evo-devo in butterfly wings. Proc Natl Acad Sci USA 99: 14262–14267.

Benirschke K (2006) Comparative Placentation. http://www.medicine.ucsd.edu/cpa.

Bolker J (1995) Model systems in developmental biology. BioEssays 17: 451–455.

Bolker J, Raff RA (1996) Developmental genetics and traditional homology. Bioessays 18: 489–494.

Bonner JT (1974) On Development: The Biology of Form. Cambridge, Mass.: Harvard University Press.

Brakefield PM, Gates J, Keys D, Kesbeke F, Wijngaarden PJ, Monteiro A, French V, Carroll SB (1996) Development, plasticity and evolution of butterfly eyespot patterns. Nature 384: 236–242.

Buss LW (1987) The Evolution of Individuality. Princeton, N.J.: Princeton University Press.

Caburet S, Cocquet J, Vaiman D, Veita RA (2005) Coding repeats and evolutionary "agility." BioEssays 27: 581–587.

Callebaut W, Rasskin-Gutman D (eds) (2005) Modularity: Understanding the Development and Evolution of Complex Natural Systems. Cambridge, Mass.: MIT Press.

Carroll SB, Grenier JK, Weatherbee SD (2005) From DNA to Diversity (2nd ed). Malden, Mass.: Blackwell Science.

Chipman AD (2001) Developmental exaptation and evolutionary change. Evol Dev 3: 299–301.

Cohn MJ, Lovejoy CO, Wolpert L, Coates MI (2002) Branching, segmentation and the metapterygial axis: Pattern versus process in the vertebrate limb. BioEssays 24: 460–465.

Colburn T, Dumanoski D, Myers JP (1996) Our Stolen Future. New York: Dutton.

Collins JP (1986) Evolutionary ecology and the use of natural selection in ecological theory. J Hist Biol 19: 257–288.

Collins JP, Zerba KE, Sredl MJ (1993) Shaping intraspecific variation: Development, ecology and the evolution of morphology and life history variation in tiger salamanders. Genetica 89: 167–183.

Daeschler EB, Shubin NH, Jenkins FA (2006) A Devonian tetrapod-like fish and the evolution of the tetrapod body plan. Nature 440: 757–763.

Darling JA, Reitzel AR, Burton PM, Mazza ME, Ryan JF, Sullivan JC, Finnerty JR (2005) Rising starlet: The starlet sea anemone, *Nematostella vectensis*. BioEssays 27(2): 211–221.

Darwin C (1859) On the Origin of Species *by Means of Natural Selection*. London: John Murray.

Davidson EH (1990) How embryos work: A comparative view of diverse modes of cell fate specification. Development 108: 365–389.

Davidson EH (2006) The Regulatory Genome: Gene Regulatory Networks in Development and Evolution. San Diego: Academic Press.

De Beer GR (1930) Embryology and Evolution. Oxford: Clarendon Press.

Edelman GM (1986) Evolution and morphogenesis: The regulator hypothesis. In: Genetics, Development, and Evolution (Gustafson JP, Stebbins GL, Ayala FJ, eds), 1–27. New York: Plenum Press.

Edelman GM (1988) Topobiology: An Introduction to Molecular Embryology. New York: Basic Books.

Emlen DJ (1994) Environmental control of horn length dimorphism in the beetle *Onthophagus acuminatus* (Coleoptera: Scarabaeidae). Proc Roy Soc Lond B256: 131–136.

Emlen DJ (1996) Artificial selection on horn length-body size allometry in the horned beetle *Onthophagus acuminatus* (Coleoptera: Scarabaeidae). Evolution 50: 1219–1229.

Emlen DJ (1997) Alternative reproductive tactics and male-dimorphism in the horned beetle *Onthophagus acuminatus* (Coleoptera: Scarabaeidae). Behav Ecol Sociobiol 41: 335–341.

Emlen DJ (2000) Integrating development with evolution: A case study with beetle horns. BioScience 50: 403–418.

Emlen DJ, Nijhout HF (2001) Hormonal control of male horn length dimorphism in the dung beetle *Onthophagus taurus* (Coleoptera: Scarabaeidae). J Insect Physiol 45: 45–53.

Extavour CG, Pang K, Matus DQ, Martindale MQ (2005) *Vasa* and *nanos* expression patterns in a sea anemone and the evolution of bilaterian germ cell specification mechanisms. Evol Dev 7: 201–215.

Finnerty JR, Pang K, Burton P, Paulson D, Martindale MQ (2004) Origins of bilateral symmetry: *Hox* and *dpp* expression in a sea anemone. Science 304: 1335–1337.

Fondon JW III, Garner HR (2004) Molecular origins of rapid and continuous morphological evolution. Proc Natl Acad Sci USA 101: 18058–18063.

Fritzenwanker J, Technau U (2002) Induction of gametogenesis in the basal cnidarian *Nematostella vectensis* (Anthozoa). Dev Genes Evol 212: 99–103.

Gilbert SF (1991) Developmental Biology (3rd ed). Sunderland, Mass.: Sinauer.

Gilbert SF (2001) Ecological developmental biology: Developmental biology meets the real world. Dev Biol 233: 1–12.

Gilbert SF (2004) Mechanisms for the environmental regulation of gene expression: Birth defects research. Embryo 72: 291–299.

Goldschmidt RB (1927) Physiologische Theorie der Vererbung. Berlin: Springer.

Gould SJ (1977) Ontogeny and Phylogeny. Cambridge, Mass.: Belknap Press of Harvard University Press.

Haeckel E (1866) Generelle Morphologie der Organismen. Berlin: Georg Reimer.

Hairston NG, Ellner SP, Geber MA, Yoshida T, Fox JA (2005) Rapid evolution and the convergence of ecological and evolutionary time. Ecol Lett 8(10): 1114.

Hand C, Uhlinger K (1992) The culture, sexual and asexual reproduction, and growth of the sea anemone *Nematostella vectensis*. Biol Bull 182: 169–176.

Hall BK, Pearson RD, Müller GB (eds) (2004) Environment, Development, and Evolution. Cambridge, Mass.: MIT Press.

Haworth KE, Islam I, Breen M, Putt W, Makrinou E, Binns M, Hopkinson D, Edwards Y (2001) Canine TCOF1: Cloning, chromosome assignment, and genetic analysis in dogs with different head types. Mammal Genome 12: 622–629.

Hayes T, Haston K, Tsui M, Hoang A, Haeffele C, Vonk A (2003) Atrazine-induced hermaphroditism at 0.1 ppb in American leopard frogs (*Rana pipiens*): Laboratory and field evidence. Environ Health Perspect 111: 568–575.

Hentschel HG, Glimm T, Glazier JA, Newman SA (2004) Dynamical mechanisms for skeletal pattern formation in the vertebrate limb. Proc Roy Soc Lond B271: 1713–1722.

Hinchliffe JR (2002) Developmental basis of limb evolution. Intl J Dev Biol 46: 835–845.

Hinchliffe JR, Johnson DR (1980) The Development of the Vertebrate Limb. Oxford: Clarendon Press.

Holland PW (1999) Gene duplication: Past, present, and future. Sem Cell Dev Biol 10: 541–547.

Hopwood N (2002) Producing development: The anatomy of human embryos and the norms of Wilhelm His. Bull Hist Med 74(1): 29–79.

Izpisúa-Belmonte JC, Tickle C, Dolle P, Wolpert L, Duboule D (1991) Expression of the homeobox *Hox-4* genes and the specification of position in chick wing development. Nature 350: 585–589.

Jernvall J, Keranen SV, Thesleff I (2000) Evolutionary modification of development in mammalian teeth: Quantifying gene expression patterns and topography. Proc Natl Acad Sci USA 97: 14444–14448.

Johannsen W (1911) The genotype conception of heredity. Amer Naturalist 45: 129–159.

Katz MJ, Lasek RJ, Kaiserman-Abramof IR (1981) Ontophyletics of the nervous system: Eyeless mutants illustrate how ontogenetic buffer mechanisms channel evolution. Proc Natl Acad Sci USA 78: 397–401.

Kirk M, Kirk DA (2004) Exploring germ-soma differentiation in *Volvox*. J Biosci 29: 143–152.

Kirschner M, Gerhart J (1998) Evolvability. Proc Natl Acad Sci USA 95: 8420–8427.

Kühn A (1941) Über eine Gen-Wirkkette der Pigmentbildung bei Insekten. Nachrichten der Akademie der Wissenschaften in Göttingen, Mathematisch-Physikalische Klasse, 231–261.

Laubichler MD, Hagen EH, Hammerstein P (2005) The strategy concept and John Maynard Smith's influence on theoretical biology. Biol Philos 20: 1041–1050.

Laubichler MD, Rheinberger HJ (2004) Alfred Kühn (1886–1968) and developmental evolution. J Exp Zool: Molec Dev Evol 302B: 103–110.

Lederberg J (1966) Remarks. Curr Top Dev Biol 1: ix–xiii.

Lindblad-Toh K, et al. (2005) Genome sequence, comparative analysis and haplotype structure of the domestic dog. Nature 438: 803–819.

Love AC (2003) Evolutionary morphology, innovation, and the synthesis of evolutionary and developmental biology. Biol Philos 18: 309–345.

Lyytinen A, Brakefield PM, Lindstrom L, Mappes J (2004) Does predation maintain eyespot plasticity in *Bicyclus anynana*? Proc Biol Sci 271: 279–283.

Magie CR, Pang K, Martindale MQ (2005) Genomic inventory and expression of *Sox* and *Fox* genes in the cnidarian *Nematostella vectensis*. Dev Genes Evol 215: 618–630.

Martindale MQ (2005) The evolution of metazoan axial properties. Nat Rev Genet 6: 917–927.

Martindale MQ, Pang K, Finnerty JR (2004) Investigating the origins of triploblasty: "Mesodermal" gene expression in a diploblastic animal, the sea anemone *Nematostella vectensis* (phylum, Cnidaria; class, Anthozoa). Development 131: 2463–2474.

Matus DQ, Thomsen GH, Martindale MQ (2006) Dorso/ventral genes are asymmetrically expressed and involved in germ-layer demarcation during cnidarian gastrulation. Curr Biol 16: 499–505.

Maynard Smith J, Burian R, Kauffman S, Alberch P, Campbell J, Goodwin B, Lande R, Raup D, Wolpert D (1985) Developmental constraints and evolution. Quart Rev Biol 60: 265–287.

McKinney ML, McNamara KJ (1991) Heterochrony: *The Evolution of Ontogeny*. New York: Plenum Press.

Metscher BD, Ahlberg PE (1999) Zebrafish in context: Uses of a laboratory model in comparative studies. Dev Biol 210: 1–14.

Müller GB (1990) Developmental mechanisms at the origin of morphological novelty: A side-effect hypothesis. In: Evolutionary Innovations (Nitecki MH, ed), 99–130. Chicago: University of Chicago Press.

Müller GB (1991) Evolutionary transformation of limb pattern: Heterochrony and secondary fusion. In: Developmental Patterning of the Vertebrate Limb (Hinchliffe JR, Hurle JM, Summerbell D, eds), 395–405. New York: Plenum Press.

Müller GB (2003) Homology: The evolution of morphological organization. In: Origination of Organismal Form (Müller GB, Newman SA, eds), 51–69. Cambridge, Mass.: MIT Press.

Müller GB (2005) Evolutionary developmental biology. In: Handbook of Evolution (vol. 2) (Wuketits F, Ayala FJ, eds), 87–115. San Diego: Wiley.

Müller GB (2007) Six memos for EvoDevo. In: From Embryology to Evo-Devo: A History of Embryology in the 20th Century (Laubichler MD, Maienschein J, eds), 499–524. Cambridge, Mass.: MIT Press.

Müller GB, Newman SA (2005) The innovation triad: An EvoDevo agenda. J Exp Zool (MDE) 304: 487–503.

Müller GB, Newman SA (eds) (2003) Origination of Organismal Form: Beyond the Gene in Developmental and Evolutionary Biology. Cambridge, Mass.: MIT Press.

Müller GB, Newman SA (eds) (2006) Evolutionary Innovation and Morphological Novelty. J Exp Zool (MDE) 304B(6), spec. iss.

Müller GB, Wagner GP (1991) Novelty in evolution: Restructuring the concept. Ann Rev Ecol Syst 22: 229–256.

National Science Foundation (NSF) (2005) An Integrated Developmental Biology: Workshop Report. Booklet 08053. Washington, D.C.: NSF.

Needham J (1933) On the dissociability of the fundamental processes in ontogenesis. Biol Rev 8: 180–223.

Neff MW, Rine J (2006) A fetching model system. Cell 124: 229–231.

Newman SA (1992) Generic physical mechanisms of morphogenesis and pattern formation as determinants in the evolution of multicellular organization. In: Principles of Organization in Organisms (Mittenthal JB, Baskin AB, eds), 241–267. Reading, Mass.: Addison-Wesley.

Newman SA (1994) Generic physical mechanisms of tissue morphogenesis: A common basis for development and evolution. J Evol Biol 7: 467–488.

Newman SA (1996) Sticky fingers: *Hox* genes and cell adhesion in vertebrate limb development. BioEssays 18: 171–174.

Newman SA, Forgacs G, Müller GB (2006) Before programs: The physical origination of multicellular forms. Intl J Dev Biol 50: 289–299.

Newman SA, Frisch HL (1979) Dynamics of skeletal pattern formation in developing chick limbs. Science 205: 662–668.

Newman SA, Müller GB (2000) Epigenetic mechanisms of character origination. J Exp Zool (MDE) 288: 304–317.

Newman SA, Müller GB (2005) Origination and innovation in the vertebrate limb skeleton: An epigenetic perspective. J Exp Zool (MDE) 304: 593–609.

Nijhout HF (2003) Development and evolution of adaptive polyphenisms. Evol Dev 5: 9–18.

Oster GF, Murray JD, Mani P (1985) A model for condrogenic condensations in the developing limb: The role of extracellular matrix and cell tractions. J Embryol Exp Morphol 78: 83–125.

Przibram H (1923) Aufbau Mathematischer Biologie. Berlin: Bornträger.

Raff RA (1996) The Shape of Life. Chicago: University of Chicago Press.

Raff RA, Wray GA (1989) Heterochrony: Developmental mechanisms and evolutionary results. J Evol Biol 2: 409–434.

Relyea RA, Mills N (2003) Predator-induced stress makes the pesticide carbaryl more deadly to gray treefrog tadpoles (*Hyla versicolor*). Proc Natl Acad Sci USA 98: 2491–2496.

Richardson MK, Jefferey JE, Tabin CJ (2004) Proximodistal patterning of the limb: Insights from evolutionary morphology. Evol Dev 6: 1–5.

Riedl R (1978) Order in Living Organisms. Chichester, U.K.: Wiley.

Salazar-Ciudad I, Jernvall J (2002) A gene network model accounting for development and evolution of mammalian teeth. Proc Natl Acad Sci USA 99: 8116–8120.

Salazar-Ciudad I, Jernvall J (2004) How different types of pattern formation mechanisms affect the evolution of form and development. Evol Dev 6: 6–16.

Salazar-Ciudad I, Newman SA, Sole RV (2001) Phenotypic and dynamical transitions in model genetic networks. I. Emergence of patterns and genotype-phenotype relationships. Evol Dev 3: 84–94.

Schlosser G, Wagner GP (eds) (2004) Modularity in Development and Evolution. Chicago: University of Chicago Press.

Shubin N, Tabin C, Carroll S (1997) Fossils, genes, and the evolution of animal limbs. Nature 388: 639–648.

Shubin NH, Alberch P (1986) A morphogenetic approach to the origin and basic organization of the tetrapod limb. Evol Biol 20: 319–387.

Sordino P, Duboule D (1996) A molecular approach to the evolution of vertebrate paired appendages. TREE 11: 114–119.

Sordino P, van der Hoeven F, Duboule D (1995) *Hox* gene expression in teleost fins and the origin of vertebrate digits. Nature 375: 678–681.

Stanojevic D, Small S, Levine M (1991) Regulation of a segmentation stripe by overlapping activators and repressors in the *Drosophila* embryo. Science 254: 1385–1387.

Stillman B, Stewart D (eds) (2005) Epigenetics. Cold Spring Harbor Symposia on Quantitative Biology 69. Cold Spring Harbor, N.Y.: Cold Spring Harbor Laboratory Press.

Stockard CR (1941) The Genetic and Endocrine Basis for Differences in Form and Behavior. Philadelphia: Wistar Institute of Anatomy and Biology.

Sultan SE (2003) Phenotypic plasticity in plants: A case study in ecological development. Evol Dev 5: 25–33.

Tabin CJ (1992) Why we have (only) five fingers per hand: *Hox* genes and the evolution of paired limbs. Development 116: 289–296.

True JR, Carroll SB (2002) Gene co-option in physiological and morphological evolution. Ann Rev Cell Dev Biol 18: 53–80.

Vila C, Savolainen P, Maldonado JE, Amorim IR, Rice JE, Honeycutt RL, Crandall KA, Lundeberg J, Wayne RK (1997) Multiple and ancient origins of the domestic dog. Science 276: 1687–1689.

Waddington CH (1942) Canalization of development and the inheritance of acquired characters. Nature 150: 563–565.

Waddington CH (1953) Epigenetics and evolution. In: Evolution (SED Symposium VII) (Brown R, Danielli JF, eds), 186–199. Cambridge: Cambridge University Press.

Waddington CH (1956) Genetic assimilation. Adv Genet 10: 257–290.

Wagner GP (2000) What is the promise of developmental evolution? Part I: Why is developmental biology necessary to explain evolutionary innovations? J Exp Zool 288: 95–98.

Wagner GP, Altenberg L (1996) Complex adaptations and the evolution of evolvability. Evolution 50: 967–976.

Wagner GP, Booth G, Homayoun-Chaichian H (1997) A population genetic theory of canalization. Evolution 51: 329–347.

Wagner GP, Chiu CH (2001) The tetrapod limb: A hypothesis on its origin. J Exp Zool (MDE) 291: 226–240.

Wake DB, Roth G (1989) The linkage between ontogeny and phylogeny in the evolution of complex systems. In: Complex Organismal Functions: Integration and Evolution in Vertebrates (Wake DB, Roth G, eds), 361–377. New York: Wiley.

Webb JF, Schilling TF (2006) Zebrafish in comparative context: A symposium. Integr Comp Biol 46: 569–576.

West-Eberhard MJ (2003) Developmental Plasticity and Evolution. Oxford: Oxford University Press.

Whyte LL (1965) Internal Factors in Evolution. New York: George Braziller.

Wilkins AS (2002) The Evolution of Developmental Pathways. Sunderland, Mass.: Sinauer.

Wilkins AS (2003) Canalization and genetic assimilation. In: Keywords and Concepts in Evolutionary Developmental Biology (Hall BK, Olson WM, eds), 23–30. Cambridge, Mass.: Harvard University Press.

Wimsatt WC (1986) Developmental constraints, generative entrenchment, and the innate-acquired distinction. In: Integrating Scientific Disciplines (Bechtel W, ed), 185–208. Dordrecht: Martinus Nijhoff.

Wolpert L (1969) Positional information and the spatial pattern of cellular differentiation. J Theor Biol 25: 1–47.

Wolpert L (1989) Positional information revisited. Development 107(suppl): 3–12.

Woltereck R (1909) Weitere experimentelle Untersuchungen über Artveränderung, speziell über das Wesen quantitativer Artunterschiede bei Daphniden. Verh Deutsch Zool Gesell 1909: 110–172.

Contributors

Giorgio A. Ascoli
Krasnow Institute for Advanced Study
George Mason University

Chandrajit Bajaj
Center for Computational Visualization
University of Texas

James P. Collins
School of Life Sciences
Arizona State University

Luciano da Fontoura Costa
Instituto de Fisica de São Carlos
University of São Paulo

Kerstin Dautenhahn
School of Computer Science
University of Hertfordshire

Nigel R. Franks
School of Biological Sciences
University of Bristol

Scott Gilbert
Department of Biology
Swarthmore College

Marta Ibañes Miguez
Department Estructura i Constituents de La Materia
University of Barcelona

Juan Carlos Izpisúa-Belmonte
Gene Expression Laboratory
Salk Institute of Biological Studies

Alexander S. Klyubin
School of Computer Science
University of Hertfordshire

Thomas J. Koehnle
Department of Neuroscience
University of Pittsburgh

Manfred D. Laubichler
School of Life Sciences
Arizona State University

Sabina Leonelli
London School of Economics and Political Science

James A. R. Marshall
Department of Computer Science
University of Bristol

George R. McGhee
Department of Geological Sciences
Rutgers University

Gerd B. Müller
Department of Theoretical Biology
University of Vienna

Chrystopher L. Nehaniv
School of Computer Science
University of Hertfordshire

Karl J. Niklas
Department of Plant Biology
Cornell University

Lars Olsson
School of Computer Science
University of Hertfordshire

Eirikur Palsson
Department of Biology
Simon Fraser University

Daniel Polani
School of Computer Science
University of Hertfordshire

Diego Rasskin-Gutman
Institute Cavanilles for Biodiversity and Evolutionary Biology
University of Valencia

Hans-Jörg Rheinberger
Max Planck Institute for the History of Science

Alexei V. Samsonovich
Krasnow Institute for Advanced Study
George Mason University

Jeffrey C. Schank
Department of Psychology
University of California

Harry B. M. Uylings
Netherlands Institute for Neuroscience

Jaap van Pelt
Netherlands Institute for Neuroscience

Iain Werry
School of Computer Science
University of Hertfordshire

Index